T0127886

World Regional and Cultural Footprints and Environmental Sustainability

Analysis of Socioeconomic Determinants

Ebenezer O. Aka, Jr.

HAMILTON BOOKS
An Imprint of
Rowman & Littlefield
Lanham • Boulder • New York • Toronto • London

Hamilton Books
4501 Forbes Boulevard
Suite 200
Lanham, Maryland 20706
UPA Acquisitions Department (301) 459-3366

Unit A, Whitacre Mews, 26-34 Stannary Street,
London SE11 4AB, United Kingdom

Printed in the United States of America

British Library Cataloging in Publication Information Available
Library of Congress Control Number: 2016954489
ISBN: 978-0-7618-6864-4 (cloth : alk. paper)
eISBN: 978-0-7618-6865-1

∞™ The paper used in this publication meets the minimum
requirements of American National Standard for Information
Sciences—Permanence of Paper for Printed Library Materials,
ANSI Z39.48-1992

For my children: Nancy, Valentine, Jennifer, Noble, and Christian,
who are the source of my joy, hope, and inspiration.
The book is devoted to them for deepening my joyous life. With Love.

Contents

List of Figures

Prologue

It's Anthropocentric Economy and Environmental Sustainability. What we do to our environment, ultimately we do to ourselves. This is our world, our responsibility, and our challenge. Healing our environment is healing ourselves.

Humans are wantonly and willfully damaging and destroying their ecological systems around the world. If emerging economies where most of the current world's population growth is taking place continue to awkwardly mimic Western consumption patterns, then sustainable development will continue to be impossible and elusive. According to Schumacher in his 1974, *Small is*

Figure P.1. World Regional Map

beautiful, Economics as People Mattered, "environmental deterioration stems from the lifestyle (Western spiritual tradition) of the modern world, which in turn arises from its basic belief." Currently, our human impact or footprint to the planet earth is enormous and intractable. The rapidification and immediatism of modern life and the ensuing climate change and environmental crisis, undoubtedly pose a universal challenge. The triple forces of rapid population growth, excessive and abusive consumption patterns, and inevitable climate change will undoubtedly exacerbate pressures on land, water access, and food security, with greater impact on the poor and low-income populations. The present and future environmental crises are unequivocally due to human's inhumanity to the environment overtime; and thus, the urgent need to join together in the fight for a livable planet. Humans in all regions and cultures should reject and eschew all forms of self-centeredness and self-absorption that lead to the present existential threat, and should be cognizant of the fact that, the social, economic, and cultural systems cannot escape the rules of abiotic and biotic nature.

We should reject the present intergenerational injustice and embrace the much needed intergenerational equity around the globe. Therefore, the urge should be for self-awareness, self-restriction, and self-direction, which should be the moral compass to our future existence as we ecologically engage planet earth for livelihood. In fact, as emphasized by William Rees, a renowned expert on ecological footprints and proponent of environmental sustainability, "human kind has remained in a state of *obligate dependence* on the productivity and life support services of the ecosphere, despite the mediating technological sophistication; and that we should limit our voracious appetite for natural capital resources, including excessive fossil fuel consumption." The invisible hands of pure market mechanisms will not adequately fix the mess we find ourselves today; we really need the visible hands of both public and private entities; and fundamental changes are urgent to avoid the impending calamities of altering the living world that it will be unable to sustain life for both human society and plant and animal kingdoms. According to Anthony Cortese, President of *Second Nature*, a nonprofit organization to promote sustainable action in higher education, "we are really the enemy as human beings and the natural environment are on a collision course, through human activities which inflict harsh and often irreversible damage on the environment." For all intents and purposes, and rather than searching for the silver bullet, such as recycling, to neutralize our damage to natural capital resources, "the largest reductions in ecological footprint can most commonly be achieved by reducing the total amount of materials consumed, rather than attempting to recycle them afterwards," according to *Earth Day Network*. Thus, reducing consumption should be a major component of the natural steps necessary for a full approach to sustainability. According to the founder of *Natural Step*, Karl-Henrik Robert, a Swedish scientist, the principles of providing a way of approaching the physical and biological metabolism of any system should be based on the actions that will: (1) reduce use of finite mineral resources; (2) reduce the use of long-lived synthetic products or molecules; (3) preserve or increase natural diversity and the capacity of ecocycles; and (4) reduce the consumption of energy and other resources.

The book, *WORLD REGIONAL AND CULTURAL FOOTPRINTS AND ENVIRONMENTAL SUSTAINABILITY: ANALYSIS OF SOCIOECONOMIC DETERMINANTS*, examines the issues of sustainability in general, and assesses the various socioeconomic determinants of ecological footprints and environmental sustainability and conservation in world's nations, regions, and cultures. It explains how sustainable development can be promoted and achieved in regional, national, and local jurisdictions. The author's objectives are to increase awareness of different causes of environmental stress and non-sustainability in general and ways to ameliorate and mitigate them. Various causes and recommendations are discussed in each nation, region, and culture represented. From each region and culture studied, the major socioeconomic determinants of ecological footprints and environmental sustainability are fleshed out using rigorous multiple regression analysis process. The study is to provide information that will help various governments and policy-makers determine if a given nation, region, and culture is on a sustainable path. It is noteworthy to understand that sustainability is a vision and a process, and not an end product; thus, it is not likely to be a state that is reached, but one toward which the world must constantly strive. Therefore, the aim is to orientate and steer local cultures in their efforts toward community development, which is a precursor to implementing sustainability. The overall objective of this book is to help current and future leaders, policy makers, development experts identify and capitalize on the opportunities afforded by the knowledge of the socioeconomic determinants, in order to properly and adequately address the issues of environmental sustainability and sustainable development. By so doing, should likely benefit the average citizens of various jurisdictions of the world, including their economies.

This book is intended to help government leaders, planners, and policy makers make a difference in ameliorating and mitigating the effects of various environmental stressors. Nonetheless, how they interpret the socioeconomic determinants and reinvent their environments will likely influence their success or failure at sustainable development and environmental conservation of their nations, regions, and cultures. Policy makers, development experts, planners, and leaders of government should choose programs and strategies that focus on specific goals, such as stemming global warming by reducing greenhouse gases in the atmosphere; environmental sustainability through consumption cutbacks, population management, resource recycling, use of adequate technology, biodiversity conservation; and poverty eradication by targeting the poor that are disproportionately affected by the negative effects of anthropogenic overuse of natural capital resources, as well as negative effects of climate change.

Recognizing that our planet earth is crying from incessant and irretrievable damages, Pope Francis in his *Encyclical on Climate Change* on September 2014, pontificated that "a true ecological approach always becomes a social approach, which must integrate questions of justice in debates on the environment, so as to hear both the cry of the earth and the cry of the poor." It is essential to emphasize that citizens everywhere must also recognize that government officials cannot do it alone; and that policing the environment should involve everyone, regardless of socioeconomic status, race, or gender. Therefore, carefully investigating policies, programs, and strategies that have been applied in different nations, regions, and cultures, which have worked (best practices) can lead to similar applications in other nations, regions, and cultures across the globe. Thus, the regional and global focus of this book is with the recognition that a global orientation is likely a precursor to understanding sustainability. We all contributed enormously to the present environmental and sustainability problems overtime, thus, we are going to solve them together, especially for the survival of coming generations. The motivation for this book arose from the inspiration on how to create more livable communities and also have a sustainable living around the world. I accept any errors in the pages of this book in trying to accomplish the above aims, directly or indirectly.

WORLD REGIONAL AND CULTURAL FOOTPRINTS AND ENVIRONMENTAL SUSTAINABILITY: ANALYSIS OF SOCIOECONOMIC DETERMINANTS does not pretend to reinvent the wheel by concentrating on calculating the various environmental footprints done elsewhere by different individuals or organizations, but focuses on elucidating the socioeconomic determinants of those environmental and ecological footprints in different countries, regions, and cultures around the world. The aim of this book is to use the strength and limitations of already existing environmental footprint accounts in different regions and cultures to address environmental sustainability issues, which is central to this present study. It is mindful that data in some countries and regions are old or missing depending on the period the research was conducted, as well as when the article was published. Nevertheless, the old or missing data episodes did not affect or negate the assumptions, theorizations, analyses, and findings of the study, concerning the stressors, aggressors, and triggers of the prevailing environmental abuses and damages in various regions and cultures. In fact in most cases, the footprints socioeconomic determinants are: carbon dioxide from voracious and omnivorous consumption of natural resources, including fossil fuels, and their greenhouse effects on global climates; large and growing absolute national and regional populations; and teeming urban populations that appropriate resources beyond their prescribed jurisdictions, even across regions and cultures. The chapters in this book seem to be a disparate lot, but they possess a collective unity concerned with environmental sustainability and footprints factors. The chapters are in journal article format with Introduction, Purpose of the

Study, Theoretical Framework of the Study, Methodology of the Study, Mode of Analysis, Hypothesis for the Study, and Variables in the Study, which are virtually the same, especially the regional studies. Articles on Biodiversity Conservation, as well as Gender Equity and Sustainable Socioeconomic Development are included to enhance and strengthen our understanding of environmental sustainability issues and problems. Articles (without any noticeable alterations or modifications) make up the volume of this book, and some had appeared in major reputable refereed national and international journals.

The book begins with the Introduction, which discusses sustainability issues and problems in general; including the Purpose of the Study, especially for the regional footprints studies; Focus of the Study (world regions represented); Significance of the Study; Theoretical Framework for the Study; Methodology for the Study; Mode of Analysis; and Variables in the Study (both dependent and independent variables in the multiple regression analysis) for the regional studies; as well as some Definitions, Descriptions, and Explanations of terms dealing with general Sustainability, Environmental Sustainability, and Ecological Footprints concepts. Chapter 2 deals with cross-cultural analysis of ecological footprint, environmental sustainability, and biodiversity conservation. Chapter 3 explores empirically the problems of biodiversity in different cultures, areas, and regions; and the relationship between biodiversity conservation and socioeconomic growth and development, especially the challenges and prospects as we endeavor to balance environmental gains (common future) with economic and social development. Chapter 4 investigates empirically the role of gender and culture in sustainable socioeconomic growth and environmental sustainability around the world. The chapter addresses the theoretical underpinning of the role of gender and culture as macroeconomic variables that affect equity, economic efficiency, and sustainability across different cultures. The following chapters (5–9) begin with historical socioeconomic and cultural backgrounds of each region covered in this study, then followed by the analysis of socioeconomic determinants of each region's ecological footprints. Thus, chapter 5 analyzes the ecological footprint factors among African countries; exposes the inequitable share of African region's ecological footprint when compared to those of other regions and cultures around the world; and also fleshes out the proximate socioeconomic determinants of African region's ecological footprint through a rigorous stepwise regression analysis of the recursive variables. Stepwise regression analysis was also performed to strengthen the results from comparative model analysis. Chapters 6, 7, 8, and 9 do the same for Organization for Economic Cooperation and Development (OECD), Latin America and Caribbean (LAC), Middle East and Central Asia, and Asia-Pacific countries, regions, and cultures respectively. Chapter 10 is a summary manuscript of the World Regional and Cultural Footprints and Environmental Sustainability. It is a complete paper by itself that summarizes the regional findings. The Epilogue gives an overview of future global sustainability in general. Some chapters have been published in reputable refereed Journals. For example, chapters 2, 3, 4, and 5 have been published in *International Journal of Environmental, Cultural, Economic and Social Sustainability*; and chapter 10 published in *National Social Science Journal*; while chapter 8 appeared in the *Proceedings* of National Technology and Social Science Conference. They granted permission for nonexclusive world rights in all languages. I am grateful for their permission.

Acknowledgments

There are a few people I am particularly indebted; who stimulated my interest on environmental sustainability for now and future generations. During several environmental conferences in the late 1990s, I was fortunate to be formally introduced to the crucial need for environmental sustainability and biodiversity conservation by Anthony Cortese, the then President of *Second Nature*, a non-profit organization to promote sustainable action in higher education. His insightful presentations during one of the sponsored workshops were very inspirational and undoubtedly stimulated (woke up) my interest. I also wish to express my gratitude to Professor William Rees, a renowned and consummate environmental sustainability proponent I met at several environmental sustainability conferences around the world, in Hawaii, USA; Vancouver, Canada; Hanoi, Vietnam; and Hiroshima, Japan, among others. He provided encouragement when I especially needed it. He was assiduous at finding and sending materials to help me in the face of his increasing workloads and pressures. I learned a lot from his books and articles. He is really my mentor, which I made him to understand. Dr. Ed Barry of The Population Institute, Washington D.C., USA, who I met during the 8th International Conference on Environmental, Cultural, Economic, and Social Sustainability at University of British Colombia, Vancouver, Canada, also provided me with valuable materials on the use and misuse of the concept of sustainability, including population and resource macro-balancing in the sustainability dialog. I thank him immensely for the materials and valuable discussions. Many thanks to the above sustainability icons, because I have enjoyed unparalleled opportunities to expand my knowledge of environmental sustainability and conservation, environmental footprints, and sustainable development concepts;

as well as the interest to proffer suggestions, recommendations and solutions on how to mitigate present and future environmental abuse and damage. I am also grateful to Morehouse College Sustainable Energy Group for research and financial support, especially to conferences to present papers. There are many others who rendered help to me in one way or another who I am thankful, and not mentioning their names is not deliberate. All errors and infelicities are mine.

Chapter One

Introduction

The earth is home to a variety of plants and animals and other living things referred to as biodiversity. Biodiversity provides many of the ecological resources and services for the anthropocentric economy, thus it is threatened by economic sectors in the aggregate (Czech et al, 2000). A fundamental conflict exists between economic growth and biodiversity conservation. To sustain livelihoods and to reduce poverty whilst conserving the earth's resources are major global challenges. Human use of the environment is the largest contributor to habitat modification and ecosystem loss (Goudie, 2000; World Resources Institute, 2000). As the human economy expands, natural capital (resources) is reallocated from non-human uses to the human economy (Czech, 2000a). The manufacturing and service sector of the human economy are responsible for the habitat and ecosystem losses. In the absence of anthropogenic threat, all natural capital is available as habitat for non-human species. Economic growth and development as a measure of gross domestic product (GDP) reflects an increase in the production and consumption of goods and services. The proportion of natural capital allocated to human economy increases with GDP, while the proportion of national capital associated to non-human habitats decreases with GDP (The Wildlife Society, 2003). Growth, according to Zovanyi (2005), constitutes unsustainable behavior because it is incapable of being continued or sustained indefinitely. Growth is blamed for such costly destructive development patterns as urban sprawl, loss of prime agricultural land and ecosystem loss, inefficient provision of public facilities and services, escalating housing prices, pervasive environmental degradation, and loss of community character. Therefore, "smart growth" as alternative to "dumb growth" (Chen, 2000; Lorentz and Shaw, 2000) should be managed to simultaneously confront social sustainability, economic sustainability, and environmental sustainability.

Sustainability refers to the ability of a system to continue and maintain a production level or quality of life for future generations. It means living in material comfort and peacefully with each other within the means of nature (Wackernagel and Rees, 1996). Sustainable development implies that a society should balance social equity, economic prosperity, and environmental integrity (Krizek and Power, 1996). Sustainable development is development that meets the needs of the present without compromising the ability of future generations to meet their needs (World Commission on Environment and Development, 1987). To balance economic and environmental goals, economic growth and development should be sustainable and imply full consideration of environmental factors.

Human societies and their places are the products of environmental, economic, political, and cultural processes, which may work across scales from the local to global. The history of civilization is also the history of ecological degradation and crisis (Chew, 2001). Of course, the world is currently utilizing nature beyond its capacity to renew and regenerate indefinitely. In fact, humankind has consistently remained in a state of "obligate dependence" on the productivity and life support services of the ecosphere, despite the mediating technological sophistication (Rees, 1990). The global modes of production and accumulation are intimately linked to environmental degradation. Market expansion through recent globalization is threatening human race with environmental disasters, and also creating conflict among three essential aims: prosperity, equity, and ecological sustainability (Jorgenson and Kick, 2003). Perhaps, this is what Anderson (2006) called "global ethical trilemma," which are production, consumption, and their consequent environmental degradation.

The pressing issues of our time are the population growth and increasing economic growth, as they interact with resource consumption rates, pushing the world toward global ecological collapse. Thus, population and consumption are the major environmental degradation factors, and resource consumption rates have increased at a faster pace in some world regions than population size in recent decades. In fact, affluence and population size (growth) have been hypothesized to be the primary drivers of human-caused environmental stressors or impacts (Earth Observatory News, 2007). Dietz et al (2007) found that increased affluence exacerbates environmental impacts and, when combined with population growth will substantially increase the human footprint on the planet. Ecological footprint is a measure of the amount of nature it takes to sustain a given population over a course of one year. It is a national-level measurement that quantifies how much land and water are required to produce the commodities consumed and assimilate the waste by them. It actually measures how consumption may affect the environment, thus the concept of ecological footprint analysis (Wackernagel and Rees, 1996; Wackernagel and Silverstein,

2000; Wackernagel et al, 2000; Jorgenson, 2003; Redefining Progress, 2005; Earth Observatory News, 2007). Ecological footprint analysis addresses the issue of human biological metabolism and humanity's industrial metabolism (Rees, 2012); and compares the humanity's ecological footprint with the earth's available capacity. It attempts to operationalize the concept of carrying capacity and sustainability. Carrying capacity is the maximum number of individuals of a defined species that a given environment can support over the long term (Hardin, 1991), or the environment's maximum persistently supportable load (Catton, 1986). In a modern capitalist world system economy, different countries, regions, and cultures have different consumption patterns, different ecological footprints, and different ecological balances.

PURPOSE OF STUDY

Limited studies exist that dealt with humanity's ecological footprint analysis tied to world's biological diversity crisis, and with major emphasis on regional, cultural, and national footprints socioeconomic determinants. Such limited footprints studies without major regional and national determinants' analysis include, among others, Burns et al, 2001; Rees, 1990; Wackernagel and Rees, 1996; and Jorgenson, 2003. This study analyzed the ecological footprints and their structural causes in different nations, regions, and cultures for ecological balance; and provided a summary of their socioeconomic determinants that were fleshed out through rigorous multiple regression analyses of their recursive variables. In effect, the study examined and explained the cross-cultural and national variations in total footprints and per capita footprints (personal planetoids) associated with ecostructural factors (socioeconomic processes) and world system human development hierarchy. Ecostructural factors are socioeconomic processes within nations that impact footprints, such as urbanization, literacy rate, gross domestic product per capita (affluence), domestic inequality (Gini Index), government social (public) spending, gender inequality index, and gender-related development index. Progress towards sustainable development can be assessed using human development index (HDI) as an indicator of well-being, and ecological footprint as a measure of demand on biosphere. The process exposes the inequitable distribution of the world's ecological footprints and at the same time heightens the concern about ecological imbalances and overshoots. The expectation is that the empirical analysis of the data and findings should yield some policy recommendations and suggestions on how to ameliorate the impacts of the detected socioeconomic determinants. The question is what factors are responsible for or are driving the total and per capita footprints among nations, regions, and cultures around the world? What type of secure environments should meet the needs of both people and natural environments in the above entities? Thus, the study proffered some solutions on how to reduce the footprints of nations and regions, which also will help to protect the national environment and promote a more equitable and sustainable society.

FOCUS OF STUDY

Nations and regions covered by the study include: African Countries, Organization for Economic Cooperation and Development (OECD) Countries, Latin America and Caribbean (LAC) Countries, Middle East and Central Asia Countries, and Asia-Pacific Countries.

SIGNIFICANCE OF THE STUDY

Our continued abuse of the earth's resources in the process of economic growth and development has generated enormous wastes, and our planet earth is currently confronted with various environment and development problems in different countries, regions, and cultures around the world. The environmental crisis poses a universal challenge. The environmental damage we are incurring today should undoubtedly affect future generations, and their sustainability is far more threatened, because of the excruciating triple forces of rapid population growth in some regions, wanton consumption patterns in some areas, and devastating global climate change effects. In effect, sustainability agenda should be a major global and local issue. This book assesses the various socioeconomic determinants of environmental footprints and environmental sustainability in world nations, regions, and cultures. Therefore, the results from this study provide the socioeconomic determinants or factors that are responsible for or are driving the per capita and total footprints or wastes in different countries, regions, and cultures of the world. The motivation for this book arose from the inspiration on how to make, particularly, a more sustainable living, and generally, more sustainable communities around the world. My motivation also arose on how to create a sustainable future. I trust that this book should in some way provide perspectives on the many vexing sustainability issues in the hearts of environmentalists, policy-makers, and development experts, which they grapple every day. I also trust that the book will become a valuable resource and inspiration in their fight for livable and sustainable communities. I accept any errors in the pages of this book, as I strive to incentive, orientate, and steer local cultures in different countries and regions in their efforts toward community development, which is a precursor to implementing sustainable development.

A major contribution of this study is the increased awareness of different causes of environmental stress and sustainability around the world, as well as possible ways to ameliorate and mitigate them. The policy makers require a clear picture of the risks their countries and regions are facing and also will face in the future, and hopefully they should use the results from this study for predictions and possible solutions. In effect, the results of this study, hopefully, will help current and future leaders, policy makers, and development experts around the globe to identify and capitalize on the opportunities afforded by the knowledge of the socioeconomic determinants, in order to properly and adequately address the issues of environmental sustainability and sustainable development. It is hoped that various guidelines in this book in some regions and cultures could help planners and policy makers in other regions and cultures to facilitate their sustainable planning and development efforts, by incorporating best practices from some programs and strategies developed somewhere. In other words, the results from this book may help communities around the world to pursue sustainable planning and development which have been determined to offer cost-effective economic, social, cultural, and environmental benefits in other areas and locations.

The study hopes to contribute to the ongoing debates and discussions on various socioeconomic determinants of environmental footprints, environmental sustainability, and sustainable development. The results from this study may also encourage and spur further footprint socioeconomic determinant research in the future at regional and cultural levels rather than at the prevailing national and local levels. If this book can sensitize and incentive a majority of individuals and policy-makers from various countries, regions, and cultures to acknowledge and recognize the indisputable and irrefutable fact that, what we do today can greatly and adversely affect future generations, then its major objective is maximally achieved. Moreover, if this book will make people and policy-makers in different countries, regions and cultures think globally but act locally, then the objectives are also served.

LIMITATIONS OF THE STUDY

1. Some national and regional analyses can be affected by many missing information or data availability, political sovereignty, threshold population requirement, and boundary issues.
2. The assumptions, proxies, and variables used in deriving the proximate footprints socioeconomic determinants in different countries, regions, and cultures are not always apparent. Social and economic information for the study may be inferred, and socioeconomic determinants are dependent on current aggregate social and economic conditions that create demand in specific regions and cultures.
3. Aggregation of information in different regions can oversimplify variable impacts.

4. The internal and external validity of the study will be enhanced if all the variables used are current and available at the same time in all countries and regions.

THEORETICAL FRAMEWORK FOR THE STUDY

Within the biosphere, everything is interconnected, including humans, thus sustainable development requires a good knowledge of human ecology. The human life-support functions of the ecosphere are maintained by nature's biological capacity which runs the risks of being depleted, the prevailing technology notwithstanding. Regardless of the humanity's mastery over the natural environment, it still remains a creature of the ecosphere and always in a state of *obligate dependency* on numerous biological goods and services (Rees, 1992, p. 123). Despite the above *dependency* thesis, the prevailing economic mythology assumes a world in which carrying capacity is indefinitely expandable (Daly, 1986; Solow, 1974). The human species has continued to deplete, draw-down, and confiscate nature's bio-capacity with reckless abandon. York et al (2003), indicated that population and affluence account for 95 percent of the variance in total footprints of countries. Thus, large human population all over the world and their excessive consumption of the scarce natural resources are responsible for the national footprint of nations. Thus, total human impact on the ecosphere is given as: population × per capita impact (Ehrlich and Holdren, 1971; Holdren and Ehrlich, 1974; Hardin, 1991). In other words, population size and affluence are the primary drivers of environmental impacts (Dietz et al, 2007). The footprints of nations provide compelling evidence of the impacts of consumption, thus the need for humans to change their lifestyles and conserve scarce natural capital. According to Palmer (1998), there are in order of decreasing magnitude, three categories of consumption that contribute enormously to our ecological footprints: wood products (53 percent), food (45 percent), and degraded land (2 percent). Degraded land includes land taken out of ecological availability by buildings, roads, parking lots, recreation, businesses, and industries. Palmer also indicated that about 10 percent or more of earth's forests and other ecological land should be preserved in more or less pristine condition to maintain a minimum base.

The environmental impacts of urban areas should be considered because a rapidly growing proportion of world's population lives in cities, and more than one million people are added to the world's cities each week and majority of them are in developing countries of Africa, Latin America, and Southeast Asia (Wackernagel and Rees, 1996). The reality is that the populations of urban regions of many nations had already exceeded their territorial carrying capacities and depend on trade for survival. Of course, such regions are running an unaccounted ecological deficit; their populations are appropriating and meeting their carrying capacity from elsewhere (Pimentel, 1996; Wackernagel and Rees, 1996; Girardet, 1996, accessed Online on 12/14/07; Rees, 1992; The International Society for Ecological Economics and Island Press, 1994; Vitouset et al, 1986; Wackernagel, 1991; WRI, 1992, p. 374). Undoubtedly, the rapid urbanization occurring in many regions and the increasing ecological uncertainty have implications for world development and sustainability. Cities are densely populated areas that have high ecological footprints which leads to the perception of these populations as "parasitic," since these communities have little intrinsic biocapacity, and instead, must rely upon large hinterlands. Land consumed by urban regions is typically at least an order of magnitude greater than that contained within the usual political boundaries or the associated built-up areas (The International Society for Ecological Economics and Island Press, 1994; Rees, 1992).

According to Rees (1992), every city is an entropic black hole drawing on the concentrated material resources and low-entropy production of a vast and scattered hinterlands many times the size of the city itself. In the same vein, Vitouset et al (1986) asserted that high density settlements "appropriate" or augment their carrying capacity from all over the globe, in the past and the future (see also Wackernagel, 1991). In modern industrial cities, resources flow through the urban system without much concern either about their origins, or about the destination of their wastes, thus, inputs and outputs are considered to be unrelated. The cities' key activities such as transportation, provision of electricity supply, heating, manufacturing and the provision of socio-economic services depend on a ready supply of fossil fuels, usually from far-flung hinterlands than within their usual political boundaries or their associated built-up areas. Cities are not self-contained entities, and their concentration of intense economic processes and high levels of consumption both increase and stimulate their demands on resources. Cities occupy only 2 percent of the world's land surface, but use some 75 percent of the world resources, and release a similar percentage of waste (Girardet, 1996). With more than half of the world's population currently in cities, the large and mega cities of the world regions should be the focus for the sustainability agenda, with enormous potential to generate change in how we use natural resources.

Like urbanization, energy footprint, created from energy use and carbon dioxide emissions, is not subject to area constraints. Energy footprint is the area of forest that would be needed to sequester the excess carbon (as carbon dioxide) that is being added to the atmosphere by the burning of fossil fuels to generate energy for travel, heating, lighting, manufacturing, recreation, among other uses. Actually, the demand for energy defines modern cities more than any other single factor. Cities contain enormous concentration of economic activities that consume enormous quantities of energy. The natural global systems of forests and oceans for carbon sequestration are not handling the human carbon contributions fast enough, thus the Kyoto Conference of early 1998. According to Suplee (1998), only half of the carbon humans generate burning fossil fuels can be absorbed in the oceans and existing terrestrial sinks. The oceans absorb about 35 percent of the carbon in carbon dioxide (Suplee, 1998), equivalent of 1.8 giga tons of carbon every year (IPCC, 2001), while the global forests under optimum management of existing forests could absorb about 15 percent of the carbon in the CO_2 produced from the burning of fossil fuels world-wide (Brown et al, 1996). The energy footprint is caused by the un-sequestrated 50 percent in the atmosphere with the potentially troubling ecological consequences, such as rapid global warming and other environmental stresses, including climate change. Carbon dioxide in the atmosphere will continue to increase unless humanity finds alternative energy sources of sufficient magnitude. It is in the humanity's best interest to get off its petroleum addiction (control and minimize fossil fuel use) and develop sustainable consumption habits.

Literacy affects the consumption of natural capital resources. Highly literate groups concentrate in urban areas where they consume more than their fair shares of biospheric resources. Literate populations generally have lower rates of domestic inequality and tend to consume more resources than their illiterate counterparts due to their higher incomes and higher standards of urban living. Furthermore, higher levels of literacy correspond with higher incomes, which allow for greater consumption (Jorgenson, 2003). This is because literate populations are subject to increased consumerist ideologies and contextual images of good life through advertising (Princen et al, 2002); what Leslie Sklair (2001) and Jennifer Clapp (2002) labeled "cultural ideology of consumerism/consumption."

METHODOLOGY FOR THE STUDY

Unit of Analysis

The units of analysis are "country" and "region" from different cultures of the world.

Samples

To test for the national variations in total footprints and per capita footprints associated with human development index, as well as their socioeconomic determinants, Comparative Model Analysis and Stepwise Regression Analysis were used as the tools of analysis. Using comparative model, samples of eligible countries out of many countries from different regions and cultures were analyzed (see Global Footprint Network, *The Ecological Footprint Atlas 2008*). Step-wise Regression Analyses were also conducted using the samples. Each table represents countries in each region of the world studied, and as also represented by Global Footprint Network, *The Ecological Footprint Atlas 2008*. These tables do not include countries that were less than one million people in 2005, as well as the dependent territories without political independence. The larger population of at least one million is chosen for the study because; population is a variable which affects the consumption rates and levels, therefore the ecological footprint. The regional tables used in the analyses are not included in this chapter. They were also included in previous regional studies by this author published elsewhere, such as *The International Journal of Environmental, Cultural, Economic, & Social Sustainability*; and *The National Social Science Journal, as well as Proceedings* of The National Social Science Association (NSSA).

Mode of Analysis

This is an explanatory study using Comparative Model that employs descriptive statistics (such as matrices, totals, averages, ratios, and percentages) as well as Step-wise Regression. Regression analysis was performed to flesh out factors that highly impact regional, national, and per capita footprints; and also to strengthen the results from comparative analysis. Potential technical problems were diagnosed that might affect the validity of the results of various analysis mode components, such as the missing data in some countries or the exclusion or omission of some member countries of a particular region not meeting the stipulated selection criteria such as minimum population threshold; non-political sovereignty (dependent territory status) during the study period; etc. The aforementioned problems did not alter the substantive conclusions of the study, especially regarding the ecological footprint accounts and environmental sustainability of different countries, regions, and cultures.

The national ecological footprints and ecological balance as dependent variables are recursively explained by the country's (region's) various independent variables. The independent variables mediated in explaining the varying levels of consumptions and wastes among different nations of the world capitalist economy. It is hypothesized that Human Development positions and Ecostructural Factors of different countries are likely to be responsible for the variations in the National and Regional Ecological Footprints and Balances. Moreover, the carbon-dioxide emission levels of different countries are included in this study to depict and emphasize the biological and industrial metabolisms (consumption and environmental impacts) of different countries, regions, and cultures. The levels reflect their ecological footprints and balances.

HYPOTHESIS FOR THE STUDY

Null Hypothesis (H_0)

None of the independent variables predicts the national and regional footprints effects. The independent variables do not influence the footprints.

Alternative Hypothesis (H_1)

Some (if not all) of the independent variables predict the national and regional footprints effects.

Variables in the Study

The *Dependent* and *Independent* variables are selected on the basis of the theoretical themes and underpinnings which indicate that national footprint (and regional footprint) as dependent variable is explained by the country's (region's) human development (HDI) hierarchical category, population size, population density, urbanization level, government social (public) spending (as percent of GDP), GDP per capita, domestic inequality (Gini Index), gender inequality index (GII), gender-related development index (GDI), and literacy rate, as the ecostructural or independent variables.

VARIABLE DEFINITIONS

Human Development Index (HDI)

The HDI is a summary measure of human development (human welfare). It measures the average achievements of a country in three basic dimensions of human development (Global Footprint Network: *Africa's Ecological Footprint-2006 Factbook*, p. 89): a long and healthy life, as measured by life expectancy at birth; knowledge, as measured by the adult literacy rate (with two-thirds weight) and the combined primary, secondary and tertiary gross enrolment ratio (with one-third weight); and a decent standard of living, as measured by GDP per capita (PPP US$). Purchasing power parity (PPP) is a rate of exchange that accounts for price difference across countries, allowing international comparisons of real output and incomes. At the PPP US$ (as used in this study), PPP US$1 has the same purchasing power in the domestic economy as $1 has in the United States of America.

Gross Domestic Product (GDP) per capita (PPP US$)

GDP is converted to US dollars using the average official exchange rate reported by the International Monetary Fund (IMF). GDP alone does not capture the international relational characteristics as does the human development hierarchy of the world economy, which accounts for a country's relative socio-economic power and global dependence position in the modern world system. It is suggested elsewhere that GDP per capita is an inadequate measure of world-system position but a more appropriate indicator of domestic affluence or internal economic development (Burns, Kentor, and Jorgenson, 2003; Jorgenson, 2003; Dietz and Rosa, 1994). The Gross Domestic Product (GDP) per capita data for this study is taken from *Human Development Report 2007/2008*, table 1, and pp. 229–232. See also *World Development Report 2005*.

Domestic Income Inequality (Gini Index)

The Gini index measures domestic income inequality of different countries, which had remained stable over a time with its impacts on other variables in the study (Bergesen and Bata, 2002; Jorgenson, 2003). Gini index measures the extent to which the distribution of income (or consumption) among individuals or households within a country deviates from a perfectly equal distribution (*Human Development Report, 2004*, p. 271)). It measures inequality over the entire distribution of income or consumption. A value of zero (0) represents perfect equality, and a value of hundred (100) represents perfect inequality. Data for domestic income inequality measured by Gini index are taken from World Bank, *World Development Report* (2005), table 2, pp. 258–259 and United Nations Development Report, *Human Development Report* 2007/2008, 15, pp. 281–284.

Gender Inequality Index (GII)

The index shows the loss in human development due to inequality between female and male achievements in reproductive health, empowerment, and labor market. It reflects women's disadvantage in those three dimensions. The index ranges from zero (0), which indicates that women and men fare equally, to one (1), which indicates that women fare as poorly as possible in all measured dimensions.

The health dimension is measured by two indicators: maternal mortality ratio and the adolescent fertility rate.

The empowerment dimension is measured by two indicators: the share of parliamentary seats held by each sex and by secondary and higher education attainment levels.

The labor market dimension is measured is measured by women's participation in the work force.

Gender Inequality Index is designed to measure the extent to which national achievements in these aspects of human development are eroded by gender inequality; also to provide empirical foundations for policy analysis and advocacy efforts. It can be interpreted as a percentage loss to potential human development due to shortfalls in the dimensions included. Countries with unequal distribution of human development also experience high inequality between women and men, and countries with high gender inequality also experience unequal distribution of human development (UNDP *Human Development Reports*). The Gender Inequality Index data for this study is taken from UNDP's *Human Development Report 2010___20th Anniversary Edition. The Real Wealth of Nations: Pathways to Human Development*. Table 4: Gender Inequality Index, pp. 156–160.

Gender-Related Development Index (GDI)

The index is one of the indicators of human development developed by the United Nations. The GDI is considered a gender-sensitive extension of the Human Development Index (HDI), which addresses gender-gaps in life expectancy, education, and income. It highlights inequalities in the areas of long and healthy life, knowledge, and decent standard of living between women and men. It measures achievement in the same basic capabilities as the HDI, but takes note of inequality in achievement between women and men (*Human Development Reports*). The methodology used imposes a penalty for inequality, such that the GDI falls when the achievement levels of both women and men in a country go down or when the disparity in basic capabilities, the lower a country's GDI compared with its HDI. The GDI is simply the HDI discounted, or adjusted downwards, for gender inequality. Thus, if GDI goes down, HDI goes down, while GII goes up. The methodology used to construct the GDI could be used to assess inequalities not only between men and women, but also between other groups such as rich and poor, young and old, etc. (*Human Development Reports*). The Gender-Related Development Index data for this study is taken from UNDP's *Human Development Report 2010___20th Anniversary Edition. The Real Wealth of Nations: Pathways to Human Development*. Table 4: Gender Inequality Index, pp. 156–160.

Urbanization Level (Urban Population as percent of Total Population)

Cities are not self-contained entities and their concentration of intense economic processes and high levels of consumption both increase and stimulate their demands on resources. The cities have limited intrinsic biocapacity which undoubtedly must rely upon large hinterlands. Land consumed by urban regions is typically at least an order of magnitude greater than that contained within the usual political boundaries or the associated built-up area (The International Society for Ecological Economics and Island Press, 1994; Rees, 1992). The data are taken from *Human Development Report 2007/2008*, table 5, and pp. 243–246.

Literacy Rate

This variable refers to the percent of a nation's population over the age of fifteen (15) that can read and write in any language of their choice. Literate population generally has low domestic inequality and tends to consume more resources than their illiterate counterpart due to their higher income and urban living. High literate groups concentrate in urban areas where they consume more than their fair shares of biospheric resources. Higher levels of literacy correspond with higher incomes, which allow for greater consumption (Jorgenson, 2003). This is because literate populations are subject to increased consumerist ideologies and contextual images of good life through advertising (Princen, 2002), what Leslie Sklair (2001) and Jennifer Clapp (2002) labeled "cultural ideology of consumerism/consumption." The data for literacy rate is taken from the *Human Development Report 2007/2008*, table 1, pp. 229–232.

Population and Population Density

Apart from consumption, many have attributed to population as driving most of the sustainability problems (Palmer, 1998; Pimentel, 1996). Likewise, Dietz et al (2007) concluded that population size and affluence are the primary drivers of environmental impacts. Population growth and increases in consumption in many parts of the world have increased humanity's ecological burden on the planet. York et al (2003) indicated that population and affluence account for 95 percent of the variance in total footprints of countries. Others also see the ensuing human impact or footprint as a product of population, affluence (consumption), and technology (i.e. I = PAT (Ehrlich and Holdren, 1971; Holdren and Ehrlich, 1974; Hardin, 1991). Population as a variable in this study is taken from, *World Development Report 2005*, table 1, and pp. 256–259.

Government Social Spending Per Capita

Spending by a government (federal, state, and local), municipality, or local authority, which covers such things as spending on healthcare, education, pensions, defense, welfare, interest, and other social services, and is funded by tax revenue, seigniorage, or government borrowing. Public expenditure exerts an effect on economic growth rate through the positive externality in the productivity of the capital stock. Investment in education and health is also investment in people and in the future (See, Education: Crisis Reinforces Importance of a Good Education, says OECD. Accessed Online on 9/23/2011, at http://www.oecd.org/document/21/0,3746. Data for this study were taken from: Human Development Report 2007/2008. *Fighting Climate Change: Human Solidarity in a Divided World.* Table 28: Gender-related Development Index, pp. 326–

329; and Human Development Report 2010___20th Anniversary Edition. *The Real Wealth of Nations: Pathways to Human Development*. Table 4; Gender Inequality Index, pp. 156–160.

COMPARATIVE MODEL ANALYSIS

The results and findings from this model of analysis in this study are discussed under the subtitles that include, "Per Capita Ecological Footprint, Biological Capacity, and Ecological Balance by World Regions," "Human Development and Affluence on Biological Capacity, Ecological Footprint and Balance by Regional Countries," and "Ecostructural Factors on Total Footprints Per Capita by Regional Countries." Please see individual national analyses and their results.

REGRESSION ANALYSIS

The tables used in the Comparative Model Analysis are also used for the Regression Analysis. Regression analysis allows the modeling, examining, and exploring of relationships and can help explain the factors behind the observed relationships or patterns. It shows the factors or independent variables that have strong correlation or association with dependent variable. The mathematical formula when applied to the explanatory variables is best used to predict the dependent variable that one is attempting to model. Each independent variable or explanatory variable is associated with a regression coefficient describing the strength and the sign of that variable relationship to the dependent variable.

In this study, the rows of the Unstandardized Coefficients of the "B" column of the Regression Model table is used to show the variables and their coefficients that exhibit strong correlation with total and per capita footprints of the nations and their regions. A regression equation might look like the one given below where Y is the dependent variable, the Xs are the explanatory variables, and the Bs are regression coefficients:

$$Y = B_0 + B_1 X_1 + B_2 X_2 + \text{-----} B_n X_n + E \text{ (Random Error Term/Residuals).}$$

The row of "unstandardized coefficients" or "Bs" gives us the necessary coefficient values for the multiple regression models or equations. See below for the Regression Analysis Terms.

Correlation or co-relation: refers to the departure of two variables from independence or they are non-independent or redundant (D'Onofrio, A., 2001/2002; Richard Lowry, 1999–2008).

Collinearity: Refers to the presence of exact linear relationships within a set of variables, typically a set of explanatory (predictor) variables used in a regression-type model. It means that within the set of variables, some of the variables are (nearly) totally predicted by the other variables [(Sundberg, R. 2002). *Encyclopedia of Environmetrics*, edited by Abdel H. El-Shaarawi and Walter W. Piegorsch (Chichester: John Wiley & Sons, Ltd), Volume 1, pp. 365–366].

Partial Correlation Coefficients (r): When large, it means that there is no mediating variable (a third variable) between two correlated variables (D'Onofrio, A., 2001/2002).

Pearson's Correlation Coefficient (r): This is a measure of the strength of the association between two variables. It indicates the strength and direction of a linear relationship between two random variables. Value ranges from - to +1; –1.0 to –0.7 Strong negative association; –0.7 to –0.3 Weak negative association; –03 to +0.3 Little or no association; +0.3 to +0.7 Weak positive association; +0.7 to 1.0 Strong positive correlation (Luke, B., "Pearson's Correlation Coefficient," Learning *From The Web.net*. Accessed Online on 5/30/2008).

Multiple "R": Indicates size of the correlation between the observed outcome variable and the predicted outcome variable (based on the regression equation).

"R^2" or Coefficient of Determination: Indicates the amount of variation (%) in the dependent scores attributable to all independent variables combined, and ranges from 0 to 100 percent. It is a measure of model performance, summarizing how well the estimated Y values match the observed Y values.

"Adjusted R^2": The best estimate of R^2 for the population from which the sample was drawn. The Adjusted R-Squared is always a bit lower than the Multiple R-Squared value because it reflects model complexity (the number of variables) as it relates to the data.

R^2 and the *Adjusted R^2* are both statistics derived from the regression equation to quantify model performance (Scott and Pratt, 2009. *ArcUser*).

Standard Error of Estimate: Indicates the average of the observed scores around the predicted regression line.

Residuals: These are the unexplained portion of the dependent variable, represented in the regression equation as the random error term (E). The magnitude of the residuals from a regression equation is one measure of model fit. Large residuals indicate poor model fit. Residual = Observed – Predicted.

ANOVA: Decomposes the total sum of squares into regression (= explained) SS and residual (= unexplained) SS.

F-test in ANOVA represents the relative magnitude of explained to unexplained variation. If F-test is highly significant (p = .000), we reject the null-hypothesis that none of the independent variables predicts the effect (scores) in the population.

The "constant" represents the intercept in the equation and the coefficient in the column labeled by the independent variables.

DEFINITION, DESCRIPTION, AND EXPLANATION OF TERMS ON SUSTAINABILITY AND ENVIRONMENTAL FOOTPRINTS CONCEPTS

Most of the definitions, descriptions, and explanations of terms are lifted from: (1) *EPA: Environmental Footprint Analysis/Sustainability Analytics/Research/ USEPA*; (2) *Earth Day Network: Ecological Footprint FAQ*; (3) Barry Cullingworth and Roger W. Caves (2014). *Planning in the USA: Policies, Issues, and Processes, 4th Edition* (London and New York: Routledge, Taylor & Francis Group); and (4) Peter Newman & Jeffrey Kenworthy (1999). *Sustainability and Cities. Overcoming Automobile Dependence* (Washington, D.C., Covelo, California: Island Press). To the above, I am heavily indebted.

Sustainability: Is a simple but powerful principle that recognizes that the natural environment is the foundation for human survival and well-being. Achieving sustainability means creating and maintaining the conditions with which people and nature can coexist in productive harmony—conditions that provide people with social, economic, and other benefits—today and in future generations. Thus, developing the science and engineering that people need in that direction is the "true north" of the national, regional, and global collective research and development efforts (*USEPA*).

Sustainability is generally viewed and depicted as a three-legged stool supported by environment, economy, and social equity, which are interrelated and should always be discussed and considered together. They require each other in order to exist and operate. According to Cullingworth and Caves (p.79), the environment leg focuses on climate, energy, air quality, loss of open space, resources conservation, air and water quality; the economy leg focuses on jobs and employment benefits, living wages, fuel prices, workforce training, market development, financial systems, and infrastructure; and the social equity leg focuses on well-being, social justice, equal opportunity, neighborhoods, diversity, discrimination, and gender equity. Thus, sustainability should have all the elements of a good quality of life: a healthy functioning natural environment; a strong economy with jobs and job security; and safe, secure communities where people have a sense of belonging and purpose and a commitment to each other. These are the things we hold in common, which together weave the fabric of sustainability. Newman and Kenworthy (p. 4) have noted that, sustainability means simply the achievement of global environmental gains along with any economic or social development. They went on to say that, the concept of sustainability emerged from a global political process that has tried to bring together, simultaneously, the most powerful needs of our time: (1) the need for economic development to overcome poverty; (2) the need for environmental protection of air, water, soil, and biodiversity, upon which we all ultimately depend; and (3) the need for social justice and cultural diversity to enable local communities to express their values in solving these issues. According to them (p. 2), the *Brundtland Report* by World Commission on Environment and Development (1987) stated four fundamental approaches to global sustainability that must be applied simultaneously for the future: (1) the elimination of poverty, especially in the Third World, is necessary not just on human grounds but as an environmental issue. Thus, Third World economic and social development is precursor to global sustainability; (2) the First World must reduce its consumption of resources and production of wastes. Thus, the primary responsibility for reducing impact on global resources lies in the rich part of the world; (3) global cooperation on environmental issues is no longer a soft option, as environmental problems will not be solved if some nations decide to hide from the necessary changes (e.g., pertaining to hazardous wastes, greenhouse gases, CFCs, loss of biodiversity, etc). Thus, a global orientation is a precursor to understanding sustainability; and (4) change toward sustainability can occur only with community-based approaches that take local cultures seriously. Change will come only when local communities determine how to resolve their economic and environmental conflicts in ways that create simultaneous improvement of both. Thus, an orientation to local cultures and community development is a precursor to implementing sustainability.

Sustainable Development: A common definition is from Brundtland Commission Report (1987, p. 43). See World Commission on Environment and Development (WCED) (1987). *Our Common Future* (Oxford: Oxford University Press). Another definition is from Carol Lancaster (2012). . . . Annual Edition: Developing World, Edited by Griffiths

"Sustainable development is development that meets the needs of the present without compromising the ability of future generations to meet their own needs." Lancaster (2012, p. 4) defined sustainable development as economic progress that does not affect the environment too harshly as we strive for the freedom to choose a fulfilling life. There are always three distinct development processes underway at the local level—economic development, community development, and ecological development. According to Newman and Kenworthy (p. 4), the economic development imperatives include, sustain economic growth, maximize private profit, expand markets, and externalize costs; the community development imperatives include, increase local self-reliance, satisfy basic human needs, increase equity, guarantee participation and accountability, and use appropriate technology; and ecological development include, respect carrying capacity, conserve and recycle resources, and reduce waste. United Nations (2012) indicates that global sustainable development efforts or actions should call for countries, regions, and cultures to renew their commitment on high-priority issues, such as jobs, energy, sustainable cities, food security and sustainable agriculture, water, oceans, disaster readiness, and continued need to close the gap between developed and developing countries, especially the eradication of poverty which represent the greatest challenge in the realization of sustainable development.

Biological Capacity (Biocapacity): This is a measure of the amount of biologically productive land and water available for human use. Biologically productive land includes areas such as crop land, forest, and fishing grounds, and excludes deserts, glaciers, and the open ocean.

Global Hectares: Are hectares with world-average productivity for all productive land and water areas in a given year. Studies that are compliant with current Ecological Footprint Standards use global hectares as a measurement unit. This makes Ecological Footprint results globally comparable. Just as financial assessments use one currency, such as dollars or Euros, to compare transactions and financial flows throughout the world.

Wastes: "Waste" from the Ecological Footprint perspective includes three different categories of materials, and each category is treated differently within Footprint accounts.

1. Biological wastes such as residues of crop products, trimmings from harvested trees, and carbon dioxide emitted from fuel wood or fossil fuel combustion are all included within Ecological Footprint accounts. A cow grazing on one hectare of pasture has a Footprint of one hectare for both creating its biological food products and absorbing its biological waste products. This single hectare provides both services, thus counting the Footprint of the cow twice (once for material production and once for waste absorption) results in double counting the actual area necessary to support the cow. The Footprint associated with the absorption of all biological materials that are harvested is thus already counted in the Footprint of those materials.
2. Waste also refers to the material specifically sent to landfills. If these landfills occupy formerly biologically productive area, then the Footprint of this landfill waste can be calculated as the area used for its long term storage.
3. Waste can also refer to toxics and pollutants released from the human economy that can not in any way be absorbed or broken down by biological processes, such as many types generally refer to the Footprint of extracting, processing, and handling these materials, but not to the Footprint of creating or absorbing these materials themselves.

Footprint: Is used generally to refer to human impact on the planet, or to a different research question. For example, the term "carbon footprint" often refers to the number of tons of carbon emitted by a given person or business during a year, or to the tons of carbon emitted in the manufacture and transport of a product. In Ecological Footprint accounts, the carbon Footprint measures the amount of biological capacity, in global hectares, demanded by human emissions of fossil carbon dioxide.

Ecological Footprint Accounts: These answer a specific research question—how much of the biological capacity of the planet is demanded by a given human activity or population? Thus, the Ecological Footprint measures the amount of biologically productive land and water area an individual, a city, a country, a region, or all of humanity uses to produce the resources it consumes and to absorb the waste it generates with today's technology and resource management practices.

Environmental Footprint Analysis: An accounting tool that measures human demand on ecosystem services required to support a certain level and type of consumption by an individual, product, or population. Through the ecological footprint analysis, it is possible to estimate the fraction (or multiples) of land/ocean area required to support a specific life-style within a specific geographic area (region, nation, state, county, city, etc, and by implication, culture). Footprint methodologies estimate the environmental impacts of anthropocentric demands. Environmental footprint methods can be classified into two broad categories of analysis: assessments that use a single unit indicator, such as carbon dioxide equivalents; and location-specific analysis, such as ecological footprint of a city.

Single-unit indicator uses different data which are converted to a single common unit, such as carbon or nitrogen; thus similar to economic tools that use currency as their single-unit indicator. Common methods available for calculating environmental footprints include: Ecological, materials, carbon, nitrogen, and water footprint analyses.

Ecological Footprint: Measures the amount of land and/or ocean required to support a certain level and type of consumption by an individual or population. This measurement is estimated by assessing the total biologically productive land and ocean areas required to produce resources consumed and mitigate the associated waste for a certain human activity or population (*USEPA*). According to *Earth Day Network*, Ecological Footprint is a resource accounting tool that measures how much biologically productive land and sea is used by a given population or activity and compares this to how much land and sea are available. Productive land and sea areas support human demands for food, fiber, timber, energy, and space for infrastructure. These areas also absorb the waste products from the human economy. The Ecological Footprint measures the sum of these areas, where ever they physically occur on the planet.

The Ecological Footprint is used widely as a management and communication tool by governments, businesses, educational institutions, and non-governmental organizations. Ecological Footprint highlights the reality of ecological scarcity, which can be disconcerting and frightening information. The existence of global overshoot suggests that human society will need to make significant changes to "business as usual" if it wants to create a sustainable future.

Ecological Footprint can be calculated for individual people, groups of people (such as a nation or region in this study), and activities (such as manufacturing a product). The ecological Footprint of a person is calculated by considering all of the biological materials consumed, and all of the biological wastes generated, by that person in a given year. These materials and wastes each demands ecologically productive areas such as cropland to grow potatoes, or forest to sequester fossil carbon dioxide emissions. All of these materials and wastes are then individually translated into an equivalent number of global hectares.

Materials Footprint: This uses material flow analysis to estimate the total material and waste generated in a well-defined system or specific enterprise (*USEPA*). This method provides several useful indicators for measuring the mass of materials entering and leaving a defined system boundary, including domestic material consumption (e.g., per capita material consumption), total materials requirements (e.g., the measure of all the material input required by a system, including direct and indirect material flows and imports), and material intensity (e.g., the ration of domestic material consumption to gross domestic product.

Carbon Footprint: This is the most developed of the footprint methods. It is a measure of the direct and indirect greenhouse gas emissions caused y a defined population, system, or activity. Carbon Footprints can be calculated by taking the inventory of six greenhouse gases identified in the Kyoto Protocol: carbon dioxide, methane, nitrous oxide, sulfur hexafluoride, perfluorocarbons, and hydrofluorocarbons. Each of these greenhouse gases can be expressed in terms of the single-unit indicator, carbon dioxide equivalents (CO_2e) or in normalized terms (e.g., CO_2e per sales dollar, land area, or production unit). CO_2e is calculated by multiplying the emissions of each greenhouse gas by their respective 100 year global warming potential.

Carbon Footprints are categorized into Scope 1 (direct greenhouse gas emissions from fuel combustion in vehicles and facilities); Scope 2 (indirect emissions from purchased electricity); and Scope 3 (other indirect greenhouse gas emissions, e.g., waste disposal, outsourced activities, business travel, emissions from leased facilities). According to *USEPA*, The World Resources Institute and the WBCSD have developed a framework for greenhouse gas accounting that is widely used and serves as the basis for international standards, such as International Organization for Standardization (ISO) 14064-1.

Nitrogen Footprint: This is a measure of the reactive nitrogen (e.g., nitrous oxides, ammonia, etc.) associated with a population or activity through agriculture, energy use, and resource consumption. Nitrogen Footprints are typically expressed in terms of mass loading per time (e.g., kg/year).

Water Footprint: This measures the total volume of freshwater that is directly or indirectly consumed by a well-defined population, business, or product. Water use can be measured by the volume of water consumed (e.g., the amount evaporated and /or polluted in a given period of time) and it is indicative of the water volume required to sustain a given population. The water footprint of a given region is the total volume of water used, direct or indirect, to produce goods and services consumed by inhabitants of a region. An internal water footprint measures the consumption within a region for goods and services, while an external water footprint measures the embodied water used outside the region for goods and services.

The Water Footprint is divided into three elements: blue (freshwater consumed from surface and groundwater sources); green (freshwater consumed from rainwater stored in the soil); and gray (the amount of polluted water, which is calculated as the volume of water needed to dilute pollutant loads to meet water quality standards. A water footprint is a hybrid method that utilizes data from a single indicator but also requires location-specific data to assess impacts from water use, which vary based on climate. For example, water-stressed or arid regions are more vulnerable to water shortages.

HOW CAN ENVIRONMENTAL FOOT ANALYSIS CONTRIBUTE TO SUSTAINABILITY?

1. It may help frame and inform sustainability discussions by providing a better understanding of the limitations of local resources to support social, economic, and environmental systems.
2. The analyses also help summarize a complex array of environmental indicators into a single or small number of values so they are more useful for decision-making.

What are the Strengths and Limits of Environmental Footprint Analysis in a Sustainability Context?

Strengths

Environmental Footprints Analysis has several advantages by providing:

1. A single index that is easy to communicate and understand.
2. A means to quickly assess and compare a variety of products, populations, or activities.
3. An easy linkage of local and global impacts.
4. A tool for exploring the relationship between different impacts.

Limits

Limitations associated with Environmental Footprints Analysis are in some cases typical of all modeling efforts.

1. Aggregation can oversimplify impacts: how the aggregation occurs is typically not included as part of the tool application.
2. The assumptions and proxies used to derive the footprint result are not always apparent.
3. Calculations can be affected by data availability and boundary issues.

Environmental Footprint Analysis is a purely environmental indicator and does not directly address social or economic issues necessary to comprehensively measure impacts. Some social and economic information may be inferred, however, as the environmental footprint is dependent on current aggregate social conditions that create demand in specific regions and cultures.

REFERENCES

Aka, Ebenezer (2006). "Gender Equity and Sustainable Socio-Economic Growth and Development," *The International Journal of Environmental, Cultural, Economic & Social Sustainability*, Volume 1, Number 5, pp. 53–71.

Andersson, Jan Otto (2006). "International Trade in a Full and Unequal World." Presented at the workshop, "Trade and Environmental Justice," Lund, 15th – 16th of February.

Bergesen, Albert and Michelle Bata (2002). "Global and National Inequality: Are They Connected?" *Journal of World-System Research*, 8: 130–44.

Brown, S., Jayant, S., Cannell, M., and Kauppi, P. (1996). "Mitigation of Carbon Emissions to the Atmosphere by Forest Management," *Commonwealth Forestry Review*, 75, 79–91.

Burns, Thomas J., Byron L. Davis, Andrew K. Jorgenson, and Edward L. Kick (2001). "Assessing the Short- and Long- Term Impacts of Environmental Degradation on Social and Economic Outcomes." Presented at the Annual Meetings of the American Sociological Association, August, Anaheim, CA.

Burns, Thomas J., Jeffery Kentor, and Andrew K. Jorgenson (2003). Trade Dependence, Pollution, and Infant Mortality in Less Developed Countries." Pp. 14–28 in *Crises and Resistance in the 21st Century World-System*, edited by Wilma A. Dunaway (Westport, CT: Pager).

Catton, W. (1986). "Carrying Capacity and the Limits to Freedom." Paper prepared for Social Ecology Session 1, XI, World Congress of Sociology, New Delhi, India (August 18).

Chen, D.T. (2000). "The Science of Smart Growth." *Scientific American*, 283 (6): 84–91.

Chew, Sing C. (2001). *World Ecological Degradation: Accumulation, Urbanization, and Deforestation 3000B.C-A.D 2000* (Walnut Creek, CA: Alta Mira).

Clapp, Jennifer (2002). "The Distancing of Waste: Over-consumption in a Global Economy," in *Confronting Consumption*, edited by T. Princen, M. Maniates, and K. Conca (Cambridge, MA: MIT Press) pp. 155–76.

Czech, B. (2000a). "Economic Growth as the Limiting Factor for Wildlife Conservation," *Wildlife Society Bulletin*, 28 (1): 4–15.

Czech, B.; P .R. Krausman; and P. K. Devers (2000). "Economic Associations Among Causes of Species Endangerment in the United States." *BioScience*, 50: 593–601.

Daly, H.E. (1986). *Beyond Growth: The Economics of Sustainable Development* (Boston, Massachusetts, USA: Beacon Press).

Dietz, Thomas and Eugene Rosa (1994). "Rethinking the Environmental Impacts of Population, Affluence, and Technology," *Human Ecology Review*, 1: 277–200.

Dietz, Thomas, Eugene Rosa, Richard York (2007). "Driving the Human Ecological Footprint." *Frontiers in Ecology and the Environment* (February).

D'Onofrio, Antonia (2001/2002). "Partial Correlation." Ed 710 Educational Statistics, Spring 2003. Accessed Online on 6/3/2008 at http://www2.widener.edu/.

Earth Observatory News (February 2007). "Human's Ecological Footprint in 2015 and Amazonia Revealed." http://earthobservatory.nasa.gov/Newsroom/MediaAlerts/2007. Accessed Online on 2/23/2007.

Education: Crisis Reinforces Importance of a Good Education, says OECD. Accessed Online on 9/23/2011, at http://www.oecd.org/document/21/0,3746.

Ehrlich and Holdren (1971). "Impacts of Population Growth," *Science*, 171, 1212–7.

Girardet, Herbert (1996). "Giant Footprints," Accessed Online on 12/14/07, at http://www.gdrc.org/uem/footprints/girardet.html.

Global Footprint Network: Africa's Ecological Footprint—2006 Factbook. Global Footprint Network, 1050 Warfield Avenue, Oakland, CA 94610, USA. http://www.footprintnetwork.org/Africa.

Global Footprint Network (2008). *The Ecological Footprint Atlas 2008*, October 28.

Goudie, A. (2000). *The Human Impact on the Natural Environment. Fifth Edition* (Cambridge, Massachusetts, USA: Harvard University Press).

Hardin, G. (1991). "Paramount Positions in Ecological Economics," in R. Costanza (ed.). *Ecological Economics: The Science and Management of Sustainability* (New York: Columbia University Press) pp. 47–57.

Holdren and Ehrlich (1974). "Human Population and The Global Environment," *American Scientist*, 62: 282–92.

IPCC (2001). Intergovernmental Panel on Climatic Change.

Jorgenson, Andrew K. (2003). "Consumption and Environmental Degradation: A Cross-National Analysis of Ecological Footprint." *Social Problems*. Volume 50, No. 3, pp. 374–394.

Jorgenson, Andrew K. and Edward L. Kick (eds.) (2003). Special Issue: Globalization and the Environment. *Journal of World-System Research*, Volume IX, Number 2, (Summer).

Krizek, K.J., Power, J. (1996). *A Planner's Guide to Sustainable Development*. Planning Advisory Service Report Number 467 (Chicago, IL: American Planning Association), 66p.

Lorentz, A., and Shaw, K. (2000). "Are You Ready to Bet on Smart Growth?" *Planning*, 66 (1): 4–9.

Lowry, Richard (1999–2008). "Subchapter 3a. Partial Correlation." Accessed Online on 6/3/08 at http://faculty.vassar.edu/lowry/cha3a.html.

Luke, Brian T. "Pearson's Correlation Coefficient." Learning From The Web.net. Accessed Online on 5/30/2008.

Palmer, A.R. (1998). "Evaluating Ecological Footprints," *Electronic Green Journal*. Special Issue 9 (December).

Pimentel, David (1996). "Impact of Population Growth on Food Supplies and Environment." American Association for the Advancement of Science (AAAS) (February, 9). See also GIGA DEATH, Accessed Online on 12/18/07, at http://dieoff.org/page13htm.

Princen, Thomas (2002). "Consumption and Its Externalities: Where Economy Meets Ecology," in *Confronting Consumption*, edited by T. Princen, M. Maniates, and K. Conca (Cambridge, MA: MIT Press) pp. 23–42.

Redefining Progress (2005). *Footprints of Nations*. 1904 Franklin Street, Oakland, California 94612. See http://www.RedefiningProgress.org.

Rees, W. (1990). Sustainable Development and the Biosphere. Teilhard Studies Number 23. American Teilhard Association for the Study of Man, or: "The Ecology of Sustainable Development." *The Ecologist*, 20 (1), 18–23.

Rees, William E. (1992). "Ecological Footprint and Appropriated Carrying Capacity: What Urban Economics Leaves Out," Environment and Urbanization, Vol. 4, No. 2, (October). Accessed Online on 12/14/07, at http://eau.sagepub.com.

Rees, William E. (2012). "Ecological Footprint, Concept of," in Simon Levin (ed.). *Encyclopedia of Biodiversity (2nd Edition)*.

Scott, Lauren and Monica Pratt (2009). "An Introduction To Using Regression Analysis With Spatial Data." *ArcUser. The Magazine for ESRI Software Users* (Spring), pp. 40–43. Lauren Scott and Monica Pratt are ESRI Geo-processing Spatial Statistics Product Engineer and ArcUser Editor respectively.

Sklair, Leslie (2001). *The Transnational Capitalist Class* (Oxford, UK: Blackwell Press).

Solow, R. M. (1974). "The Economics of Resources or the Resources of Economics," *American Economics Review*, Vol. 64, pp. 1–14.

Sundberg, Rolf (2002). *Encyclopedia of Environmetrics*, edited by Abdel H. El-Shaarawi and Walter W. Piegorsch (Chichester: John Wiley & Sons, Ltd), Volume 1, pp. 365–366.

Suplee, D. (1998). "Unlocking the Climate Puzzle." *National Geographic*, 193 (5), 38–70.

The international Society for Ecological Economics and Island Press (1994). "Investing in Natural Capital: The Ecological Approach to Sustainability." Accessed Online on 12/18/07, at http://www.dieoff.org/page13.htm.

The Wildlife Society (TWS) (2003). *The Relationship of Economic Growth to Wildlife Conservation*, Technical Review 03-1 (Bethesda, MD: Wildlife Society).

United Nations Development Program (2004). *Human Development Report 2004 (HDR). Cultural Liberty in Today's Diverse World* (New York, N.Y: UNDP).

United Nations Development Program (2007/2008). *Human Development Report 2007/2008. Fighting Climate Change: Human Solidarity in a Divided World.*

United Nations Development Program (2010). *Human Development Report 2010. The Real Wealth of Nations: Pathway to Human Development*, p. 187.

Vitousek, P., P. Ehrlich, A. Ehrlich and P. Matson (1986). "Human Appropriation of the Products of Photosynthesis," *Bioscience*, Vol. 36, pp. 368–374.

Wackernagel, M. (1991). "Using 'Appropriated Carrying Capacity' as an Indicator: Measuring the Sustainability of a Community." Report for the UBC Task Force on Healthy and Sustainable Communities. UBC School of Community and Regional Planning, Vancouver, Canada.

Wackernagel, Mathis and William Rees (1996). *Our Ecological Footprint: Reducing Human Impact on the Earth* (Gabriola Island, B.C., Canada: New Society Publishers).

Wackernagel, Mathis, Alejandro C. Linares, Diana Deumling, Maria A.V. Sanchez, Ina S.L. Falfan, and Jonathan Loh (2000). *Ecological Footprints and Ecological Capacities of 152 Nations: The 1996 Update* (San Francisco, CA: Redefining Progress).

Wackernagel, Mathis and Judith Silverstein (2000). "Big Things First: Focusing on the Scale Imperative with the Ecological Footprint." *Ecological Economics*, 32: 391–4.

World Bank (2005). *World Development Report 2005. A Better Investment Climate for Everyone* (New York, N.Y: A Co-publication of the World Bank and Oxford University Press).

World Commission on Environment and Development (WCED) (1987). *Our Common Future* (Oxford: Oxford University Press).

World Resources Institute (WRI) (1992). *World Resources, 1992–1993* (New York: Oxford University Press).

World Resources Institute (2000). *World Resources, 2000–2001: People and Ecosystems; The Fraying Web of Life* (Washington, D.C., USA: World Resources Institute).

York, Richard, Eugene A. Rosa, and Thomas Dietz (2003). "Footprints on the Earth: The Environmental Consequences of Modernity." *American Sociological Review*, 68: 279–300.

Zovanyi, G. (2005). "Urban Growth Management and Ecological Sustainability Confronting the 'Smart Growth Fallacy,'" in *Policies for Managing Urban Growth and Landscape Change. A Key to Conservation in the 21st Century*. Published by North Central Research Station, Forest Service U.S. Department of Agriculture, St. Paul, MN 55108.

Chapter Two

Ecological Footprint, Environmental Sustainability, and Biodiversity Conservation

A Cross-Cultural Analysis

INTRODUCTION

The modern world economy has experienced rapid expansion in recent years, made possible by the rapid technological growth. Human societies and their places are the products of environmental, economic, political and cultural processes which may work across scales from the local to the global. The history of civilizations is also the history of ecological degradation and crisis (Chew, 2001). The global modes of production and accumulation are intimately linked to environmental degradation (Jorgenson and Kick, 2003). The human economy is related to nature and all economic activities are dependent on the ecosystem. Market expansion through recent globalization is now threatening the human race with environmental disasters (Jorgenson and Kick, 2003), as well as creating the apparent conflict among three essential aims: prosperity, equity, and ecological sustainability. The latter, perhaps, is what Andersson (2006) called "the global ethical trilemma," which are production and consumption and their consequent environmental degradation. In other words, the pressing issues of our time are the population growth and increasing economic growth as they interact with resource consumption rates, pushing us toward global ecological collapse.

It is a known fact that the present anthropocentric economic development efforts around the world have utilized nature beyond its capacity to renew and regenerate indefinitely. Some countries, especially the more developed, consume natural resources at levels higher than the global ecological system can continue to support. Likewise and paradoxically, according to Jorgenson (2005), countries with lower levels of consumption (i.e., less developed) also experience high rates of domestic environmental degradation, due to deforestation, organic water pollution, and growing carbon dioxide emissions (see also Clapp, 2002; Princen, 2002; Jorgenson and Burns, 2004). This is a possibility because, many social scientists have argued that material goods consumed in the more developed countries have disastrous effects on the environment in other regions of the world (Burns et al, 2001; Hornborg, 2001; Clapp, 2002; Conca, 2002; Tucker, 2002).

In a human development hierarchy of the modern world economy or world system, different countries have different consumption patterns, different ecological footprints (EF), and different ecological balances. A recent study showed that human population size and affluence are the main drivers of human-caused environmental impacts or stressors (Earth Observatory News, February 6, 2007). Increased affluence exacerbates the environmental impacts and, when combined with population growth, substantially increases the human footprints on the planet (Dietz et al, 2007). Ecological footprint (EF) is a measure of the amount of nature it takes to sustain a given population over a course of one year. It is a national-level measurement that quantifies how much land and water are required to produce the commodities consumed and assimilate the wastes generated by them. It actually measures how consumption may affect the environment, thus the concept of ecological footprint analysis (Wackernagel and Rees, 1996; Wackernagel and Silverstein, 2000; Wackernagel et al, 2000; Jorgenson, 2003; Redefining Progress, 2005; Earth Observatory News, 2007). Therefore, ecological footprint analysis (EFA) compares the humanity's ecological footprint with the earth's available biological capacity. It attempts to operationalize the concept of carrying capacity and sustainability. Carrying capacity is the maximum number of individuals of a defined species that a given environment can support over the long term (Hardin, 1991), or the environment's maximum persistently supportable load (Catton, 1986). Sustainability refers to the ability of a system to continue and maintain a production level or quality of life for future generations. Thus, sustainability means living in material comfort and peacefully with each other within the means of nature (Wackernagel and Rees, 1996, p. 32). According to Rees (1990), despite our technological sophistication, humankind remains in a state of "obligate dependence" on the productivity and life support services of the ecosphere.

Population and consumption are major players, and in fact, the shrinking carrying capacity may soon become the most important single issue confronting humanity. At present, both the human population and average consumption are increasing while the total area of productive land and stocks of natural capital are fixed or in decline. Many studies have dealt with the environmental and ecological degradations of the present anthropocentric economic development efforts, especially their effects on environmental sustainability and biodiversity conservation (Burns et al, 2001; Chew, 2001; Conca, 2002; Clapp, 2002; Princen, 2002; Tucker, 2002; Andersson, 2006). However, very few of the previous studies dealt with the ecological footprint analysis which has had explicit anthropocentric stance that had failed to capture the world's biological diversity crisis (Wackernagel and Rees, 1996; Jorgenson, 2003). Likewise, a few essays and studies of anthropogenic-caused environmental destruction exist that analyzed the structural causes of ecological footprints (Jorgenson, 2003).

PURPOSE OF STUDY

This study draws from the human development hierarchy theoretical perspective to analyze the structural causes of national-level per capita ecological footprints in the world economy. It analyzes the ecological footprints of different nations, regions, and cultures for ecological balance. It examines the cross-cultural or national variations in total footprints and per capita footprints associated with ecostructural factors (or socioeconomic processes) and human development hierarchy of the world system. The paper also strives to explain various forms of anthropogenic-caused environmental and ecological degradation and depletion using consumption or per capita ecological footprints. The paper strives to show that affluence and increased population size and growth level increase the consumption of natural capital, thus, higher per capita footprints. It exposes the inequitable distribution of the world's ecological footprints and heightens the concern about ecological imbalances and overshoots. After the examinations and analyses of data, the paper proffers some policies and solutions to reduce the per capita footprint of nations and at the same time protect biodiversity.

In order to achieve the above objectives, the study will analyze and articulate the effects of population, affluence (GDP per capita), and ecostructural factors such as urbanization, literacy level, and domestic income inequality (Gini index) on the ecological footprints associated with national human development hierarchy of the world economy. This supports the work of Jorgenson (2003) who drew from world-systems and ecostructuralist theoretical perspectives to analyze the structural causes of per capita footprints. He examined the cross-national variation of per capita footprints within and between "zones" of the world-economy, namely, the core (high consumption and politically powerful countries), semi-periphery, high periphery, and low periphery (low consumption and politically weak countries). In this study, the author examined the cross-cultural variations of total footprints and per capita footprints within and between different human development zones or hierarchies of the world economy. According to the United Nations Development Program (UNDP, e.g., *Human Development Report 2004*), these zones are: High human development, Medium human development, and Low human development. These zones almost correspond with Jorgenson's zones respectively, especially in terms of level of development and national representatives.

The purpose of the ecological footprint analysis is to illustrate the possibility of overshoot (Wackernagel and Silverstein, 2000) a condition in which a given population outgrows or out-consumes its ecological carrying capacity (Harper, 2001, p. 20). Ecological footprint analysis has many advantages as a measure of consumption (Wackernagel and Rees, 1996; Wackernagel et al, 2000; Jorgenson, 2003; Redefining Progress, 2005): (a) Its use does not require researchers to know what specific region of the world the resources came from; (b) It provides a common unit of measurement that allows for comparisons of various types of impacts (Wackernagel et al, 2000); (c) The ecological footprint is a helpful concept useful for teaching, which shows how humans are over-exploiting nature and depleting on a continuous basis the nature's capital (Wackernagel and Rees, 1996); (d) Ecological Footprint Accounts allow national governments and their agencies to evaluate risks and formulate better policy, thus, it is a tool for national security; (e) The calculations of ecological footprints will impress the world community and help politicians, businesses, engineers, and the public-at-large to find new and exciting paths towards sustainable development (Redefining Progress, 2005); and (f) Ecological Footprint Account transforms "sustainability" from a vague concept into a measurable goal. It is a reliable, comprehensive method to evaluate progress toward sustainability. It is a 21st century tool measuring the drive toward sustainability in the 21st century (Redefining Progress, 2005).

THEORETICAL FRAMEWORK FOR THE STUDY

National ecological footprints are the manifestations of the interactions between socio-cultural and ecological systems over time. Relationships exist between social, political, economic inequalities and environmental problems among different nations and globally. To understand the theory behind footprints inequalities among nations and regions, perspectives are incorporated from diverse fields such as world systems analysis, environmental history, environmental sociology, environmental anthropology, and ecological economics (Hornborg, 2005).

We live in an unequal world system where the consumption levels of the higher human development countries are substantially higher than those of the lower human development. The high human development countries are politically and economically powerful, highly urbanized, with higher levels of per capita income and literacy, as well as lower levels of domestic income inequality. They consume biospheric resources at levels above the global bio-capacity per capita, which could lead to potentially catastrophic outcomes for all societies (Jorgenson, 2003). According to Hedenius and Azar (2005), if the quintile of the world population that lives in the richest countries is compared to that of the quintile living in the poorest countries, the following ratios become apparent income in terms of market exchange rates (MER) is more than 70 times higher; income in terms of purchasing power parity exchange rates (PPP) is more than 10 times higher; release of carbon dioxide is 22 times higher; consumption of electricity is 35 times higher; and consumption of paper is 89 times higher. Since the 1960s the income disparity in terms of MER has grown like wildfire from the ratio of around 25:1 to now around 75:1, and in absolute terms the PPP-gap has increased from $9,200 per capita in 1960 to $23,000 in 2000 (see Hedenius and Azar, 2005, pp. 3 and 4).

Affluence and population size (i.e., growth) have been hypothesized to be the primary drivers of human-caused environmental stressors or impacts (E O News, 2007). It is found that, over the last half century or more, resource consumption rates have increased at a faster pace than population size. Thus, the growing rates of consumption of specific resources (e.g., energy, paper, etc.) constitute a major threat to natural resources or capital. In fact, Dietz et al (2007) found that increased affluence exacerbates environmental impacts and, when combined with population growth will substantially increase the human footprint on the planet. From the above analyses, a country's position in a world-system or world economy, its affluence (per capita income), population size and growth rate, and urbanization level, all prove to be significant predictors of per capita ecological footprint (Wallerstein, 1974, 1979; Kick, 2000; Jorgenson, 2003; Andersson, 2006; Hedenius and Azar, 2005; EO News). Nevertheless, the Earth Observatory News (EO News) (February 6, 2007) shows human population size and affluence as the main drivers of human-caused environmental stressors, while urbanization, economic structure and age of population have little effect.

The human development hierarchy of the world capitalist economy is asymmetric, exploitative, disjunctive and unsustainable. In the theory or model of unequal exchange, the high human development countries dominate and exploit the low and weak human development countries through trade and other control mechanisms (e.g., military and political controls). Often the gains from trade are unequally or asymmetrically distributed. The exchange between the development hierarchies is often non-equivalent in terms of market prices, labor and biomass inputs, as well as ecological footprints. Thus, what one country gains, the other loses disjunctively, which aggravates the income-and development-gaps (Andersson, 2006). According to the ecological and environmental economists, the unequal exchange is unsustainable especially, in terms of ecological footprints, rucksack (displaced environmental impacts) and eMergy (energy memory) (Hornborg, 2001). Energy memory (eMergy) is used to estimate the amount of energy that has been invested in a product and can be applied to approximate the extent to which a population or territory (e.g., a nation) is a net importer or exporter of invested energy (Odum, 1988; Odum and Arding, 1991). This entails social exploitation and increasing unsustainability at the periphery or low human development countries as the core and advanced countries or high human development countries run massive unaccounted ecological deficits with the rest of the world (Rees, 1996). Disjunctively the high human development countries import "eXergy" (e.g., biomass) from low human development countries as a prerequisite for the development of the industrial "technomass," ultimately resulting in ecological and socioeconomic impoverishment of the later (Hornborg, 2001; Andersson, 2006). The unequal exchange of energy or eXergy is essential and central in analyzing the underdevelopment of periphery, low human development areas (Bunker, 1985). Perhaps, this is what Foster and Clark (no date) call ecological imperialism, which is the curse of capitalism. Ecological imperialism is the growth of the center of the system (core) at unsustainable rate, through the thoroughgoing ecological degradation of the periphery, which is currently generating a planetary-scale set of ecological contradictions, imperiling the entire biosphere (Foster and Clark, no date; Crosby, 1986; Foster, 2000). According to Foster and Clark, ecological imperialism presents itself most obviously in the following ways: the pillage of the resources of some countries by others and the transformation of the whole ecosystems upon which states and nations depend; massive movements of population and labor that are interconnected with the extraction and transfer of resources; the exploitation of ecological vulnerabilities of societies to promote imperialist control; the dumping of ecological wastes in ways that widen the chasm between center and periphery; and over all, the creation of a global "metabolic rift" that characterizes the relation of capitalism to the environment, and at the same time limits capitalist development.

Studies have linked a country's per capita consumption level to its position in the human development hierarchy of the world system. For example, Wallerstein (1974; 1979), Chase-Dunn (1998; 2002), Kentor (2000), Kirk (2000), and Jorgenson (2003) stressed the importance of the world capitalist economy for socioeconomic processes in all the nations of the "core," "periphery," and "semi-periphery." In this study, it is equally true for the "high human development," "medium human development," and "low human development" (*Human Development Report 2004*; for classification of countries, see p. 279). Socioeconomic processes include domestic inequality, urbanization, and literacy rates, which according to Jorgenson (2003) are variables that partially mediate the effect of world-system position (in this study, human development hierarchy), which affect per capita ecological footprints. Significance in the capitalist world-system in terms of success and position is based on a combination of relative military power, economic power, and global dependence (Chase-Dunn 1998, 2002; Kentor 2000; Jorgenson, 2003). In this paper, Kentor's national-level index scores are used, which measure relative position in the modern world-system (Kentor, 2000). His multidimensional measure of world-system position combines level of capital intensiveness, production size, trade size, global capital control, military expenditures, military exports, global military control, export commodity concentration, foreign capital dependence, and military dependence. To increase his sample size, he created a new measure that included total Gross Domestic Product (GDP), GDP per capita, and military expenditures, which correlate with the original index at 0.98 (see Jorgenson, 2003, p. 382). Among the geopolitical units there exist power-dependent linkages wherein the core nations dominate the semi-periphery and periphery nations in global production and consumption by virtue of their domestic and international economic strength. The core nations include the major powers of Western Europe, North America, and Japan. The periphery and to a lesser extent, the semi-periphery are once colonized countries that suffer from domestic economic weaknesses, even in recent globalization, related to their relative positions of dependency in the world-system (Lindeke, 1995; Kirk, 2000). The core nations are associated with economic diversity, wealth, public well-being, robust personal health, clean environmental circumstances (Burns et al, 1997), and continuing educational and human capital expansion (Kirk, 2000). On the other hand, the periphery nations are associated with commodity exports of primary products, abject poverty accompanied by debilitating population growth, growing environmental degradation (Burns et al, 1997; Ward, 1993), technological and education disadvantage, and greater percentage of labor force engaged in agrarian activities. The periphery suffers from economic stagnation and backwardness as well as lack of mobility in the world-system (Wallerstein, 1974; Kentor, 1998; Kirk, 2000; Jorgenson, 2003). According to Kirk (2000), semi-peripheral nations (e.g., much of the Middle East, Latin America, and Asia) occupy an intermediate position in the global system and have characteristics of both the core and the periphery (see also Terlouw, 1993; Arrighi and Grangel, 1986). The middle position of the periphery is manifested by its economic domination over the periphery, especially in terms of the exchange of finished goods for raw material; and its economic dependence upon the core, especially in investment. Peripheral countries are rapidly industrializing and experiencing transitions in national institutions and human capital outcomes, moderately well-educated and technically advantage, and increasing alarming environmental outcomes (Burns et al, 1997; Wallerstein, 1979; Kirk, 2000).

The human development index in this study is a summary measure of human development in an anthropocentric world economy. It is an improvement on core-periphery hierarchy of the world-system positions as it is based on national and regional socio-economic well-being of the world human economy rather than solely on economilitary growth. The economic and social forms of development are coupled in the human development index, which is a good proxy for development dynamics. Besides, there is parity among the human development and core-periphery hierarchies of the world economy as they mirror each other in their world system positions, as well as in their structurally-caused national ecological footprints. Thus, the argument here is that the national per capita footprints measure the national consumption levels that reflect the national human development positions in the world system.

According to the *Human Development Report (2004)*, human development index (HDI) is "a composite index measuring average achievement in three basic dimensions of human development, namely, long and healthy life, as measured by life expectancy at birth; knowledge, as measured by the adult literacy rate; and a decent standard of living, as measured by Gross Domestic Product (GDP) per capita" (for details on how the index is calculated, see Technical note 1, pp. 258–259). Countries in the human development aggregates, excluding the United Nations member countries for which the HDI cannot be computed, are classified as high human development, medium human development, and low human development in the world economy or world system (HDR, 2004, p. 279). In the human development hierarchy of the world economy, the high human development countries are countries with HDI of 0.800 and above, found mainly in North America, Western Europe and Australia (55 countries or areas). According to the World Bank (2005), these are mostly high income countries of the Organization for Economic Co-operation and Development (OECD) with GNP per capita of $9,386 or more. The medium development countries are countries with HDI of 0.500–0.799, found in Eastern Europe, South-East Asia, South America and North Africa (86 countries or areas). These are the countries Kick (2000) referred to as "semi-core" and "semi-periphery" (Arrighi and Drangel, 1989; Wallerstein, 1979) with the obvious theoretical and technical problems of treating them as monolithic bloc in the world system (Nemeth and Smith, 1985; Smith and White, 1992; Smith, 1993; Smith and Timberlake, 1997). They are the lower middle income (LMC) and upper middle income (UMC) countries with per capita GNP of between $766 and $9,385 (World Bank, 2005, p. 255). The low human development countries are countries with HDI below 0.500, found mainly in Sub-Saharan Africa except the country of South Africa, Gabon, Ghana, which are included in the Medium human development (36 countries or areas). They are the low income countries (LIC) with per capita GNP of $765 or less (World Bank, 2005, p. 255).

METHODOLOGY FOR THE STUDY

Sample

To test national-regional ecological footprints, environmental sustainability, and biodiversity conservation with respect to human development positions in the world capitalist economy or world system, the author analyzed a sample of 58 high, medium, and low development countries listed in tables 1–6, representing a worldwide population. Each table represents the first 20 countries selected from each human development category, except where the country's population is less than a million people (HDR, 2004, pp. 152–155). The larger population is chosen for the study because; population is a variable which affects the consumption rates and levels, therefore the ecological footprints. This is in tune with Dietz et al (2007) that restricted data to countries of at least one million people when calculating basic forms of consumption, including crops, meat, energy, and living space. Missing data for the formerly socialist but newly independent countries of the Eastern bloc, as well as some countries at war prevented analysis for them, especially regarding world system positions, domestic inequality (Gini index), and ecological footprints per capita (personal planetoid). With these exceptions, the author detected no other problems in the generalizability of the results due to missing cases (see Kick, 2000).

Mode of Analysis

This is an explanatory study using comparative model and descriptive statistics such as totals, ratios and percentages. Potential technical problems are diagnosed such as the missing data mentioned above, which did not alter the substantive conclusions, especially regarding the ecological footprint accounts and environmental sustainabilities of different countries. The national ecological footprint, as the dependent variable, is explained recursively by the country's world system position, human development hierarchical category, population size, GDP per capita, domestic inequality, urbanization, and literacy rate as the independent variables. The independent variables helped in explaining the varying levels of consumption among different human development hierarchies of the capitalist world economy or world system. The carbon dioxide emission levels (*World Development Report, 2005*, table 1, pp. 256–257) were included to depict the consumption and environmental degradation impacts of different countries, which reflect their national footprint levels.

Variables

Dependent and independent variables were selected on the basis of the theoretical themes and underpinnings described above. The dependent variable, ecological footprint, is used in most studies like this such as Wackernagel and Rees, 1996; Wackernagel and Silverstein, 2000; Wackernagel et al, 2000; Jorgenson, 2003; and Redefining Progress, 2005. The measures of the dependent and independent variables employed in the present study are as follow:

Dependent Variable

The *ecological footprints* or consumption level is measured by the ecological footprint per capita or personal planetoid, taken from Wackernagel et al, 2000, and Redefining Progress (2005).

Independent Variables

Population

This is the total population in 2002 which refers to medium-variant projections, taken from *World Development Report 2004*, table 5, pp. 152–155. It is suggested that population growth and increasing economic growth which interact with resource consumption rates would likely push the world toward ecological collapse.

Gross Domestic Product (GDP) Per Capita (PPP US $)

The PPP (Purchasing power parity) is a rate of exchange that accounts for price differences across countries, allowing international comparisons of real output and incomes. At the PPP US$ rate, $1 has the same purchasing power in the domestic economy as $1 has in the United States (for definitions see, WDR, 2004, p. 274). The *Gross domestic product (GDP) per capita* for this study is taken from *World Development Report 2004*, table 1, pp. 139–142. It is suggested elsewhere that GDP per capita is an inadequate measure of world-system position but a more appropriate indicator of domestic affluence or internal economic development (Burns, Kentor, and Jorgenson, 2003; Jorgenson, 2003; Dietz and Rosa, 1994). GDP per capita also does not capture the political-military attributes of a country's position in the core-periphery hierarchy of the modern world system (Chase-Dunn, 1998, 2002; Kentor, 2000; Jorgenson, 2003). Increased affluence exacerbates the environmental impacts and, when combined with population growth, substantially increases the human footprints on the planet (Dietz et al, 2007).

Country's World System Position/Human Development Category (HDI)

The world-system position has been found to have direct and significant effect on per capita ecological footprints, and tend to support other arguments regarding core/periphery differences in consumption (Bunker, 1985; Burns et al, 2001; Hornborg, 2001; Jorgenson, 2003). The human development category in the world-economy is likely to have significant effect on the per capita ecological footprints as well as consumption levels since it mirrors core-periphery hierarchy of the world economy or world system. Data for both world-system positions and human development categories are taken from Kentor (2000) and *Human Development Report*, 2004 respectively.

Domestic Income Inequality

The Gini Index measures domestic income inequality of different countries, which had remained stable over a time with its impacts on other variables in this study (Bergesen and Bata, 2002; Jorgenson, 2003). Gini Index measures the extent to which the distribution of income (or consumption) among individuals or households within a country deviates from a perfectly equal distribution (*Human Development Report*, 2004, p. 271). It measures inequality over the entire distribution of income or consumption. A value of zero (0) represents perfect equality, and a value of hundred (100) represents perfect inequality. Data for domestic income inequality measured by Gini Index are taken from World Bank, *World Development Report* (2005, table 2, pp. 258–259) and *Human Development Report* (2004, 14, pp. 188–191).

Urbanization Level

This represents the mid-year percent of a nation's total population residing in urban areas. Urbanization measures the effective consumption of living space, which when combined with increasing affluence and population, exacerbates environmental impacts that substantially increase the human footprint on the planet. The data are taken from *Human Development Report* (2004, table 5, pp. 153–155).

Literacy Rate

This variable refers to the percent of a nation's population over the age of fifteen (15) that can read and write in any language of their choice. Literate population generally has low domestic inequality and tends to consume more resources than their illiterate counterpart due to their higher income and urban living. The data for 2002 is taken from the *Human Development Report* (2004, table 1, pp. 139–142).

RESULTS AND DISCUSSION

The results and findings in this research are discussed under subtitles that include, "Population and Affluence on Total Ecological Footprint of Countries," "World-System Positions and Ecostructural Factors on Per Capita Ecological Footprints," and "Environmental Sustainability and Biodiversity Conservation."

POPULATION AND AFFLUENCE ON TOTAL ECOLOGICAL FOOTPRINTS OF COUNTRIES

Tables 2.1 to 2.3 present the cross-national comparisons for both dependent (total footprint) and some of the independent variables (human development category or position, population, and GDP per capita). The tables show that a coun-

Table 2.1. Population, Affluence, Total Ecological Footprints, and Ecological Balance in High Human Development Countries

High Human Development Country (High Income OECD)	*Population (Millions) 2002*[1]	*GDP Per Capita 2002*[1] *(US$)*	*HDI 2002*[1]	*Total Footprint Per Capita 2001*[2]	*Ecological Balance 2001*[2]	*Carbon-Dioxide Emission Mil. Tons 2003*
Norway	4.5	36,600	0.956	93.13	–44.24	49.9
Sweden	8.9	26,050	0.946	66.76	–40.38	46.9
Australia	19.5	28,260	0.946	79.05	31.16	344.8
Canada	31.3	29,480	0.943	83.03	2.92	435.9
Netherlands	16.1	29,100	0.942	69.09	–61.13	138.9
Belgium	10.3	27,570	0.942	68.87	–61.69	102.2
United States	291.0	35,750	0.939	108.95	–88.58	5601.5
Japan	127.5	26,940	0.938	53.21	–44.44	1184.5
Ireland	3.9	36,360	0.936	65.42	–38.40	42.2
Switzerland	7.2	30,010	0.936	67.40	–59.03	39.1
United Kingdom	59.1	26,150	0.936	62.56	–52.11	567.8
Finland	5.2	26,190	0.935	44.48	–12.33	53.4
Austria	8.1	29,220	0.934	51.13	–41.19	60.8
France	59.8	26,920	0.932	65.82	–54.54	362.4
Denmark	5.4	30,940	0.932	61.84	–45.56	44.6
New Zealand	3.8	21,740	0.926	48.54	36.18	32.1
Germany	82.4	27,100	0.925	52.21	–43.77	785.5
Spain	41.0	21,460	0.922	50.68	–40.24	282.9
Italy	57.5	26,430	0.920	41.51	–33.46	428.2

Iceland and Luxembourg not included due to their populations are less than one million

1. *Human Development Report 2004*
2. Wakernagel et al (2000). *Ecological Footprints and Ecological Capacities of 152 Nations: The 1996 Update* (San Francisco, CA: Redefining Progress). See also Redefining Progress (2005). *Footprint of Nations.* Accessed Online on 4/3/2006, at http://www.rprogress.org/education. Note that ecological footprint is measured in global hectares. The enormous share of ecological overshoot attributable to carbon dioxide emissions is made explicit in total footprint per capita (Redefining Progress, 2005).
3. *World Development Report (WDR) 2005.* Table 1. Key Indicators of Development, pp. 256–257.

Table 2.2. Population, Affluence, Total Ecological Footprints, and Ecological Balance in Medium Human Development Countries

Medium Human Development Country	*Population (Millions) 2002*[1]	*GDP Per Capita 2002*[1] *(US$)*	*HDI 2002*[1]	*Total Footprint Per Capita 2001*[2]	*Ecological Balance 2001*[2]	*Carbon Dioxide Emission (Mill. Tons) 2000*[3]
Bulgaria	8.0	7,130	0.796	33.65	–22.88	42.3
Russian Federation	144.1	8,230	0.795	48.35	–12.42	1435.1
Libya	5.4	7,570	0.794	39.50	31.60	58.86[a]
Malaysia	24.0	9,120	0.793	35.48	–18.38	144.4
Panama	3.1	6,170	0.791	22.32	0.51	6.3
Belarus	9.9	5,520	0.790	35.17	–22.94	59.2
Albania	3.1	4,830	0.781	9.90	–1.61	2.9
Bosnia Herzegovina	4.1	5,970	0.781	21.20	–13.07	19.3
Venezuela	25.2	5,380	0.778	28.80	–9.05	157.7
Romania	22.4	6,560	0.778	28.94	–20.71	86.3
Ukraine	48.9	4,870	0.777	40.46	–30.94	342.8
Brazil	176.3	7,770	0.775	14.11	15.05	307.5
Colombia	43.5	6,390	0.773	11.17	7.47	58.5
Thailand	62.2	7,010	0.768	15.95	–6.27	198.6
Saudi Arabia	23.5	12,650	0.768	68.10	–44.91	374.3
Kazakhstan	15.5	5,870	0.766	35.66	–1.05	121.3
Jamaica	2.6	3,980	0.764	25.90	–14.41	10.8
Philippines	78.6	4,170	0.753	8.55	–0.56	77.5
Peru	26.8	5,010	0.752	7.06	23.05	29.5
Turkey	70.3	6,390	0.751	16.25	–7.17	221.6

a. *Human Development Report 2004*, p. 208

1. *Human development Report 2004*.
2. Wakernagel et al (2000). *Ecological Footprints and Ecological Capacities of 152 Nations: The 1996 Update* (San Francisco, CA: Redefining Progress). See also Redefining Progress (2005). *Footprint of Nations*. Accessed Online on 4/3/2006, at http://www.rprogress.org/education. Note that ecological footprint is measured in global hectares. The enormous share of ecological overshoot attributable to carbon dioxide emissions is made explicit in total footprint per capita (Redefining Progress, 2005).
3. *World Development Report (WDR) 2005*. Table 1. Key Indicators of Development, pp. 256–257.

Table 2.3. Population, Affluence, Total Ecological Footprints, and Ecological Balance in Low Human Development Countries

Low Human Development Countries	*Population (Millions) 2002*[1]	*GDP Per Capita 2002*[1] *(US$)*	*HDI 2002*[1]	*Total Footprint Per Capita 2001*[2]	*Ecological Balance 2002*[2]	*Carbon Dioxide Emission (Mi. Tons) 2000*[3]
Pakistan	149.9	1,940	0.497	4.69	2.39	104.8
Togo	4.8	1,480	0.495	3.83	5.23	1.8
Rep. of Congo	3.6	980	0.494	3.46	52.44	1.8
Uganda	25.0	1,300	0.493	3.55	6.44	1.5
Zimbabwe	12.8	2,400	0.491	9.03	7.32	14.8
Kenya	31.5	1,020	0.488	3.87	8.83	9.4
Nigeria	120.9	860	0.466	5.84	3.20	36.1
Haiti	8.2	1,610	0.463	2.15	5.03	1.4
Senegal	9.9	1,580	0.437	5.98	8.48	4.2
Tanzania	36.3	580	0.407	3.39	12.42	4.3
Cote d'Ivoire	16.4	1,520	0.399	3.84	10.68	10.5
Zambia	10.7	840	0.389	2.89	27.51	1.8
Angola	13.2	2,130	0.381	6.41	38.30	6.4
Chad	8.3	1,020	0.379	4.21	28.75	0.1
D. Rep. Congo	51.2	650	0.365	—	—	2.7
Ethiopia	69.0	780	0.359	1.56	6.97	5.6
Mozambique	18.5	1,050	0.354	1.49	22.50	1.2
Mali	12.6	930	0.326	3.03	22.19	0.6
Burkina Faso	12.6	1,100	0.302	3.05	9.42	1.0

1. *Human development Report 2004.*
2. Wakernagel et al (2000). *Ecological Footprints and Ecological Capacities of 152 Nations: The 1996 Update* (San Francisco, CA: Redefining Progress). See also Redefining Progress (2005). *Footprint of Nations.* Accessed Online on 4/3/2006, at http://www.rprogress.org/education. Note that ecological footprint is measured in global hectares. The enormous share of ecological overshoot attributable to carbon dioxide emissions is made explicit in total footprint per capita (Redefining Progress, 2005).
3. *World Development Report (WDR) 2005.* Table 1. Key Indicators of Development, pp. 256–257.

try's total footprint, ecological balance, and carbon dioxide emission (waste) are a function of its human development position in the world economy. This supports the findings of Jorgenson (2003), which show that world-system position has significant and positive effect ($R^2 = 0.32$) on per capita ecological footprints. From the tables it became apparent that the higher the national population and GDP per capita (affluence), the higher the imbalanced consumption, total footprint, carbon dioxide emission (waste), and also the lower the ecological balance, and vice versa. In fact, York et al (2003) indicated that population and affluence account for 95 percent of the variance. Dietz et al (2007) concluded that population size and affluence are the primary drivers of environmental impacts. They projected 20 nations that will have the largest ecological footprint in 2015, with the United States, China, and India topping the list. They further indicated that the greatest absolute increase will occur with China and India, where both population and economic growth represent 37 percent of the increase in global human footprint (see also E O News, 2007).

According to table 2.1, among the high human development countries (high income OECD), large countries such as U.S., Japan, United Kingdom, France, Germany, and Italy, have large GDP per capita (affluence), high carbon dioxide emissions (waste), high total footprints, and low ecological balance that are incredibly negative. Apart from Australia, Canada, and New Zealand that have positive ecological balance, the rest of the countries under this human development hierarchy have negative balance. Under this cultural group (high human development index), United States of America is the worst offender with population of 291 million people; GDP per capita of $35,750; total footprint of 108.95, ecological balance of –88.58 (Redefining Progress, 2005); and carbon dioxide emission of 5,602 million tons (see *World Development Report*, 2005, pp. 256–257). According to Wackernagel et al (2000), the carbon dioxide portion accounts for approximately 50 percent of the United States' total per capita footprints. Table 2.2 constitutes the Medium human development countries, which exhibit intermediate impacts of the indicators. This cultural group (medium human development index) in the hierarchy of world economy has variable impacts that are not as high as those of the high human development but also not as low as those of the low human development (see table 2.3). This is not a monolithic bloc in terms of structure and indicator impacts but resembles what Kick (2000), Wallerstein (1979), and Arrighi and Drangel (1989) referred to as "semi-core" and "semi-periphery." Although most of the countries under this category have lower total footprints, lower carbon dioxide emissions, and higher ecological balance, a majority of them still have negative ecological balance (except Libya, Brazil, Colombia, Peru, and Panama). The worst faired country in this category is the Russian Federation, followed by Saudi Arabia. Russian Federation has population of 144.1 million people; per capita income of $8,230; total footprint of 48.35; ecological balance of –12.42; and carbon dioxide emission of 1435.1 million tons. Saudi Arabia has a population of 23.5 million people; GDP per capita of $12,650; total footprint of 68.10, ecological balance of –44.91; and carbon dioxide emission of 374.3 million tons. Table 2.3 shows the Low human development countries or what Wallerstein (1979), Wackernagel and Rees (1986), Kick (2000), and Jorgenson (2003) called the periphery in the word capitalist economy or world-system position. Apart from Pakistan and Haiti, the rest are Sub-Saharan African countries that were once colonies of the core or high human development countries of Western Europe. The Low human development countries have relatively GDP per capita, thus low consumption levels and low total footprints, as well as low carbon dioxide emissions (waste). The high carbon dioxide emissions found in Pakistan and Nigeria could be accounted for by their probable high resource consumptions due in part by their large populations. In the world economy and trade, this region possesses more extractive economies and tends to be raw-material oriented, thus the extant positive ecological balance. According to Jorgenson (2003), Bunker (1985), Bornschier and Chase-Dunn (1985), and Boswell and Chase-Dunn (2000), this region has disarticulated markets, and higher levels of dependent industrialization and underdevelopment.

WORLD-SYSTEM POSITIONS, ECOSTRUCTURAL FACTORS, AND PER CAPITA ECOLOGICAL FOOTPRINTS OF COUNTRIES

Human development hierarchies, world-system positions, and ecostructural factors have significant effects on per capita ecological footprints of nations. The above hierarchies, positions, and factors lead to per capita footprints differences and variations among different nations. Tables 2.4, 2.5, and 2.6 show that indicators which include urbanization, domestic inequality, literacy factors constitute the structural causes of ecological footprints per capita. This supports the work of Jorgenson (2003) who argued that the effect of world-wide position on per capita ecological footprints is both direct and indirect, mediated by its effects on domestic income inequality, urbanization, and literacy. Ecological footprints per capita reflect the consumption patterns of different nations in the world economy with the attendant environmental degradations.

Table 2.4 shows that the higher the human development, the higher the position in world-system, as well as the higher rates of urbanization, higher literacy rates, higher ecological footprints per capita, and lower domestic inequality. These promote domestic consumption of biospheric resources. The urban regions, especially in this cultural region, are the most powerful cities in the global city-system, and contain key markets for material goods that require material resources in their production (Chase-Dunn and Jorgenson, 2003; Sassen, 1991;

Table 2.4. World System Position, Ecostructural Factors, and Per Capita Footprints in High Human Development Countries

High Human Development Countries	*World-System Position 2000*[1]	*Urbanization Level 2002*[2]	*Literacy Rates (Percent) 2002*[2]	*Domestic Inequality or Gini Index 2002*[3]	*Ecological Footprint Per Capita 2001*[4]
Norway	1.25	77.6	99.0	25.8	6.13
Sweden	1.54	83.3	99.0	25.0	7.53
Australia	1.53	91.6	99.0	35.2	8.49
Canada	2.5	80.1	99.0	33.1	7.66
Netherlands	1.54	65.4	99.0	32.6	5.75
Belgium	1.41	97.2	99.0	25.0	5.88
United States	16.96	79.8	99.0	40.8	12.22
Japan	5.44	65.3	99.0	24.9	5.94
Ireland	—	59.6	99.0	24.9	—
Switzerland	2.17	67.6	99.0	33.1	6.63
United Kingdom	3.15	89.0	99.0	36.0	6.29
Finland	1.18	61.0	99.0	26.9	8.83
Austria	1.29	65.8	99.0	30.0	5.45
France	3.62	76.1	99.0	32.7	7.27
Denmark	1.44	85.2	99.0	24.7	9.88
New Zealand	0.69	85.8	99.0	36.2	9.54
Germany	3.98	87.9	99.0	28.3	6.31
Spain	1.17	76.4	97.7	32.5	5.50
Italy	2.68	67.3	98.5	36.0	5.51
Greece	0.20	60.6	97.3	35.0	5.58

1. Kentor, Jeffery (2000). *Capital and Coercion: The Economic and Military Processes That Have Shaped the World-Economy, 1800–1990* (New York: Garland Press). Note, ecological footprint is measured in global hectares. The unsustainable use of fisheries, croplands, and the enormous share of ecological overshoot attributable to carbon dioxide emissions are not made explicit (Redefining Progress, 2005).
2. *Human Development Report 2004*, tables 1 and 5, pp. 139–142 and 153–155 respectively.
3. *Human Development Report 2004*, 14, pp. 188–191; and *World Development Report 2005*, table 2, pp. 258–259.
4. Wackernagel et al (2000). *Ecological Footprints and Ecological Capacities of 152 Nations: The 1996 Update* (San Francisco, CA: Redefining Progress). See also Redefining Progress (2005). *Footprint of Nations*. Accessed Online on 4/3/2006, at http://www.rprogress.org/education.

Smith and Timberlake, 2001). Higher levels of urbanization correspond to higher levels of consumption because biospheric resources are consumed at higher levels of urban areas, which generally contain high literate groups. Higher levels of literacy correspond with higher incomes, which allow for greater material consumption (Jorgenson, 2003). This is because literate populations are subject to increased consumerist ideologies and contextual images of good life through advertising (Princen et al, 2002), what Leslie Sklair (2001) and Jennifer Clapp (2002) labeled "cultural ideology of consumerism/consumption." The arch-type of this cultural group is the United States that ranks highest on the world-system positions and highest in the ecological footprints per capita or personal planetoid. Generally, countries in the high human development category have high consumption levels and high ecological footprint per capita, when compared with countries in medium and low human development.

Table 2.5 shows a cultural group that is non-monolithic in terms of factor attributes. The countries in this group are at the intermediate levels of human development, world-system position and ecological footprint per capita. Some countries in this group are potentially upwardly mobile in the world-system, capable of increasing both total and per capita levels of consumption (Arrighi and Drangel, 1986; Boswell and Chase-Dunn, 2000; Burns et al, 2001; Chase-Dunn, 1998; Chase-Dunn and Hall, 1997). Compared to high and low human development countries, the medium human development countries are moderately well-educated, experiencing transitions in national institutions and human capital outcomes (Kick, 2000), as well as technically advantaged which is leading to some alarming environmental outcomes (see table 2.2, carbon-dioxide emissions column). Some countries are moderately urbanized while some and not. Likewise, some countries have low domestic income inequality while some have high inequality, these affect consumption levels which are reflected in their levels of ecological footprint per capita. According to Jorgenson (2003, p. 387), the relatively high level of variation in the semi-periphery supports the theoretical characterization of the multidimensional structural and cultural heterogeneity of semi-peripheral countries (Boswell and Chase-Dunn, 2000; Chase-Dunn, 1998, 2000; Chase-Dunn and Hall, 1997). In this category, Russian Federation and Saudi Arabia are outliers regarding their very high ecological footprints when compared to other countries in this group. The high footprints of the Russian Federation could be accounted for by its large population, military, and economy which enhance its consumption of biospheric resources, while the Saudi Arabia's relatively high per capita footprint could be accounted for by its overall level of oil production (extraction and refinement of fossil fuels) and refineries, coupled with its relatively small population and greater per capita use of fossil fuels (Jorgenson, 2003, p. 387; Redefining Progress, 2005). This is equally true for Kuwait where environmental impact is high (due to extraction and refinement of fossil fuels), especially in terms of overall footprint per capita (Podobnik, 2002; Redefining Progress, 2005; Jorgenson, 2003).

Table 2.5. World-System Position, Ecostructural Factors and Per Capita Footprints in Medium Human Development Countries

Medium Human Development Countries	*World-System Position 2000*[1]	*Urbanization Level 2002*[2]	*Literacy Rates (Percent) 2002*[2]	*Domestic Inequality or Gini Index 2002*[3]	*Ecological Footprint Per Capita 2001*[4]
Bulgaria	—	69.4	98.6	31.9	—
Russian Federation	9.29	73.3	99.6	45.6	5.36
Libya	—	86.0	81.7	—	—
Malaysia	–0.58	63.3	88.7	49.2	3.68
Panama	–1.09	56.8	92.3	56.4	2.35
Belarus	—	70.5	99.7	30.4	—
Albania	—	43.2	98.7	28.2	—
Bosnia Herzegovina	—	43.9	94.6	26.2	—
Venezuela	–0.09	87.4	93.1	49.1	2.88
Romania	–0.82	54.5	97.3	30.3	3.49
Ukraine	—	67.2	99.6	29.0	—
Brazil	0.33	82.4	86.4	59.1	2.6
Colombia	–0.62	76.0	92.1	57.6	1.9
Thailand	–0.61	31.6	92.6	43.2	2.7
Saudi Arabia	0.76	87.2	77.9	—	6.15
Kazakhstan	—	55.8	99.4	31.3	—
Jamaica	—	52.1	87.6	37.9	—
Philippines	–1.07	60.2	92.6	46.1	1.42
Peru	–1.07	73.5	85.0	49.8	1.33
Turkey	–0.52	65.8	86.5	40.0	2.73

1. Kentor, Jeffery (2000). *Capital and Coercion: The Economic and Military Processes That Have Shaped the World-Economy, 1800–1990* (New York: Garland Press). Note, ecological footprint is measured in global hectares. The unsustainable use of fisheries, croplands, and the enormous share of ecological overshoot attributable to carbon dioxide emissions are not made explicit (Redefining Progress, 2005).
2. *Human Development Report 2004*, tables 1 and 5, pp. 139–142 and 153–155 respectively.
3. *Human Development Report 2004*, 14, pp. 188–191; and *World Development Report 2005*, table 2, pp. 258–259.
4. Wackernagel et al (2000). *Ecological Footprints and Ecological Capacities of 152 Nations: The 1996 Update* (San Francisco, CA: Redefining Progress). See also Redefining Progress (2005). *Footprint of Nations*. Accessed Online on 4/3/2006, at http://www.rprogress.org/education.

Table 2.6. World-System Position, Ecostructural Factors and Per Capita Footprints in Low Human Development Countries

Low Human Development Countries	*World-System Position 2000*[1]	*Urbanization Level 2002*[2]	*Literacy Rates (Percent) 2002*[2]	*Domestic Inequality of Gini Index 2002*[3]	*Ecological Footprint Per Capita 2001*[4]
Pakistan	–1.16	33.7	41.5	33.0	1.09
Togo	–1.54	34.45	59.6	—	0.82
Rep. of Congo	–1.56	53.1	82.8	—	1.15
Uganda	–1.49	12.2	68.9	43.0	0.88
Zimbabwe	–1.38	34.5	90.0	56.8	1.45
Kenya	–1.45	38.2	84.3	44.5	1.15
Nigeria	–1.34	45.9	66.8	50.6	1.31
Haiti	–1.49	36.9	51.9	—	0.78
Senegal	–1.44	48.9	39.3	41.3	1.06
Tanzania	–1.55	34.4	77.1	38.2	1.02
Cote d'Ivoire	–1.44	44.4	49.7	45.2	0.95
Zambia	–1.52	35.4	79.9	52.6	1.21
Angola	–1.53	34.9	42.0	—	0.82
Chad	–1.59	24.5	45.8	—	0.75
D. Rep. Congo	–1.23	31.2	62.7	—	0.69
Ethiopia	–1.57	15.4	41.5	30.0	0.85
Mozambique	–1.50	34.5	46.5	39.6	0.76
Mali	–1.57	31.6	19.0	50.5	0.86
Burkina Faso	–1.56	17.4	12.8	48.2	0.90

1. Kentor, Jeffery (2000). *Capital and Coercion: The Economic and Military Processes That Have Shaped the World-Economy, 1800–1990* (New York: Garland Press). Note, ecological footprint is measured in global hectares. The unsustainable use of fisheries, croplands, and the enormous share of ecological overshoot attributable to carbon dioxide emissions are not made explicit (Redefining Progress, 2005).
2. *Human Development Report 2004*, tables 1 and 5, pp. 139–142 and 153–155 respectively.
3. *Human Development Report 2004*, 14, pp. 188–191; and *World Development Report 2005*, table 2, pp. 258–259.
4. Wackernagel et al (2000). *Ecological Footprints and Ecological Capacities of 152 Nations: The 1996 Update* (San Francisco, CA: Redefining Progress). See also Redefining Progress (2005). *Footprint of Nations*. Accessed Online on 4/3/2006, at http://www.rprogress.org/education.

Table 2.6 shows the low human development countries or the peripheral countries with low levels of urbanization, literacy rates, and high levels of domestic inequality, as depicted by the Gini index. Lower levels of literacy and urbanization correspond to lower levels of income, which allow for smaller material consumption and vice versa. Findings here parallel the theorization above, which emphasizes the low consumption levels and the accompanying low ecological footprint per capita (personal planetoid) and high ecological balance based on this cultural group's position in the world system. As mentioned before, apart from Pakistan, these are sub-Saharan African countries that were once colonized, where the processes of underdevelopment, economic stagnation, and dependent industrialization limit their populations to consuming a greater share of unprocessed bioproductive resources. The high domestic income inequality or intra-inequality has a negative effect on per capita consumption levels (Bornschier and Chase-Dunn, 1985; Kick, 2000; Kentor, 2001; and Jorgenson, 2003) that yield both low total and per capita ecological footprints.

ENVIRONMENTAL SUSTAINABILITY AND BIODIVERSITY CONSERVATION

The ecological footprints of nations manifest the degree of their environmental and ecological degradations based on their consumption patterns in the world system and economy. The ecological footprint accounts in effect measure the degree of sustainability prevailing in the nations as they engage in socioeconomic development. Currently, national ecological deficits are becoming an ever-increasing liability to the competitive position of national economies. Humanity's increasing demands on nature world-wide have strained the environmental utilization space or ecospace, which makes meeting the national environmental debt and intergenerational equity virtually impossible. Sustainability refers to the ability of a system to continue and maintain a production level or quality of life for future generations (Redefining Progress, 2005). Thus, it involves ensuring that future generations will have the means to achieve a quality of life equal to or better than today's. Intergenerational equity is the principle of equity or fairness between people alive today and future generations. The implication is that unsustainable production and consumption by today's society will degrade the ecological, social, and economic basis for tomorrow's society. Ecological deficit or overshoot occurs when human consumption and waste production exceed the capacity of the earth to create new resources and absorb the waste. During overshoot, the earth's ability to support future life declines as natural capital is liquidated to support current resource use in the human economy. Natural capital to the environmental goods and services, can be divided into renewable (e.g., new trees) and non-renewable (e.g., fossil fuels). Environmental utilization space or ecospace is the capacity of the environment to support human activities by regenerating renewable resources and absorbing waste, which are determined by the patterns and level of economic activity (http://www.iisd.org/susprod/principles.htm). Environmental debt is the cost of restoring previous environmental damage as well as the cost of recurring restoration measures. Environmental debt continues to rise and the burden is transferred to future generations, unless measures are taken to alleviate the environmental degradation. Some environmental damage and biodiversity destruction such as species extinction is not restorable. What is really needed is eco-efficiency, which is the more efficient use of materials and energy in order to reduce economic costs and environmental impacts.

In anthropocentric economy, humans are reshaping the environment and using up many parts faster than nature can renew itself through urbanization and other practices. According to the Commission for Environmental Cooperation (2001), transformation of the landscape, including habitat loss and alteration, has become the primary threat to biodiversity. Human use of the environment is the largest contributor to habitat modification and ecosystem loss (Goudie, 2000; World Resources Institute, 2000; E O News, 2007). For example, E O News (2007) indicates that the Amazon Basin, one of the world's most important bioregions, harboring rich array of plant and animal species and offering a wealth of goods and services to society, has shown how large scale forest clearings cause decline in biodiversity and the availability of forest products. Ecological science research indicates that deforestation cause also collateral damage to the surrounding forest through enhanced drying of the forest floor, increased frequency of fires, and lowered productivity. According to E O News (2007), the loss of healthy forests can degrade key ecosystem services, such as carbon storage in biomass and soils (carbon sequestration function), the regulation of water balance and river flow, the modulation of regional climate patterns, and the amelioration of infectious diseases (see http://www.esa.org).

The above supports the theorization and assertion that the global environmental stress is caused primarily by the increase in resource consumption (affluence) and human population size and growth. Modernization and developmentalism are based on Western rationalism, which encourages the belief that nature is an infinitely exploitable domain (McMichael, 1993, p. 51), through thought, logic and calculation or what Peet (1999, p. 68) called "scientific-technological rationalism." Thus, in the human development hierarchy of the world economy or world system, the high human development countries of the North or Western Europe, which are more affluent consume more resources than the medium- and low- human development countries of the South or the developing countries. The environmentalists saw the excessive and greedy consumption of natural resources as inevitably linked to economic development. Najam (1996) argues that consumption control among developed countries (the core or high human development countries) is a major key to avoid continuing environmental crises. Nevertheless, some landmark studies have linked global environmental concerns to poverty and malnutrition in developing countries (the periphery or the low and moderate human development countries), as well as to world political structures or world systems (Meadows et al, 1972, 1992; Hoff, 1998, p. 7; Asafu-Adjaye, 2000). They argue that there is a causal link between increase in resource consumption, poverty, rapid population growth, and global environmental degradation. Nevertheless, the medium- and low-human development countries have refused to take responsibility for the earth's damage by resisting the implication that population in the South is responsible for global environmental stress, but insisting that over the last half century or more, resource consumption rates have increased at a faster pace than population size. They argue that it is more efficient to focus on consumption concerns, and that the effect of population on the environment is related to consumption (Harf and Lombardi, 2004, pp. 129–130). They believe that the high human development countries (the North and core countries) must reduce their consumption of resources and reduce wastes or total ecological footprints. For example, the average American consumes natural biospheric resources at a rate 50 times that of an average Indian, and the poorest groups in abject poverty across the world consume 500 times less (Newman and Kenworthy, 1999, p. 3).

From the findings of this research, it is indicative that the high human development countries or the core countries are at the higher end of contributing to the global ecological footprints and environmental stress and degradation. However, whether it is the high development countries (the core) or the medium- and low- development countries (the periphery) are responsible for or the main drivers of human-caused environmental stressors, man-made or anthropocentric forces lead to species decline and damage to biodiversity such as (Myers, 1994; WRI, 1996; Hunter et al, 2003; Harf and Lombardi, 2004; Aka, 2007): (1) habitat loss and fragmentation due to human encroachment and habitat destruction. This tends to accelerate with the increase in a country's population density; (2) biological invasion, especially through the introduction of non-indigenous species; (3) pollution and human induced climate change; (4) over-harvesting and hunting (intense overexploitation); (5) deliberate extermination; and (6) overestimating and undervaluing of available natural resources and biodiversity. Thus, this is, believing in the limitlessness of natural biospheric resources, as well as feigning ignorance of the values of biodiversity.

POLICY RECOMMENDATIONS

The trio of world human development dynamics, world-system dynamics, and population dynamics are important predictors of environmental outcomes, including those related to ecological footprints. The present paper has shown how the capitalist world economy impacts the global ecological system. What is needed is the development and implementation of more informed international policies and practices that will reduce the negative impacts of production and accumulation on the biosphere and human populations living throughout the world, particularly in noncore, peripheral, low human development regions (Jorgenson and Kick, 2003). It is mentioned above that ecological imperialism, which means the worst forms of ecological destruction in terms of pillage of resources, the disruption of sustainable relations to the earth, and the dumping of wastes, fall on the periphery or low human development countries than the core or high human development countries (Foster and Clark, no date; Andersson, 2006).

It has been shown that most so-called "advanced" or high human development countries are running massive unaccounted ecological deficits with the use of the planet (Rees, 1996). Rees (1996) concluded that not all countries can be net importers of carrying capacity, and the material standards of the wealthy cannot be extended sustainably to even the present world population using prevailing technology. For world-wide sustainability, Orton (1999) suggested that industrialized countries need to reduce their impact upon the earth to about one tenth of what it is at the present time. Furthermore, sustainability may well depend on such measures as greater emphasis on equity in international relationships, significant adjustments to prevailing terms of trade, increasing regional self-reliance, and policies to stimulate a massive increase in the material and energy efficiency of economic activities (Rees, 1996; Jorgenson, 2003). Expanding trade and dominant technologies are allowing humanity dangerously to overshoot long term global carrying capacity. Therefore, global terms of trade must be reexamined to ensure that it is equitable, socially constructive, and confined to true ecological surpluses. In effect, prices must reflect ecological externalities, and benefits of growth from trade should flow to those who need them most.

Ecological footprint analysis supports the argument that to be sustainable, economic growth must be much less material and energy intensive than at present as found in the core or high human development countries (Pearce, 1994). According to Wackernagel and Rees (1996, p. 144), one hopeful strategy involves massive improvements in the efficiency of economic activity so that growth in consumption of goods and services is "decoupled" from growth in the use of energy and material. This should permit an increase in consumption to be accompanied by a decrease in resource use. It is suggested that this "dematerialization" of economic goods and services must proceed faster than economic growth to produce the necessary reduction in humanity's total load on the ecosphere. The above notwithstanding, growth is a pressing moral imperative for all human development hierarchies of the world economy or world system.

Thus, growth uses up a lot of natural biospheric resources, increases ecological footprints, and encourages biodiversity loss.

There is need to address the underlying causes of biodiversity loss in all the human development categories. For example, in high human development countries of Europe and North America, the voracious appetite for natural resources should be reduced and managed (Aka, 2007). Another avenue advocated by Goklany and Sprague (1992) is to improve the efficiency and productivity of all activities that use land while ensuring those activities that are environmentally sound. These are based on the fact that global population growth and rise in per capita consumption are inevitable in the future. In the medium- and low-human development countries, there is every need to address the underlying causes of biodiversity loss, which include poverty, lack of property right and tenure regimes that at times make them prone to consfisticating both regulated and unregulated natural biospheric resources (Asafu-Adjaye, 2000; Ukanga and Afoaku, 2005). The above negative issues raised in all the human development categories increase their total and per capita ecological footprints. For what can be done to reduce the size of personal footprint or increase biocapacity include, among others, supporting policies and companies that promote renewable energy, locally produced organic food, energy and resource conservation, and recycling (Redefining Progress, 2005).

CONCLUSIONS

In a human development hierarchy of the world economy or world system, different countries have different consumption patterns, different ecological footprints, and different ecological balance. Increase affluence exacerbates the environmental impacts and, when combined with population growth, substantially increases the human footprints on the planet. Ecological footprint is the amount of nature it takes to sustain a given population over a course of one year. It is a national-level measurement that quantifies how land and water are required to produce the commodities consumed and assimilate the waste generated by them. It attempts to operationalize the concept of carrying capacity and sustainability. Sustainability refers to the ability of a system to continue and maintain a production level or quality of life for future generations.

Many studies have dealt with the environmental and ecological degradations of the present anthropocentric economic development efforts, especially their effects on environmental sustainability and biodiversity conservation (Burns et al, 2001; Chew, 2001; Clapp, 2002; Conca, 2002; Princen, 2002; Tucker, 2002; Andersson, 2006). However, very few studies have dealt with the ecological footprint analysis that captured the world's biological diversity crisis (Wackernagel and Rees, 1996; Jorgenson, 2003). The purpose of this study is to analyze the ecological footprints of different nations, regions, and cultures for ecological balance. It examines the cross-cultural or national variations in total footprints and per capita footprints associated with ecostructural factors and human development hierarchy of the world economy or world system. The paper exposes the inevitable distribution of the world's ecological footprints and heightens the concern about ecological imbalances and overshoots, as well as biodiversity crisis.

To achieve the above objectives, the study analyzed and articulated the effects of population, affluence (GDP per capita), and ecostructural factors (or socio-economic processes) such as urbanization, literacy level, and domestic income inequality (Gini index) on the ecological footprints based on the national human development hierarchies of the world economy. The human development hierarchies of the world economy or world-system include, high human development, medium human development, and low human development (see *Human Development Report*, 2004). This study supports the work of Jorgenson (2003) who drew from world-systems and ecostructuralist theoretical perspectives to analyze the structural causes of national-level per capita ecological footprints within and between "zones" of the world-economy, namely, the core, semi-periphery, high periphery, and low periphery countries.

To understand the theory behind footprints inequalities among nations and regions, perspectives are incorporated from diverse fields as world system analysis, environmental history, environmental sociology, environmental anthropology, and environmental economics (Hornborg, 2005). The present society lives in an unequal world economy and world system, where the consumption levels of the higher human development countries are substantially higher than those of lower human development. The higher human development countries are politically more powerful, highly urbanized, with higher levels of per capita income (consumption) and literacy, as well as lower levels of domestic income inequalities. They consume biospheric resources at levels above the global bio-capacity per capita which could lead to potentially catastrophic outcomes for all societies (Jorgenson, 2003). Affluence and population size (i.e., growth) have been hypothesized to be the primary drivers of human-caused environmental stressors or impacts (E O News, 2007). This study concurred with the above assumptions and findings. Moreover, resource consumption rates have recently increased at a faster pace than population size (Dietz et al, 2007). From the above discussions, it was found that, a country's position in a world system, its affluence (GDP per capita), population size and growth, urbanization level, domestic income inequality (Gini index), and literacy level, all proved to be significant predictors of per capita ecological footprint (Wallerstein, 1994, 1979; Kick, 2000; Jorgenson, 2003; Hedenius and Azar, 2005; Andersson, 2006; E O News, 2007).

The human development hierarchy of the world capitalist economy is asymmetric, exploitative, disjunctive, and unsustainable. In the theory of unequal exchange, the high human development countries dominate and exploit the low human development and weak countries through trade and other control mechanisms. Often the gains from trade are unequally or asymmetrically distributed within the development hierarchy. The exchange is often non-equivalent in terms of market prices (terms of trade), labor and biomass inputs, as well as ecological footprints. Thus, what one country gains, the other loses disjunctively, which aggravates and exacerbates the income- and development-gaps (Andersson, 2006), constituting what Foster and Clark (no date) call "ecological imperialism."

To test for regional-national ecological footprints, environmental sustainability, and biodiversity conservation with respect to human development positions in the world capitalist economy or world system, the author analyzed a sample of 58 high, medium, and low development countries that are each, at least, one million in population. This was an explanatory study using comparative model and descriptive statistics such as totals, ratios, and percentages. The national ecological footprint (total or per capita), as the dependent variable, is explained by the independent variables that are recursive which include, the country's world system position, human development category, population size, GDP per capita (affluence), domestic inequality (Gini index), urbanization, and literacy rate. The independent variables helped in explaining the varying levels of consumption among different human development hierarchies of the capitalist world economy or world system. The carbon-dioxide emission levels (*World Development Report*, 2005, table 1, pp. 256–257) were included to depict the consumption and environmental degradation impacts of different countries, which reflect their national footprint levels.

The results and findings show that a country's total footprint, ecological balance, and carbon-dioxide emissions (waste) are a function of its population size and human development position in the world economy (see tables 2.1–2.3). This supports the findings of Jorgenson (2003) which showed that world-system position has significant and positive effect ($R^2 = 0.32$) on per capita ecological footprints. It became apparent from the analyses that, the higher the national populations and GDP per capita (affluence), the higher the *imbalanced* consumption, total footprint, carbon-dioxide emission, and the lower the ecological balance, and vice versa. The above is particularly true among high human development countries that sustain large populations (see table 2.1) such as, the United States, Japan, United Kingdom, France, Germany, and Italy. The medium human development category of the world economy is not a monolithic bloc and exhibits intermediate and variable impacts that are not as high as those of the high human development but also not as low as those of the low human development category (see table 2.2). Table 2.3 shows the low human development category countries or what Wallerstein (1979), Wackernagel and Rees (1996), Kick (2000), and Jorgenson (2003) called the periphery. Most of the representative countries of this cultural group are sub-Saharan African countries that were once colonies of the core or high human development countries of Western Europe. According to Bunker (1985), Bornschier and Chase-Dunn (1985), Boswell and Chase-Dunn (2000), and Jorgenson (2003), this region has disarticulated markets, higher levels of dependent industrialization, and underdevelopment. They possess more extractive economies and tend to be raw-material oriented. They have relatively low GDP per capita, low consumption levels, low total ecological footprints, high ecological balance, and low carbon-dioxide emissions (waste).

Human development hierarchies, world-system positions, and ecostructural factors have significant effects on per capita footprints of nations. Tables 2.4–2.6 show that indicators which include, urbanization, domestic inequality, and literacy factors constitute the structural (or socioeconomic processes) causes of ecological footprints per capita. This supports the work of Jorgenson (2003) who argued that the effect of world-wide position on per capita ecological footprints is both direct and indirect, mediated by its effects on domestic income inequality, urbanization, and literacy. Table 2.4 shows that the higher the human development, position in the world-system, rates of urbanization, literacy rates, and low domestic inequality (Gini index), the higher the ecological footprints per capita. Generally, countries in the high human development category have high consumption levels and high ecological footprint per capita, when compared to those of the medium and low human development countries (see tables 2.5 and 2.6).

The ecological footprint of nations manifest the degree of their environmental and ecological degradations based on their consumption patterns in the world system. Ecological footprint accounts, in effect, measure the degree of sustainability prevailing as nations engage in socioeconomic development. Humanity's increasing demands on nature world-wide have strained the environmental utilization space or ecospace, which make the meeting of national environmental debt and intergenerational equity virtually impossible. The results are environmental damage and biodiversity destruction, which in some cases make species extinction not restorable, thus eco-efficiency is needed to reduce economic costs and environmental negative impacts. According to E O News (2007), the loss of healthy forests (deforestation) can degrade key ecosystem services, such as carbon storage in biomass and soils (carbon sequestration), the regulation of water balance and river flow, the modulation of regional climate patterns, and the amelioration of diseases. Environmentalists have seen that the excessive and greedy consumption of natural resources and high ecological footprints of nations are inevitably linked to economic development. Man-made or anthropocentric forces have led to species and biodiversity damages such as (Myers, 1994; WRI, 1996; Hunter et al, 2003; Harf and Lombardi, 2004; Aka, 2007): habitat loss and fragmentation; biological invasion; pollution and human induced climate change; over-harvesting and hunting; deliberate extermination; and over-estimating and undervaluing of available natural resources and biodiversity.

The trio of world human development dynamics, world system dynamics, and population dynamics are important predictors of environmental outcomes that include ecological footprints. It has been shown that most advanced or high development countries are running massive unaccounted ecological deficits with the use of planet (Rees, 1996). Najam (1996) argued that consumption control among developed countries (the core or high human development countries) is a major key to avoid continuing environmental crises. Aka (2007) suggests that the voracious appetite of the high human development countries of Europe and North America for natural resources should be reduced and managed. For world-wide sustainability, Orton (1999) suggested that industrialized countries need to reduce their impact upon the earth to about one tenth of what it is at the present time. On the other hand, global environmental concerns have been linked to world political structure, poverty, and malnutrition in the periphery or low and moderate human development countries (Meadows et al, 1972, 1992; Hoff, 1998; Asafu-Adjaye, 2000; Ukanga and Afoaku, 2005). They argue that there is a causal link between increase in resource consumption, poverty, and rapid population growth, and global environmental degradation.

It is suggested that sustainability will depend on such measures as greater emphasis on equity in international relationships, significant adjustments to prevailing terms of trade, increasing regional self-reliance, and policies to stimulate a massive increase in the material and energy efficiency of economic activities (Pearce, 1994; Rees, 1996; Jorgenson, 2003). Goklang and Sprague (1992) argue that the efficiency and productivity of all activities that use land should be improved, while ensuring that they are environmentally sound. In the medium- and low- human development countries, there is the need to address the underlying causes of biodiversity loss, which include lopsided income distribution and poverty, lack of property right and tenure regimes, lack of adequate rural infrastructure, health and education services, and lack of employment opportunities (Asafu-Adjaye, 2000; Ukanga and Afoaku, 2005). For what can be done to reduce the size of personal footprint or increase biocapacity include, among others, supporting policies and companies that promote renewable energy, locally produced organic food, energy and resource conservation, and recycling (Redefining Progress, 2005).

REFERENCES

Aka, Ebenezer (2007). "Biodiversity, Conservation and Sustainable Socio-economic Development." *The International Journal of Environmental, Cultural, Economic and Social Sustainability*, Volume 2, Number 5, pp. 51–66.

Andersson, Jan Otto (2006). "International Trade in a Full and Unequal World." Presented at the workshop, "Trade and Environmental Justice," Lund, 15th–16th of February.

Arrighi, Giovanni and Jessica Drangel (1986). "The Stratification of the World Economy: An exploration of the Semi-Peripheral Zone." *Review* 10: 9–74.

Asafu-Adjaye, John (2000). "Biodiversity Loss and Economic Growth: A Cross-Country Analysis." Paper presented at the Western Economic Asociation International Annual Conference in Vancouver, BC, Canada.

Bergesen, Albert and Michelle Bata (2002). "Global and National Inequality: Are they Connected?" *Journal of World-System Research*, 8: 130–44.

Bornschier, Volker and Christopher Chase-Dunn (1985). *Transnational Corporations and Underdevelopment* (New York: Pager).

Boswell, Terry and Christopher Chase-Dunn (2000). *The Spiral of Capitalism and Socialism: Toward Global Democracy* (Boulder, CO: Lynne Reinner Publisher).

Bunker, Stephen G. (1985). *Underdeveloping the Amazon: Extraction, Unequal Exchange, and the Failure of the Modern State* (Urbana: University of Illinois Press).

Burns, Thomas J., Byron L. Davis, and Edward L. Kick (1997). "Position in the World-System and National Emissions of Greenhouse Gases." *Journal of World-Systems Research*, 3: 432–465.

Burns, Thomas J., Byron L. Davis, Andrew K. Jorgenson, and Edward L. Kick (2001). "Assessing the Short- and Long- Term Impacts of Environmental Degradation on Social and Economic Outcomes." Presented at the Annual Meetings of the American Sociological Association, August, Anaheim, CA.

Burns, Thomas J., Jeffery Kentor, and Andrew K. Jorgenson (2003). Trade Dependence, Pollution, and Infant Mortality in Less Developed Countries." Pp. 14–28 in *Crises and Resistance in the 21st Century World-System*, edited by Wilma A. Dunaway (Westport, CT: Pager).

Catton, W. (1986). "Carrying Capacity and the Limits to Freedom." Paper prepared for Social Ecology Session 1, XI, World Congress of Sociology, New Delhi, India (August 18).

Chase-Dunn, Christopher (1998). *Global Formation: Structures of the World-Economy* (Lanham, MD: Rowman & Littlefield).

Chase-Dunn, Christopher (2002). "World-Systems Theorizing." Pp. 589–612 in *Handbook of Sociology Theory*, edited by Jonathan H. Turner (New York: Kluwer Academic/ Plenum Publishers).

Chase-Dunn, Christopher and Thomas D. Hall (1997). *Rise and Demise: Comparing World-Systems* (Boulder, CO: Westview).

Chase-Dunn, Christopher and Andrew K. Jorgenson (2003). "System of Cities." Pp. 140–144 in *Encyclopedia of Population*, edited by Geoffreey Mcnicoll and Paul Demeny (New York: McMillan).

Chew, Sing C. (2001). *World Ecological Degradation: Accumulation, Urbanization, and Deforestation 3000B.C-A.D 2000* (Walnut Creek, CA: Alta Mira).

Clapp, Jennifer (2002). "The Distancing of Waste: Over-consumption in a Global Economy." Pp. 155–76 in *Confronting Consumption*, edited by T. Princen, M. Maniates, and K. Conca (Cambridge, MA: MIT Press).

Commission for Environmental Cooperation (2001). *The North American Mosaic: A State of the Environment Report*. Commission for Economic Cooperation, Montreal, Quebec, Canada.

Conca, Ken (2002). "Consumption and Environment in a Global Economy." Pp. 133–54 in *Confronting Consumption*, edited by Thomas Princen, Michael Maniates, and Ken Conca (Cambridge, MA: MIT Press).

Crosby, Alfred W. (1986). *Ecological Imperialism: The Biological Expansion of Europe* (Cambridge: Cambridge University Press).

Dietz, Thomas and Eugene Rosa (1994). "Rethinking the Environmental Impacts of Population, Affluence, and Technology." *Human Ecology Review*, 1: 277–300.

Dietz, Thomas, Eugene Rosa, Richard York (2007). "Driving the Human Ecological Footprint." *Frontiers in Ecology and the Environment* (February).

Earth Observatory News (February 2007). "Human's Ecological Footprint in 2015 and Amazonia Revealed." http://earthobservatory.nasa.gov/Newsroom/MediaAlerts/2007. Accessed Online on 2/23/2007.

Foster, Bellamy John (2000). *Marx's Ecology: Materialism and Nature* (New York: Monthly Review Press).

Foster, John Bellamy and Brett Clark (no date). "Ecological Imperialism: The Curse of Capitalism."

Goklany, I.H. and M.W. Sprague (1992). Policy Analysis, Sustaining Development and Biodiversity: Productivity, Efficiency, and Conservation. Cato Policy Analysis. No. 175 (August 6): 18p. http://www.cato.org/pubs/pas/pa-175html. Accessed Online on 12/27/2005.

Goudie, A. (2000). *The Human Impact on the Natural Environment. Fifth Edition* (Cambridge, Massachusetts, USA: Harvard University Press).

Hardin, G. (1991). "Paramount Positions in Ecological Economics," in R. Costanza (ed.). *Ecological Economics: The Science and Management of Sustainability* (New York: Columbia University Press) pp. 47–57.

Harf, James E. and Lombardi, Mark Owen (2004). *Taking Sides. Clashing Views on Controversial Global Issues. Second Edition* (Guilford, Connecticut: McGraw-Hill).

Harper, Charles (2001). *Environment and Society: Human Perspectives on Environmental Issues* (Upper Saddle River, NJ: Prentice Hall).

Hedenius, Fredrick and Christian Azar (2005). "Estimates of trends in global income and resource inequalities." *Ecological Economics*, 55, 351–364.

Hoff, Marie D. (1998). *Sustainable Community Development Studies in Economic, Environmental, and Cultural Revitalization* (London, Washington, D.C., : Lewis Publishers).

Hornborg, Alf (2001). *The Power of the Machine: Global Inequalties of Economy, Technology, and Environment* (Walnut Creek: Alta Mira Press).

Hornborg, Alf (2005). "Fair Trade? Non-monetary Measures of Global Resource Flow." Paper to be presented and discussed on Workshop, 27th of October 2005. Division of Human Ecology, Lund University.

Human Development Report, 2004. Cultural Liberty in Today's Diverse World. Published by the United Nations Development Program (UNDP), 1 UN Plaza, New, York, New York, 10017, USA.

Hunter, Lori, John Beal, and Thomas Dickinson (2003). "Integrating Demographic and GAP Analysis Biodiversity Data: Useful Insight?" *Human Dimensions of Wildlife*, 8: 145–157.

Jorgenson, Andrew K. (2003). "Consumption and Environmental Degradation: A Cross-National Analysis of Ecological Footprint." *Social Problems*. Volume 50, No. 3, pp. 374–394.

Jorgenson, Andrew K. (2005). "Unpacking International Power and the Ecological Footprints of Nations: A Quantitative Cross-National Study." *Sociological Perspectives*. Volume 48, Issue 3, pp. 383–402.

Jorgenson, Andrew K. and Edward L. Kick (eds.) (2003). Special Issue: Globalization and the Environment. *Journal of World-System Research*, Volume IX, Number 2, (Summer).

Jorgenson, Andrew K. and Tom Burns (2004). "Globalization, Environment, and Infant Mortality: A Cross-National Study." *Humboldt Journal of Social Relations*, 28: 7–52.

Kentor, Jeffery (2000). *Capital and Coercion: The Economic and Military Processes that Have Shaped the World-Economy, 1800–1990* (New York: Garland Press).

Kick, Edward L. (Summer 2000). "World-System Position, National Political Characteristics and Economic Development." *Journal of Political and Military Sociology*. http://www.findarticles.com/p/articles/mi-qa3719. Accessed Online on 4/4/07.

Lindeke, William (1995). "Democratization in Namibia: Soft State, Hard Choices." *Studies in Comparative International Development*, 30: 3–29.

McMichael, A.J. (1993). *Planetary Overload: Global Environmental Change and the Health of the Human Species* (Cambridge: Cambridge University Press).

Meadows, D.H., D.L Meadows, J. Randers, and W.W. Behrens (1972). *The Limits to Growth* (New York, NY: Universe Books).

Meadows, D.H., D.L. Meadows, and J. Randers (1992). *Beyond the Limits: Confronting Global Collapse, Envisioning A Sustainable Future* (Post Mills, Vermont, USA: Chelsea Green).

Myers, N. (1994). "Population and Biodiversity," in F. Graham-Smith (ed.), *Population: The Complex Reality* (London: North American Press) pp. 117–136.

Najam, Adil (1996). "A Developing Countries' Perspective on Population, Environment, and Development." *Population Research and Policy Review*, Volume 15, No. 1 (February).

Nemeth, Roger and David Smith (1985). "International Trade and World-System Structure: A Multiple Network Analysis." *Review*, 8: 577–590.

Newman, Peter and Jeffery Kenworthy (1999). *Sustainability and Cities. Overcoming Automobile Dependency* (Washington, D.C., Covelo, California: Island Press).

Odum, H.T. (1988). "Self-Organization, Transformity, and Information." *Science*, 242:1132–1139.

Odum, H.T. and J.E. Arding (1991). Emergy Analysis of Shrimp Mariculture in Ecuador. *Working Paper*, prepared for Coastal Resources Center, University of Rhode Island.

Orton, David (July 1999). "Commentary on Our Ecological Footprint: Reducing Human Impact on the Earth" by Mathis Wackernagel and William Rees, published by New Society Publishers, 1996. Accessed Online on 2/23/07, at http://home.ca.inter.net/~greenweb/Ecological_Footprint.html.

Pearce, D. (1994). "Sustainable Consumption through Economic Instruments." Paper prepared for the Government of Norway Symposium on Sustainable Consumption, Oslo, 19–20 January.

Peet, Richard (1999). *Theories of Development* (New York, London: The Guilford Press).

Podobnik, Bruce (2002). "Global Energy Inequalities: Exploring the Long-Term Implications." *Journal of World-System Research*, 8: 252–75.

Princen, Thomas (2002). "Consumption and Its Externalities: Where Economy Meets Ecology." Pp. 23–42 in *Confronting Consumption*, edited by T. Princen, M. Maniates, and K. Conca (Cambridge, MA: MIT Press).

Redefining Progress (2005). *Footprints of Nations*. 1904 Franklin Street, Oakland, California 94612. See http://www.RedefiningProgress.org.

Rees, W. (1990). Sustainable Development and the Biosphere. Teilhard Studies Number 23. American Teilhard Association for the Study of Man, or: "The Ecology of Sustainable Development." *The Ecologist*, 20 (1), 18–23.

Rees, William E. (1996). "Revisiting Carrying Capacity: Area-Based Indicators of Sustainability." *Population and Environment: A Journal of Interdisciplinary Studies*, Volume 17, Number 3 (January). http://www.dieoff.org/page110.htm. Accessed Online on 2/23/07.

Sassen, Saskia (1991). *The Global City: New York, London, and Tokyo* (Princeton, NJ: Princeton University).

Sklair, Leslie (2001). *The Transnational Capitalist Class* (Oxford, UK: Blackwell Press).

Smith, David A. (1993). "Technology and the Modern World System: Some Reflections." *Science, Technology and Human Values*, 18: 186–195.

Smith, David and Douglas White (1992). Structure and Dynamics of the Global Economy: Network Analysis of International Trade, 1965–1980." *Social Forces*, 70: 857–893.

Smith, David and Michael Timberlake (1997). "Urban Political Economy." Pp. 109–28 in *The Urban World*, edited by John Palen (New York: McGraw-Hill).

Smith, David A. and Michael Timberlake (1997). "World City Networks and Hierarchies, 1977–1997: An Empirical Analysis of Global Air Travel Links." *American Behavioral Scientist*, 44: 1656–78.

Terlouw, C.P. (1993). "The Elusive Semiperiphery: A Critical Examination of the Concept of Semiperiphery." *International Journal of Comparative Sociology*, 34: 87–102.

Tucker, Richard (2002). "Environmentally Damaging Consumption: The Impact of American Markets on Tropical Ecosystems in the Twentieth Century." Pp. 177–96 in *Confronting Consumption*, edited by Thomas Princen, Michael Maniates, and Ken Conca (Cambridge, MA: MIT Press).

Ukanga, Okechukwu and Afoaku, Osita G. (2005). *Sustainable Development in Africa. A Multifaced Challenge* (Trenton, NJ: African World Press).

Wackernagel, Mathis and William Rees (1996). *Our Ecological Footprint: Reducing Human Impact on the Earth* (Gabriola Island, B.C., Canada: New Society Publishers).

Wackernagel, Mathis and Judith Silverstein (2000). "Big Things First: Focusing on the Scale Imperative with the Ecological Footprint." *Ecological Economics*, 32: 391–4.

Wackernagel, Mathis, Alejandro C. Linares, Diana Deumling, Maria A.V. Sanchez, Ina S.L. Falfan, and Jonathan Loh (2000). *Ecological Footprints and Ecological Capacities of 152 Nations: The 1996 Update* (San Francisco, CA: Redefining Progress).

Ward, Kathryn (1993). "Foreign Debt and Economic Growth in the World System." *Social Science Quarterly*, 74: 703–720.

Wallerstein, Immanuel (1974). *The Modern World-System* (New York: Academic Press).

Wallerstein, Immanuel (1979). *The Capitalist World Economy* (New York: Cambridge University Press).

World Bank (2005). *World Development Report, 2005. A Better Investment Climate for Everyone* (New York, NY: A Co-publication of the World Bank and Oxford University Press).

World Resource Institute (WRI) (1996). *World Resources 1996–1997. A Guide to the Global Environment. The Urban Environment.* A Joint Publication by: The World Resources Institute, The United Nations Environment Program, The United Nations Development Program, and The World Bank (New York, Oxford: Oxford University Press).

World Resources Institute (WRI) (2000). *World Resources, 2000–2001: People and Ecosystems: The Fraying Web of Life* (Washington, D.C., USA: World Resources Institute).

York, Richard, Eugene A. Rosa, and Thomas Dietz (2003). "Footprints on the Earth: The Environmental Consequences of Modernity." *American Sociological Review*, 68: 279–300.

Chapter Three

Biodiversity, Conservation, and Sustainable Socio-Economic Growth and Development

Problems and Prospects

INTRODUCTION

The earth is home to a variety of plants and animals and other living things. Biodiversity refers to variety and variability among living organisms and the ecological complexes in which they occur (Dovie, 2003, p. 39). Diversity can be defined as the number of different items and relative frequencies [Office of Technology Assessment (OTA), 1987a; Noss, 1990]. Biological diversity exists at several levels, from the finer scale of genetic diversity to the coarse scale reflecting variation across ecosystem (OTA, 1987b; WRI, 1994). At all levels it is important to the composition, structure, and functioning of ecosystem. Moreover, it provides many of the ecological services for the anthropocentric economy.

Attempt to sustain livelihoods and to reduce poverty whilst conserving the earth's resources are major global challenges. Challenges to biodiversity conservation have become a major topic of academic and public debate (Czech, 2003). There is an agreement, especially among conservationists (e.g., conservation biologists, ecologists, ecological economists, etc.) that a fundamental conflict exists between economic growth and biodiversity conservation (The Wildlife Society, 2003). The evidence of negative effects of economic growth on biodiversity is quite robust whether in developed countries (Czech, 2003; Czech et al, 2000; Czech and Daly, 2004; Daly, 2002) or in developing countries (Asafe-Adjaye, 2000; Ukanga and Afoaku, 2005). The adverse effects depend on the type and composition of growth. For example, in more developed countries of Europe and the United States of America, the growth pressure on biodiversity is more by increased urbanization and rise in per capita consumption. In developing countries, the pressure is more by rapid population growth and inappropriate farm practice such as slash-and-burn, since agricultural output still remains a higher component of total output. Furthermore, in developing countries the underlying causes of biodiversity loss include poverty, lack of property rights and tenure, and lack of employment opportunities (Asafu-Adjaye, 2000).

Economic growth, which is a measure of gross domestic product (GDP), is the synthesis of population and consumption and represents an increase in the production and consumption of goods and services. According to The Wildlife Society (2003), economic growth entails the reallocation of natural resources from the "economy of nature" and its nonhuman species to the anthropocentric or human economy. Thus, biodiversity is threatened by economic sectors in the aggregate (Czech et al, 2000). A social concern is that the unfettered pursuit of profit by business firms will lead to the degradation of the environment and the overuse of resources (Czech and Pister, 2005). Human use of the environment is the largest contributor to habitat modification and ecosystem loss (Goudie, 2000; Harrison and Pearce, 2000; World Resources Institute, 2000). Growth has been blamed for such diverse problems as the costly destructive development pattern associated with urban sprawl, loss of prime agricultural land and ecosystem loss, inefficient provision of public facilities and services, escalating housing prices, pervasive environmental degradation, and loss of community character. Zovanyi (2005), postulates that growth constitutes unsustainable behavior because it is incapable of being continued or maintained indefinitely. Thus growth should be managed to simultaneously confront social sustainability, economic sustainability and environmental sustainability. Perhaps this is what is called "smart growth" as an alternative to "dumb growth" represented by sprawl and its destructive forces (Chen, 2000; Lorentz and Shaw, 2000).

Although the concept of sustainable development is elusive and poorly understood (Goklany and Sprague, 1992; Daly, 1996), it is defined by the World Commission on Environment and Development (1987) as "development that meets the needs of the present without compromising the ability of future generations to meet their needs." To those who believe that current path of development is clearly unsustainable because the planet is about to choke on humanity's wastes and there is not enough land to meet everyone's demand, "sustainable development" implies virtually no additional growth and development (e.g., Catton, Jr., 1980; Hubert, 1987; Wackernagel, 2002; Czech and Daly, 2004; Attarian, 2004; Feinman, 2005) or negative growth and contraction (e.g., Goldsmith, 1979). This faction is inherently suspicious of technological change or progress and economic growth. To those inclined toward balancing economic and environmental goals, "development" implies economic growth, and "sustainable" implies full consideration of environmental factors (e.g., Goklany and Sprague, 1992). The sustainable development movement postulates that a sustainable society must balance social equity, economic prosperity, and environmental integrity (Krizek and Power, 1996). In the same way, the Millennium Development Goals and the Convention on Biological Diversity 2010 (CBD 2010) goals directly imply that we need to find ways to develop the global economy while also protecting the environment and the welfare of those whose health and economic well-being are most dependent on the ecosystem services supplied by the natural environment.

Conserving biodiversity has become high and dynamic since the signing of the Biodiversity Convention at the first Earth Summit in Rio de Janeiro, Brazil in 1992 (UNEP, 1993). The concerned environmental or ecological societies include, among others, Ecological Society of America (ESA), The Wildlife Society (TWS), the US Society for Ecological Economics (USSEE), Society for Conservation Biology (SCB), a global community of conservation professionals, and American Fisheries Society (AFS). Their goals, devotions, and mission statements are dovetailed toward biodiversity conservation. For example, The Wildlife Society in its Technical Review 03-1 identified a "fundamental conflict between economic growth and wildlife conservation" (The Wildlife Society, 2003); and the U.S Society for Ecological Economics adopted a position in 2003 that identifies a "fundamental conflict between economic growth and ecosystem health" (http://www.ussee.org/about). The North American Section of the Society for Conservation Biology has adopted a position on economic growth to include major points such as, "recognition of the conflict between economic growth and conservation, a goal of replacing the growth economy with a steady state economy that does not breach ecological carrying capacity, understanding the difference between economic development and growth, and support for prosperous transition to steady state economics for all nations" [Society for Conservation Biology (SCB), 2005]. In the case of the Ecological Society of America, it is devoted to ensuring "the appropriate use of ecological science in environmental decision making by enhancing communication between the ecological community and policy makers" (see, http://www.frontiersinecology.org).

Nevertheless, the mainstream economists and policy makers around the world tend to ignore the aforementioned problems as they typically opine that there is no practical limit to economic growth. In agreement with those above, the economic triangle of politicians, bureaucrats, and government agencies (Czech, 2000) that dominate policy arenas, such as in the United States of America has claimed that there is no conflict between economic growth and environmental protection. They argue that technological progress, which is referred to as invention and innovation, is a reconciler of economic growth and biodiversity conservation (Lomborg, 2001; Czech, 2003). Thus with rising productive efficiency (due to technological progress), more is produced with a given amount of resource input. They further argue that at higher levels of socioeconomic growth and development, structural changes towards information-intensive industries and services, coupled with increased environmental awareness, enforcement of environmental regulations, better technology and higher environmental

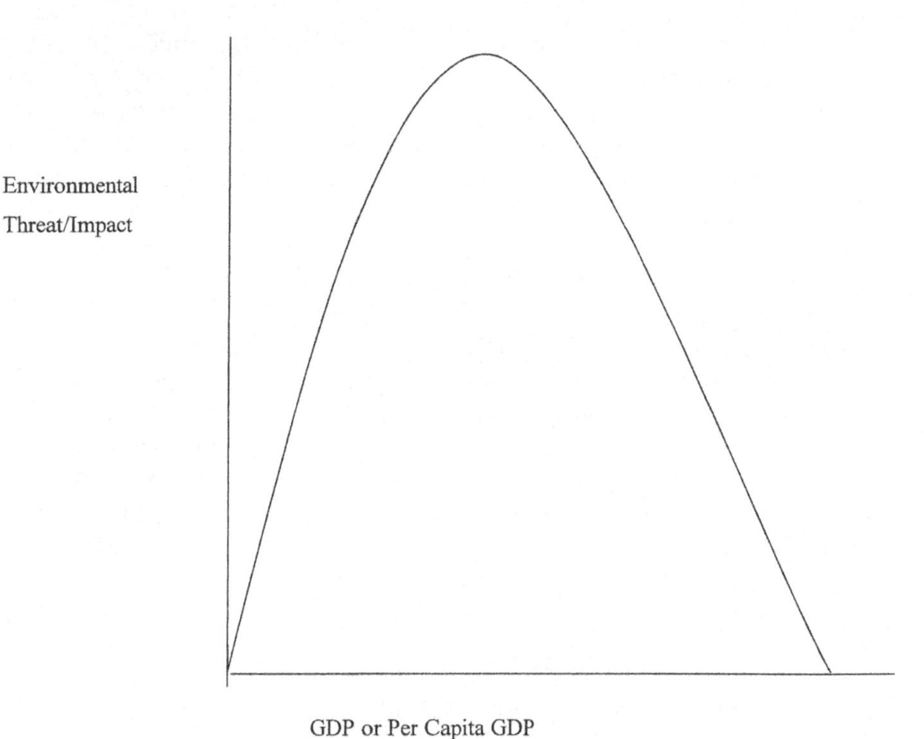

Figure 3.1. Environmental Kuznets Curve

expenditures, and would result in leveling off and gradual decline of environmental degradation (Stern, 2004). This phenomenon will lead to what is known as the environmental Kuznets curve (EKC) (see figure 3.1).

The environmental Kuznets curve is the hypothesis that the environment deteriorates during the early phases of economic growth then recovers after a threshold of growth is achieved (Barbier, 1997). The EKC implies that socio-economic growth would be the means to eventual environmental improvement.

There are many values associated with biodiversity as the earth is home to a variety of plants, animals, and other living things. These are environmental, economic, and aesthetic and spiritual values (Hunter et al, 2003; WRI, 1996).

1. Environmental value contributes to ecosystem maintenance, balance, and vitality. Species diversity plays a vital role in the functioning of ecosystems, especially in trophic and niche dynamics.
2. Economic value contributes to material and physical sustenance. Species diversity provides a host of wild and domestic plant, fish, and animal products used for medicines (pharmaceuticals), cosmetics, industrial products, fuel and building materials, and food, among other things.

3. Aesthetic and Spiritual values feed the human craving for intellectual, cognitive, and spiritual meaning and satisfaction (Kellert, 1984, 1996; Aka, 2006). In some cultures certain plant and animal species have religious and spiritual meanings and functions. They are revered by the people and spared from entering into the human economy, thereby staying with the economy of nature.

PURPOSE OF STUDY

The paper explores theoretically and empirically the problems of biodiversity in different cultures, areas, and regions. The relationship between biodiversity conservation and socioeconomic growth is also explored, especially the challenges and prospects as we endeavor to balance environmental gains (common future) with economic and social development. The paper highlights ecological footprints of the human economy as it liquidates the natural capital resources, such as in the United States of America. Solutions are proffered on how to conserve biodiversity in the face of anthropogenic threats, and at the same time make socioeconomic growth sustainable.

THEORETICAL ARGUMENT FOR THE STUDY

Ecological economists, conservation biologist, and other environmental conservationists believe that a fundamental conflict exists between economic growth and biodiversity conservation. Theoretical and empirical arguments suggest that economic growth is the primary challenge to biodiversity conservation and ultimately to human economic sustainability (Czech et al, 2000). This theoretical construction rests upon the principles of carrying capacity, niche breadth, competitive exclusion, and trophic levels (The Wildlife Society, 2003).

Carrying Capacity

The ecosystem contains populations of human and non-human species that have limits and are finite. Nebel and Wright (2000) defined carrying capacity as the maximum population of a given organism that a particular environment or habitat can sustain or support without degradation of the habitat over the long term. In the human economy, economic carrying capacity is a function of population size and per capita consumption (Daily and Ehrlich, 1992). Economic growth entails increase in population and per capita consumption that occurs through the relocation of natural capital (water, forests, minerals, etc.) from the economy of nature to human economy. Neoclassical economists and the public regularly assume that perpetual economic growth is possible and that such growth can occur without interfering with the need of biodiversity or people (Willers, 1994; Czech, 2000b).

Some economists posit that there is no carrying capacity imposed upon humans because they have the ability to modify their environments and to protect themselves from decimating factors such as hunting, pollution, and severe weather (Leopold, 1933; Simon, 1996). Thus as resources become scarce, humans find substitutes for those resources, and invention and innovation or technological progress lead to increasing efficiency in the economic production process (Ehrlich and Ehrlich, 1990; Hawken, 1993; Simon, 1996). Nevertheless, ecological economists have criticized the theory's unlimited economic growth (Daly, 1996; 1997; Erickson and Gowdy, 2000; Boulding 1996; Constanza et al, 1997) and an infinite expansion of production and consumption of goods and services (Ehrlich, 1994). The above criticisms had been noted by Meadows et al (1972) who axiomatically stated that "infinite growth is impossible in a finite system," and stated that global growth would soon be reached. In short, they stated in no distant past (Meadows et al, 1992) that some of those limits had already been reached as reflected by unsustainable resource use and pollution generation. According to the Wildlife Society (2003), ecologists generally share the perspective that the world's resources are limited, and when certain limits are reached, both biodiversity and people suffer (Daily and Ehrlich, 1992; Meadows et al, 1972; 1992; Pulliam and Haddad, 1994; Harrison and Peace, 2000).

Niche Breadth and Competitive Exclusion

Niche breadth and competitive exclusion are two ecological principles that underlie the conflict between economic growth and biodiversity conservation. According to the Wildlife Society (2003), the niche pertains to the breadth of habitats used and the extent of interactions with other species (Hutchinson, 1978). Competitive exclusion is the principle that if two species compete for the same resources, then one species can succeed only at the expense of the other (Ricklefs and Miller, 2000). Niche breadth is generally correlated with intelligence and body size, both of which determine options a species has in exploiting its environment (Czech and Krausman, 2001). The niche breadth of a species is clearly correlated with the number of species affected by that species. Humans are large-bodied species with the highest intelligence known with unprecedented niche expansion and reside in all regions of the earth and in all types of ecosystems (Czech, 2002a).

Expanding economies and large populations tend to divert resources from other species for the exclusive use of humans. Technological progress broadens the human niche (Kingdon, 1993), and economic growth is the process of filling the broadened niche (Czech, 2003). The result is the conversion of natural capital to human economy leaving a dangerous ecological footprint. Ecological footprint is a resource management tool that measures how much land and water area (the biologically productive land and sea surfaces) a human population required to produce the resources it consumes and to absorb its waste under prevailing technology. In order to live, humans consume what nature offers. Today, according to the Global Footprint Network, humanity's ecological footprint is over 23 percent larger than what the planet can regenerate (http://www.footprint network.org/gfn_sub.php?). Currently humans have moved into ecological overshoot rather than ecological balance or deficit. Humans maintain this overshoot by liquidating the planet's ecological resources. In a sustainable world, society's demand on nature is in balance with nature's capacity to meet that demand.

Trophic Levels

Trophic levels refer to the nutritional organization of an ecosystem (Ricklefs and Miller, 2000). A trophic level is a step in the transfer of food or energy within a chain (The Wildlife Society, 2003). Several trophic levels exist within a system, which include producers, primary consumers, and secondary consumers. The primary consumers consume producers, and secondary consumers consume other consumers. In the "economy of nature" producers are plants that produce their food via photosynthesis, and in the "human economy" agriculture and extractive sector (i.e., logging, mining, ranching, and harvesting wildlife) constitute the economy's foundation or its producer trophic level (Czech, 2000b). In the human economy, consumer trophic levels are represented primarily by the manufacturing sector. Service sectors are represented by recreation and urbanization (The Wildlife Society, 2003). The manufacturing and service sectors of the human economy are responsible for the habitat and ecosystem losses. The carrying capacity of a habitat depends largely upon the biomass and productivity of consumable species residing at lower trophic levels. The underlying conflict between economic growth and biodiversity is illustrated by a consideration of natural capital (see figure 3.2). In the absence of anthropogenic threat, all natural capital is available as habitat for non-human species. But as the scale of the human economy expands, natural capital is re-allocated from nonhuman uses to the human economy (Czech, 2000a).

METHODOLOGY FOR THE STUDY

This is an exploratory study that is cross-sectional in approach. Several indicators which inform the nature and impact of anthropogenic threat on the environment, thereby, biodiversity, are discussed. Such indicators include, among

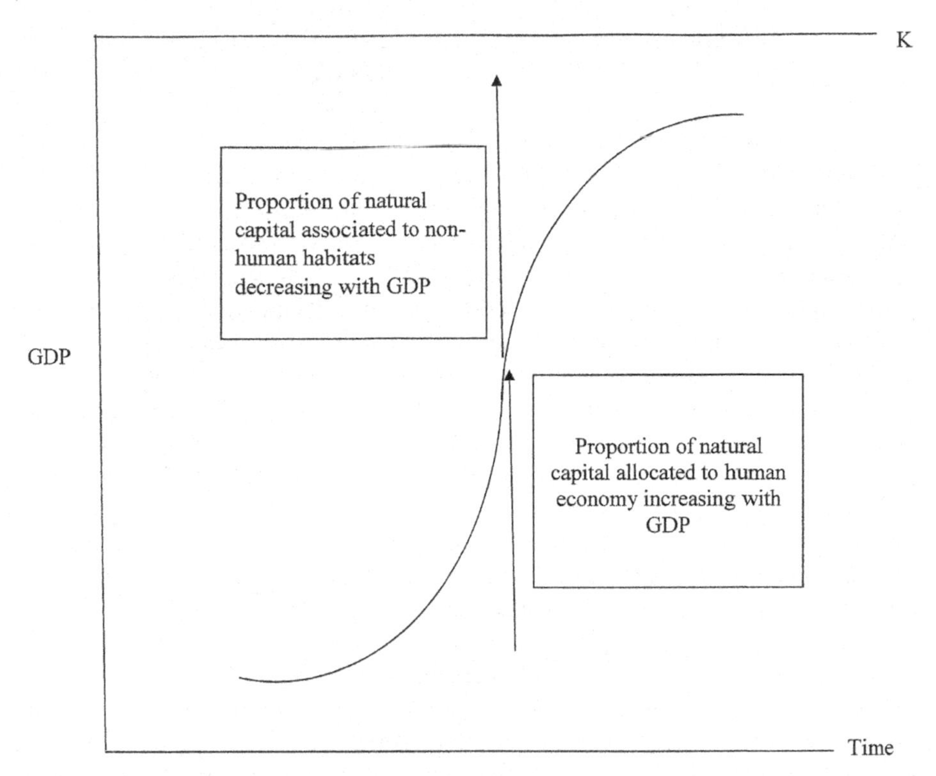

Figure 3.2. Allocation of Natural Capital to Human Economy. GDP= Gross Domestic Product; K= Carrying capacity. Source: The Wildlife Society (2003). The Relationship of Economic. Growth to Wildlife Conservation. Technical Review 03-1.

others, ecosystem loss by region, species endangerment by region, species threat by region, and economic growth (GDP). The regions include Asia, Latin America, Europe, North and Central America, South America, former Soviet Union, Oceania, USA, and Africa. In the U.S. case, several indicators that cause species endangerment include, among others, interactions with nonnative species, urbanization, agriculture, harvest, logging, road presence and construction, disease and vandalism. These indicators are the basis for assessing the impact of human economy on the natural capital or resources as human economy liquidates the natural capital or resources. The secondary data are collected from (a) World Resources Institute, 1996–1997. A Guide to the Global Environment. The Urban Environment; and (b) The Wildlife Society (2003). Technical Review 03-1. Data analyses for the study are performed by comparing the absolute and relative numbers of each species by region. Policy recommendations for the study are based on the findings of the analysis, and the conclusions are based on the findings of the study.

BIODIVERSITY CONSERVATION PROBLEMS: EMPIRICAL EVIDENCE

Empirical evidence for the fundamental conflict between economic growth and biodiversity conservation lies in the trends and causes of species endangerment, ecosystem loss, and reduction of ecosystems and habitats. Urban sprawl has been implicated as the leading cause of habitat loss and species endangerment (Czech et al, 2000; Zovanyi, 2005). Urban sprawl is characterized as relatively low-density, noncontiguous, automobile-dependent, residential and nonresidential development that convert and consume relatively large amounts of farmland and natural areas (Burchell et al, 1998). Sprawl has been identified as the most significant factor affecting forest ecosystems (Wear and Greis, 2002) and has been linked to an array of economic and social costs, such as higher costs for providing public infrastructure, e.g., roads, utilities, and less cost-efficient transit (Burchell et al, 1998). Urban sprawl has supporters whose arguments are often based on private property rights and consumer sovereignty. However, public discussion of sprawl has been largely opposed to it by viewing sprawl as a problem and undesirable, which should be avoided and stopped (Gillham, 2002; Fan et al, 2005).

Humans are reshaping the environment and using up many parts of the world faster than nature can renew itself. According to the Commission for Environmental Cooperation (2001), transformation of the landscape, including habitat loss and alteration, has become the primary threat to biodiversity. Human use of the environment is the largest contributor to habitat modification and ecosystem loss (Goudie, 2000; Harrison and Pearce, 2000; World Resources Institute, 2000). Land use change has critical effects on wildlife habitats through activities that result in fragmentation and destruction. Three broad types of land-altering human activities are generally recognized as contributing most to ecosystem loss: (i) urbanization, (ii) agriculture, and (iii) resource extraction, including forest management (The Wildlife Society, 2003). Urbanization and agricultural practices endanger ecosystems by replacing them directly. Forest management generally endangers ecosystems when severe modification such as degradation occurs.

Some natural and man-made forces lead to species decline. Such forces include (Myers, 1994; WRI, 1996; Hunter et al, 2003; Harf & Lombardi, 2004):

1. Habitat loss and fragmentation due to human encroachment (especially in developed nations) of habitat destruction. This tends to accelerate with the increase in a country's population density.
2. Biological invasion, especially through the introduction of non-indigenous species.
3. Pollution and human induced climate change.

4. Over-harvesting and hunting (intense overexploitation). For example, over-grazing and over-collection of firewood, which is a function of growing population, are degrading 14 percent of the world's remaining large areas of virgin forest (Harf & Lombardi, 2004, p.138).
5. Deliberate extermination.
6. Overestimating and undervaluing of available natural resources and biodiversity. That is, believing in the limitlessness of natural resources as well as feigning ignorance of the values of biodiversity.

Table 3.1 shows percent of regional coastlines under low, moderate, or high potential threat in 1995. Threat ranking depicts potential risk to coastal ecosystems from development-related activities. From the table, 34 percent of the world coastlines is potentially under high threat. The highest threat is found in Europe (70 percent), followed by Asia (52 percent), and Africa (38 percent). The lowest coastline threat is found in the former Soviet Union (12 percent), and then followed by North and Central America (7 percent). Coastal development contributes to habitat loss, which includes (WRI, 1996, p.254) the conversion of mangroves and other wetlands as a result of urbanization and agricultural expansion, the building of shoreline stabilization structures such as breakwaters, mining, and oil drilling, and dredging and filling.

THE U.S. CASE

In the United States, economic growth is a foremost national effort and has been a primary goal of the public and polity (Collins, 2000; Czech, 2000b). Many mission statements of key federal agencies reiterate the primacy of economic growth. Such agencies include U.S. Department of Commerce (2002), U.S. Department of Treasury (2002), U.S. Agency for International Development (2002), and the Federal Reserve System, the Central Bank of the United States (2002). For example, the goal of the Department of Treasury is to "promote domestic economic growth," and the Federal Reserve System noted that "economic growth is the driving force for the United States economy," while the Agency for International Development advocates that "broad-based economic growth is the most effective means of bringing poor, disadvantaged, and marginalized groups into the mainstream of the economy." U.S. economy is the largest in the world, accounts for nearly one-fourth of global gross product (Eves et al, 1998), and its gross domestic product (GDP) ranks first in the world (Knox and Agnew, 1998). The GDP measures the monetary value of goods and services produced within a nation's borders.

The United States is the third most populous country in the world (National Research Council, 2000) with population of 281,421,906 people in 2000 (U.S.

Table 3.1. Percent of Regional Coastlines under Low, Moderate, or High Potential Threat, 1995

	Percent of Coastline Under Potential Threat[a]		
Region	*Low*[b]	*Moderate*[c]	*High*[d]
Africa	49	14	38
Asia	31	17	52
North and Central America	71	12	17
South America	50	24	26
Europe	14	16	70
Former Soviet Union	64	24	12
Oceania	56	20	24
World	**49**	**17**	**34**

a, b. World Resources (1996). *World Resources, 1996–1997. A Guide to the Global Environment. The Urban Environment*. A joint publication by The World Resources Institute, The United Nations Environment Program, The United Nations Development Program, and The World Bank (New York, Oxford: Oxford University Press), p. 250.

c, d. Bryant, Dirk et al. (1995). "Coastlines at Risk: An Index of Potential Indicator Development-Related Threats to Coastal Ecosystems," World Resources Institute (WRI) Brief (WRI, Washington, D.C.), pp. 5–6.

Census Bureau, 2000), only second and third to India and China respectively. The United States is a consumer society based on continuous growth of consumer expenditure (Schor, 1997). A strong relationship (R^2=98.4%) exists between species endangerment and economic growth in the United States (The Wildlife Society, 2003, p.13). The country comprises of less than 5 percent of the world's population (Smith, 1999) but consumes 30 percent of its resources, and in the last 50 years its per capita resource use rose 45 percent overall (Suzuki, 1998). It is noted that a baby born in the U.S. consumes approximately 20 times the resources as a baby born in less developed country (Mckibben, 1998). The United States is among the rich nations which "sky-high" consumption patterns have remained a major threat to the global environment. The economy of the United States depends heavily on fossil fuel use as it consumes more than 20 million barrels of oil daily which is a quarter of all global consumption. The country is a leading producer of carbon dioxide, a leading emitter of greenhouse gases, accounting for more than a quarter of the world's total. These gases are thought to be responsible for global climate change (Suzuki, 1998; Smith, 1999).

Anthropogenic threat to natural ecosystems is very severe in North America, particularly in the United States. Humans are reshaping the environment and using up many parts, faster than nature can renew itself, through urbanization and other practices. U.S. contains the widest span of biome types ranging from rain forests to arctic tundra and from coral reefs to great lakes, of any country in the world. However, for many years, species (more than 200,000 described) have declined (Mattheissen, 1959) and some have gone extinct (Czech and Krausman, 2001). In the U.S. study nearly all the species threat and endangerment are because of human activity (see table 3.2). Economic activities in the study include urbanization, agriculture, mineral extraction, outdoor recreation, logging, industry, among others. Indirect economic activities used in the study include disease and interactions with nonnative species. Roads (their construction, presence, and maintenance) that are required for urbanization, mining, agriculture were found to be associated with more other causes of species endangerment than any other cause. Urbanization and agriculture that are associated with more cases of species endangerment do these through habitat destruction (Czech et al, 2000; Wilcove et al, 1998).

BIODIVERSITY CONSERVATION PROSPECTS

It is a known fact that fundamental conflict exists between economic growth and biodiversity conservation. However, technological progress has been proffered as a reconciler of economic growth (increase in GDP) and biodiversity conservation (Lomborg, 2001), especially in militating against adverse environmental impacts associated with economic growth. A relationship exists between GDP and environmental quality, which is depicted by the environmental Kuznets curve (EKC) (See figure 3.1). The EKC is named for Kuznets (1955) who hypothesized that income inequality first rises then falls as economic development proceeds. It implies that economic growth would be the means to eventual environmental improvement. The EKC concept emerged in the early 1990s (Grossman and Krueger, 1991; Shafik and Bandyopadhyay, 1992) and popularized in 1992 by the World Bank Development Report (IBRD, 1992). As Stern (2004) indicated, the World Bank's 1992 report argued that "the view that greater economic activity inevitably hurts the environment is based on static assumption about technology, tastes, and environmental investment" (p.38).

It further stated that as incomes rise, the demand for improvements in environmental quality will increase, as will the resources available for investment (p.39; see also Beckerman, 1992, p.491; Lomborg, 2001; Cole, 2003a; Panayotou, 1993). According to Stern (2004), there are proximate causes of the EKC relationship which include, scale, changes in economic structure (product or output mix), changes in technology, and changes in input mix. There are also underlying causes such as environmental regulation, awareness, and education, which can only have effect via the proximate variables. The scale effect occurs if there were no changes in the structure or technology of the economy, pure growth in the scale of the economy would result in growth in pollution and other environmental impact (See figure 3.1). The output or product mix implies that different industries have different pollution intensities.

During the course of economic development the output-mix changes, for example, the earlier phases of development may see a shift away from agriculture toward heavy industry that increases emissions. The later stages of development

Table 3.2. Causes of Endangerment for American Species Classified as Threatened or Endangered by the U.S. Fish and Wildlife Service

Cause (Human Activities)	*Number of Species Endangered by Cause*	*Estimated Number of Species Endangered by Cause*
Interactions with nonnative species	305	340
Urbanization	275	340
Agriculture	224	260
Outdoor recreation and tourism development	186	200
Domestic livestock and ranching activities	182	140
Reservoirs and other running water diversions	161	240
Modified fire regimes and silviculture	144	80
Pollution of water, air, or soil	144	140
Mineral, gas, oil, and geothermal extraction or exploration	140	140
Industrial, institutional, and military activities	131	220
Harvest, intentional and incidental	120	220
Logging	109	80
Road presence, construction, and maintenance	94	100
Loss of genetic variability, inbreeding depression, or hybridization	92	240
Aquifer depletion, wetland draining or filling	77	40
Native species interactions, plant succession	77	160
Disease	19	20
Vandalism (destruction without harvest)	12	0

Source: Czech, B. and P. R. Krausman (1997). "Distribution and Causation of Species Endangerment in the United States." *Science* 277: 1116–1117.

may see a shift from the more resource intensive and extractive heavy industrial sector toward services and light manufacturing that have lower emissions per unit output. The input mix involves the substitution of less environmentally damaging inputs and vice versa, e.g., substituting natural gas for coal and low sulfur coal in place of high sulfur coal. The improvement in the state of technology (or technique effect) involves changes in productivity, in terms of using less, ceteris paribus, of polluting inputs per unit output, and emissions specific changes in process result in lower emissions per unit input.

Some studies have shown that there is little evidence for a common inverted U-shaped pathway that countries follow as income rises. Some found that EKC does not exist (Perman and Stern, 2003; Copeland and Taylor, 2004), especially when diagnostic statistics and specification tests are done. Research has also shown that some pollutants do not obey the Kuznets curve pattern, especially the new pollutants or toxics such as carcinogenic chemicals, carbon dioxide, etc., that are replacing the traditional pollutants such as SO_2 and NO_x emissions (Dasgupta et al, 2002). With "new toxics," EKC shows a monotonic increase of emissions in income. That is, they found that emissions of most pollutants and flows of waste are monotonically rising with income

Specific factors contribute to economic productivity. As noted by Denison (1985), such factors include increases in knowledge, education, allocation improvement, and economies of scale. Technological or technical progress is said to have occurred when increases in knowledge and education result in invention and innovation that lead to more efficient production (Pearce, 1992). In the native, technological progress refers to invention and innovation. In economics, technological progress occurs when more is produced with a given amount of resource input which is synonymous with rising productive efficiency. The various economic sectors have a "trophic structure" as in the economy of nature (Czech, 2003), and technological progress has long been characterized the dynamics of this trophic structure. There are two basic types of technological progress: extractive and end-use (Wils, 1998). The purpose of extractive technology is the extraction of natural capital, which is not conducive to biodiversity conservation because it reallocates more natural capital from the economy of nature to human economy. End-use technological progress increases productive efficiency without necessitating a reallocation of natural capital from the economy of nature to human economy, e.g., the development of more efficient engines. Efficiency has been described as just a tool for lessening the ecological impacts of production and consumption and like any other tool, can be used or abused. Czech (2003), specified that efficiency is no substitute for frugality, and economic growth is no substitute for a steady-state economy. Frugality induces efficiency, but efficiency does not induce frugality (Daly, 2002; Czech, 2003). It is seen from the above discussions that technological progress may not reconcile the fundamental conflict between economic growth and biodiversity conservation. Instead, the macroeconomic implication is that a steady-state economy with stable population and per capita consumption (i.e., stable GDP) is required for biodiversity conservation and economic sustainability.

There has been a growing argument that if a steady-state economy is adopted, it would deny all nations the means of creating more technology and wealth. Such measures are believed to be poor social and economic policy and would be counterproductive which would likely cause more environmental harm than good, and might even bring about the very catastrophe environmentalists strive to avert (Goklany and Sprague, 1992). No wonder the United States government refused to sign the Convention on Biodiversity Diversity, which would reduce the incentives for research and development (R&D), the very technologies that would help meet the competing demands made on land by humans and other species.

As the world population is likely to increase and there is the likelihood that the average economic wellbeing of that population will improve, there should be a continued technological progress to maintain and improve the quality of life of all the societies of the world. According to Goklany and Sprague (1992), the only way to feed, clothe, and shelter the greater world population that the future will inevitably bring, while limiting deforestation and loss of biodiversity and carbon-dioxide sinks, is to increase, in environmentally sound manner, the productivity of all activities that use land. Such increases in production are possible only within a legal, economic, and institutional framework that relies on free markets, fosters decentralized decision-making, respects individual property rights, and rewards entrepreneurship. This is also true for developing countries (see Asafu-Adjaye, 2000).

POLICY RECOMMENDATIONS AND PROSPECTS

There is no magic or existing blue print for adequate biodiversity conservation, because the problem is intractable. Nonetheless, we need to protect the future and the planet which is the only habitat we have. An approach to sustainable biodiversity conservation should take into account the biological, socio-cultural, and economic parameters. There is the need to take a global view of the consequences of actions that affect the productivity of land and support a more comprehensive, careful, and objective analysis of the benefits and costs of such actions. The recognition of the role and responsibilities of all stakeholders is likely to offer the best guarantee for sustainable biodiversity conservation.

1. The best hope for a world that is facing severe pressure on its land base is to devise strategies that should lead to sustainable development, conserve biological diversity, and combat global deforestation. In a world or ecological system that population growth and rise in per capita consumption are inevitable, the strategies should be directed toward improving the efficiency and productivity of all activities that use land while ensuring that those activities are environmentally sound (Goklany and Sprague, 1992). Habitat liquidation is a prerequisite of economic growth. Population and per capita consumption parameters are primary in determining economic scale, and it is the economic activities of the population that impact the ecosystem.
2. It is argued that the technology nostrum creates an augmented carrying capacity which rest on capital consumption rather than the profit. It is chimerical as well as "phantom carrying capacity" that is fated to be temporary (Catton, 1980; Hubbert, 1987; Attarian, 2004). Instead of allowing growth to come to a halt by itself, society should seek purposefully to achieve a steady-state economy (Czech, 2003; Daly, 1991; Goldsmith, 1991; Czech and Daly, 2004; Feinman, 2005; Attarian, 2004) or an "equilibrium society" (Meadows et al, 1972; 1992).
3. Maintaining consistent Conservation Policy Goals at local, regional, and national levels. For example, (a) educational programs in primary, secondary,

and higher institutions to capture the attention of the young individuals who are the future custodians of the planet, e.g., Conservation or General Biology curriculum should include materials on the conflict between economic growth and biodiversity conservation [Society for Conservation Biology (SCB), 2005]. The SCB also noted that concepts central to the national conservation policy should be developed and featured prominently in "media talking points," societies' and organizations' publications, conferences, workshops, and meetings; (b) protection of biodiversity/managing biodiversity. Protection and management actually call for the institution of frugality and thrift in the conservation of nature. It basically merges development with conservation agendas, which should be socially and culturally determined, thus the challenges of human pressure (development) and the need to provide for the livelihoods of multitudes of poor people (WRI, 1996; Dovie, 2003; Hunter et al, (2003). This also entails the community based natural resource management (CC) or community-based conservation, which calls for local participation in nature protection. This is natural resource or diversity protection by, for, and with local communities e.g., in Thailand, Ethiopia, and Norway (IBCR, 2001; Daugstad, 2006; Trisurat, 2006). Perhaps, this may provide the much-needed meaningful balance for meeting local needs; and (c) growth management policies and programs. Growth management is a tool for implementing planning and land use control. Without repudiating growth, it is advanced to address the ills associated with future growth (McHarg, 1994; Nelson and Duncan, 1995; Kelly, 2004; Zovanyi, 1998, 2005).

4. There is the need to address the underlying causes of biodiversity loss, especially in developing countries, which include poverty, lack of property rights and tenure regimes, lack of adequate rural infrastructure, health and education services, and lack of employment opportunities. In more developed countries of Europe and North America, the voracious appetite for natural resources should be reduced and managed. With their not-so-much growing population, they consume more than their fair share, especially the United States of America. Women are the backbone of many households in developing countries, especially in Asia and African countries, and they are highly affected by unemployment and poverty. In the gender division of labor, women are often more aware of the environmental degradation because they make up a significant proportion of agricultural workforce. They are heavily involved in food production from planting to cultivation, harvesting, at times in most unsustainable ways. Thus the empowerment of women for gender equity, poverty reduction, and environmental sustainability is urgently needed. Empowerment is guaranteed or achieved through education, employment, and political representation (Aka, 2005/2006).
5. Cultural policy should be one of the main components of endogenous and sustainable development policy (Aka, 2005/2006). It has been noted that the

destabilization and sheer destruction of cultures are traceable to unsustainable environmental and economic policies. Sustainable development and the flourishing of culture, especially the preservation of indigenous knowledge culture, are interdependent (UNDP, 2004; UNESCO, 1998). Different cultures have different ways of preserving their heritage, including natural resources and biodiversity.

6. The use of indigenous knowledge or community knowledge to preserve nature and biodiversity. Indigenous knowledge can be called "local knowledge," "vernacular knowledge/science," "citizen science," "gray literature" or "anecdotal information" (Warren, 1992). They help guide development along a path that is both ecologically and socially sound, creating new models of development that help think globally but act locally. Indigenous knowledge provides the bases for grassroots decision-making, much of which takes place at the community level through indigenous organizations where problems are identified and solutions to them are determined. Rural communities have the indigenous knowledge base that is sophisticated in understanding of their environment, e.g., in Niger and Sierra Leone, among others. There is a need in all countries to create National Indigenous Resource Centers for systemizing the local resource base. Their functions are to record, codify, and store indigenous knowledge (intellectual property) that might be exchanged among cultures, regions, and nations, as found already in Netherlands, Ghana, Nigeria, Indonesia, Mexico, Philippines, Kenya, etc. (Warren, 1992).
7. Create Institutes or Centers for Biodiversity as vehicles to endorse, sign, and implement international agreement and obligations on biodiversity, e.g., as found in Ethiopia; Hague and Leiden, Netherlands; Ibadan, Nigeria; Silong, Cavite, Philippines; Cape Coast, Ghana; Bandung, Indonesia; Puebla, Pue, Mexico; Nairobi, Kenya; and Ames, Iowa, United States of America. Their general objective is to undertake conservation study, research, and promote the development and sustainable utilization of the country's biodiversity (Warren, 1992; IBCR, 2001). It also includes the consideration of the potential multiple uses that wildlife and biodiversity can provide, such as recreation, food, scientific, cultural, economic, and ecological functions (FAO, 2003).

These biological diversity conservation efforts play a role in in-situ conservation with emphasis on local and national needs and values.

CONCLUSIONS

There is an agreement, especially among conservationists such as conservation biologists, ecologists, and ecological economists that fundamental conflict exists between economic growth and biodiversity conservation. There is a strong correlation between increase in economic growth or gross domestic product (GDP) and biodiversity loss. The negative effects of economic growth on biodiversity are quite robust whether in developed or developing countries. Theoretical and empirical arguments suggest that economic growth is the primary challenge to biodiversity conservation and ultimately to human economic sustainability.

The paper explores theoretically and empirically the problems of biodiversity in different cultures, areas, and regions. The challenges and prospects are explored, as the society endeavors to balance environmental gains with economic growth and social development. The ecological footprints of human economy are highlighted in the United States and other countries as it liquidates the natural capital resources. The human or anthropogenic use of the environment is the largest contributor to habitat modification and ecosystem loss. Theoretical construction rests upon the principles of carrying capacity, niche breadth, competitive exclusion, and trophic levels. Empirical evidence lies in the trends of species endangerment, ecosystem loss, and reduction of ecosystems and habitats. Urban sprawl is implicated as the leading cause of habitat loss and species endangerment. The United States is among the rich nations which "sky-high" consumption patterns have remained a major threat to the global environment.

The study is exploratory and cross-sectional in approach. Several indicators that inform the nature and impacts of anthropogenic threat to biodiversity and conservation include, among others, ecosystem loss by region, species endangerment by region; and species threat by region. The regions include Asia, Latin America, Europe, North and Central America, South America, former Soviet Union, Oceania, USA and Africa. The United States of America was studied as a special case by using several indicators that cause species endangerment, such as urbanization, interaction with non-native species, agriculture, harvest, logging, road presence and construction, disease and vandalism. Most of the secondary data were collected from World Resource Institute (1996–1997); and The Wildlife Society (2003), Technical Review 03-1. Analyses of data indicate that an inverse relationship exists between the human economy and natural capital resources, as well as a direct relationship between human economy and biodiversity loss.

Economic growth, which is the synthesis of increase in population and rise in per capita consumption, diverts resources from other species for the exclusive use of humans, leaving a negative and dangerous ecological footprint (Czech, et al, 2000; Burchell et al, 1998; Wear and Greis, 2002; Fan et al, 2005; Goudie, 2000; World Resources Institute, 2000; The Wildlife Soceity, 2003; Czech, 2003; Wackernagel et al, 2002; Asafu-Adjaye, 2000; Czech and Daly, 2004; Goklany and Sprague, 1992). A major disagreement exists on how to balance economic growth and biodiversity conservation, and still maintain a sustainable development. A major school of thought believes and pontificates that technological progress is the reconciler of the environmental impacts of human economy on the ecological systems and the growth of the economy or the gross domestic product (GDP), thus the belief on the ecological Kuznets curve (EKC) (Grossman and Krueger, 1991; IBRD, 1992; Lomberg, 2001; Stern, 2004; Pearce, 1992; Panyotou, 1993; Beckerman, 1992). Some do not believe that EKC exists, especially with the new toxics (CO_2 and carcinogenic chemicals), even if it exists it is for the old pollutants such as SO_2 and NO_x (Perman and Stern, 2003; Copeland and Taylor, 2004; Dasgupta et al, 2004). They found out that EKC shows a monotonic increase in income. Technological progress pontificators believe that increases in knowledge and education result in invention and innovation that lead to more efficient production (Pearce, 1992; Goklany and Sprague, 1992). According to Goklany and Sprague (1992), technological progress in the last century has enabled the United States to feed a much larger population a healthier diet without significant total land conversion for agriculture, thereby protecting natural habitats and biological diversity. Goklany and Sprague (1992, p.15), argue that the consequences of low-input sustainable agriculture (LISA) is that more land would be devoted to crops than should be otherwise. Their stance is that equal emphasis should be on both high output and the sustainability criteria, and what is needed is not LISA but high output sustainable agriculture (HOSA).

The second school of thought argues that the technological nostrum creates an augmented carrying capacity which rests on natural economy's capital consumption. It further indicates that this is chimerical as well as a "phantom carrying capacity" that is fated to be temporary (Catton, 1980; Hubbert, 1987; Attarian, 2004). Instead of allowing growth to come to halt by itself, society should seek purposefully to achieve a "steady-state economy' (Czech, 2005; Daly, 1991; Czech and Daly, 2004; Feinman, 2005; Attarian, 2004) or purposefully a decline in economic growth (Goldsmith, 1979) or an "equilibrium society" (Meadows et al, 1972; 1992). They argue that the technological progress, which characterizes the dynamics of the trophic structure, is to extract more of the natural capital which is not conducive to biodiversity conservation. The extractive technology reallocates more natural capital from the economy of nature to the human economy. The end-use technological progress increases productive efficiency, but efficiency has been described as just a tool for lessening the ecological impacts of production and consumption, and like any other tool, can be used or abused. Czech (2003) specified that efficiency is no substitute for frugality, and economic growth is no substitute for a steady-state economy. Frugality induces efficiency, but efficiency does not induce frugality (Daly, 2002; Czech, 2003). They called for a limit to technological progress (R&D) which is financed by increase in economic growth or GDP.

It is a fact that economic growth has negative effects on biodiversity or ecosystem conservation, but curtailing economic growth in the presence of ever increasing population and per capita consumption seems to be unattainable, unsustainable, counterproductive, and utopian. On the other hand, one should be cognizant of the fact that there is a limit to infinite technological progress and economic growth in a finite world system (Meadows et al, 1972; 1992). Growth constitutes unsustainable behavior because it is incapable of being continued or maintained indefinitely (Zovanyi, 2005). Thus, while recognizing that biodiversity conservation problem is intractable, and there is no magic or existing blue print for adequate conservation while maintaining economic growth and well-being, the policy recommendations in this study call for:

1. Improving the efficiency and productivity of all activities that use land while ensuring that those activities are environmentally sound (Goklany and Sprague, 1992). This is based on the fact that global population growth and rise in per capita consumption are inevitable in the future.
2. Economic growth liquidates natural resources and biodiversity, thus erodes the nature's carrying capacity, technological progress notwithstanding. Instead of alloying growth to come to a halt by itself, society should seek purposefully to achieve a steady-state economy (Czech and Daly, 2004; Goldsmith, 1979; Feinman, 2005) or what Meadows et al (1972, 1992) called "equilibrium society." Currently, the universality and practicality of this development alternative are in doubt, but should be pondered.
3. Maintain a consistent conservation policy goals at local, regional, and national levels through: (a) educational programs at all levels to capture the attention of the youthful individuals who are the future custodians of the planet (SCB, 2005); (b) growth management policies and programs to implement land use planning, management, and control, thus, "smart growth" instead of the existing "dumb growth" (McHarg, 1994; Nelson and Duncan, 1995; Kelly, 2004; Zovanyi, 1998, 2005); and (c) protecting and managing biodiversity through the institution of frugality and thrift in the conservation of nature, while providing for the livelihoods of the putative beneficiaries (WRI, 1996; Dovie, 2003; IBCR, 2001; Daugstad, 2006).
4. Addressing the underlying causes of biodiversity loss, especially in developing countries that include poverty, lack of property rights and tenure, lack of adequate rural infrastructure, health and education services, lack of employment (Asafu-Adjaye, 2000; Ukanga and Afoaku, 2005), and empowerment of women, who are closest to nature, for environmental sustainability, through education, employment, and political representation (Aka, 2005/2006); and curtailing of the voracious appetite of the more developed countries for natural resources is also needed, at least for intergenerational equity.
5. Cultural policy that is one of the main components of endogenous and sustainable development policy. Sustainable development and the flourishing of culture, especially the preservation of indigenous knowledge culture, are interdependent (UNESCO, 1998; UNDP, 2004; Aka, 2005/2006). Different

cultures have different ways of preserving their heritage, including natural resources and biodiversity.

6. The use of indigenous knowledge or community knowledge, local knowledge, vernacular knowledge/science, citizen science, gray literature, and anecdotal information to preserve nature and biodiversity (Warren, 1992). They help guide development along a path that is both ecologically and socially sound, creating new models of development that are on the path of thinking globally but acting locally
7. Creating Institutes or Centers for Biodiversity as vehicles to endorse, sign, and implement international agreement and obligations on biodiversity. The objective is also to undertake conservation, study, research, and promote the development and sustainable utilization of the country's biodiversity (Warren, 1992; IBCR, 2001).

Therefore, by recognizing that infinite growth is impossible in a finite world (Meadows et al, 1972), we need to be more frugal and efficient in any production process that uses land in order to protect the future and the planet.

REFERENCES

Aka, Ebenezer O. (2005/2006). "Gender Equity and Sustainable Socio-Economic Growth and Development: A Cross-Cultural Analysis." The International Journal of Environmental, Cultural, Economic & Social Sustainability, Volume 1, 32p.

Aka, Ebenezer O. (2006). "Biodiversity, Conservation, and Sustainable Socioeconomic Development: Problems and Prospects." Paper presented at the Second International Conference on Sustainable Heritage Development: Environmental, Cultural, Economic and Social Sustainability, Hanoi and Halong Bay, Vietnam, 9–12 January, 2006.

Asafu-Adjaye, John (2000). "Biodiversity Loss and Economic Growth: A Cross-Country Analysis." Paper presented at the Western Economic Association International Annual Conference in Vancouver, BC, Canada.

Attarian, John (2004). "The Steady-State Economy: What It Is, Why We Need It." NPG. Accessed Online on January 2, 2006. 12p. http://www.npg.org/forum_series/steady-state.html.

Barbier, E.B. (1997). Environ. Dev. Econ, 2, 369.

Beckerman, W. (1992). "Economic Growth and the Environment: Whose Growth? Whose Environment?" World Development, 20, 481–496.

Boulding, K. E. (1996). "The Economics of the Coming Spaceship Earth," pages 3–14 in H. Jarrett, editor. Environmental Quality in a Growing Economy (Baltimore, Maryland, USA: Johns Hopkins University Press).

Burchell, R. W.; Shad, N. A.; Listokin, D.; et al (1998). The Cost of Sprawl-Revisited. TCRP Report 39, Transit Cooperative Research Program, Transportation Research Board, National Research Council (Washington, D.C.: National Academy Press) 268 p. Available Online: http://gulliver.trb.org/publications/tcrp_rpt_39-a.pdf.

Catton, Jr., William R. (1980). Overshoot: The Ecological Basis of Revolutionary Change (Urbana and Chicago: University of Illinois).

Chen, D.T. (2000). "The Science of Smart Growth." Scientific American, 283 (6): 84–91.

Cole, M. M. (2003a). "Environmental Optimists, Environmental Pessimists and the Real State of the World," an article examining the "Real State of the World," by Bjorn Lomborg, Economic Journal, 113, 362–380.
Collins, R. M. (2000). More: The Political Economy of Growth in Postwar America (Oxford, United Kingdom: Oxford University Press).
Commission for Environmental Cooperation (2001). The North American Mosaic: A State of the Environment Report. Commission for Economic Cooperation, Montreal, Quebec, Canada.
Constanza, R.; J. Cumberland; H. Daly; R. Goodland; and R. Norgaard (1997). An Introduction to Ecological Economics (Boca Raton, Florida, USA: St. Lucie Press).
Copeland, B. R., and Taylor, M. S. (2004). "Trade, Growth and the Environment," Journal of Economic Literature, 42, 7–71.
Czech, B. (2000a). "Economic Growth as the Limiting Factor for Wildlife Conservation," Wildlife Society Bulletin, 28 (1): 4–15.
Czech, B. (2000b). Shoveling Fuel for a Runaway Train: Errant Economists, Shameful Spenders, and A Plan to Stop Them All (Berkeley, USA: University of California Press).
Czech, B. (2002a). "The Importance of Ecological Economics to Wildlife Conservation: An Introduction." Wildlife Society Bulletin, 28: 2–3.
Czech, B. (2003). "Technological Progress and biodiversity Conservation: A Dollar Spent, A Dollar Burned." Conservation Biology, Volume 17, No. 5, (October) pp. 1455–1457.
Czech, B.; P .R. Krausman; and P. K. Devers (2000). "Economic Associations Among Causes of Species Endangerment in the United States." BioScience, 50: 593–601.
Czech, B.; and P. R. Krausman (2001). The Endangered Species Act: History, Conservation Biology, and Public Policy (Baltimore, Maryland, USA: Johns Hopkins University Press).
Czech, B.; and H. E. Daly (2004). "In My Opinion: The Steady- State Economy—What It Is, Entails, and Connotes." Wildlife Society Bulletin 2004, 32 (2): 598–605.
Czech, B.; and P. Pister (2005). "Economic Growth, Fish Conservation, and The American Fisheries Society: Introduction to A Special Series," Fisheries, 30 (1): 38–40.
Daily, G. C.; and p. R. Ehrlich (1992). "Population, Sustainability, and Earth's Carrying Capacity: a framework for estimating population sizes and lifestyles that could be sustained without undermining future generations." BioScience, 42: 761–771.
Daly, H. E. (1991). Steady-State Economics, 2nd Edition (Washington, D.C.: Island Press).
Daly, H. E. (1996). Beyond Growth: The Economics of Sustainable Development (Boston, Massachusetts, USA: Beacon Press).
Daly, H. E. (1997). Steady-State Economics: The Economics of Biophysical Equilibrium and Moral Growth (San Francisco, California, USA: W.H. Freeman).
Daly, H. (2002). "Sustainable Development: Definitions, Principles, Policies." (Washington, D.C.: World Bank).
Daugstad, Karoline (2006). "Local Participation in Nature Protection: Magic or Mistake?" Paper presented at The Second International Conference on Sustainable Heritage Development: Environmental, Cultural, Economic and Social Sustainability, Hanoi and Halong Bay, Vietnam, 9–12, January 2006.
Dasgupta, S.; Laplante, B.; Wang, H.; and Wheeler, D. (2002). "Confronting the Environmental Kuznets Curve," Journal of Economic Perspectives, 16, 147–168.
Denison, E. F. (1985). Trends in American Economic Growth, 1929–1982 (Washington, D.C., USA: Brookings Institution Press).
Dovie, Delali B.K. (2003). "Determining Livelihood and Disputing Biodiversity: Whose Dilemma?" Ethics, Place and Environment, Vol. 6, No. 1, (March), pp. 27–41.
Ehrlich, P. R. (1994). "Ecological Economics and the Carrying Capacity of Earth," pages 38–56, in A.M. Jansson, M. Hammer, C. Folke, and R. Constanza, editors. Investing in Natural Capital: the Ecological Economics Approach to Sustainability (Washington, D.C., USA: Island Press).

Ehrlich, P.R.; and A.H. Ehrlich (1990). The Population Explosion (New York, USA: Simon and Schuster).

Erickson, J.D.; and J.M. Gowdy (2000) "Resource Use, Institutions, and Sustainability: A Tale of Two Pacific Island Cultures." Land Economics, 76: 345–354.

Eves, R.; A. Gilbert; S. Higgins; C. Howard; S. Jones; D. Martin; D. McKelvey; H. Nelder; N. Wiseman; and S. Wright, compliers (1998). The Economist Pocket World in figures. 1999 Edition (New York, USA: John Wiley & Sons).

Fan, D.P.; David N. Bengston; Robert S. Potts, and Edward G. Goetz (2005). The Rise and Fall of Concern about Urban Sprawl in the United States: An Updated Analysis. Published by: North Central Research Station Forest Service, U.S. Department of Agriculture. 1992 Folwell Avenue, St. Paul, MN 55108. Available Online: http:/www.ncrs.fs.fed.us.

FAO (2003). "The State of Tropical Forest Management: Forest Management for Conservation and Protection." Food and Agricultural Organization (FAO) of the United Nations (New York, N.Y 10017, USA). Accessed Online on December 27, 2005, at: http://www.fao.org/DOCREP/003/X4110e/X4110E05.htm.

Federal Reserve (2000). The 87th Annual Report, Board of Governors of the Federal Reserve System, Washington, D.C. Available Online at: http:/www.federalreserve.gov/boarddocs/Rpt Congress/annual00/ar00.pdf.

Feinman, R.D. (2005). "Planning for a Steady State (No Growth) Society," Accessed Online on January 2, 2006. 4p. at: http://robertdfeinman.com/society/no_growth.html.

Gillham, O. (2002). The Limitless City: A Primer On the Urban Sprawl Debate (Washington, D.C.: Island Press), 309p.

Goklany, I.H., and M.W. Sprague (1992). Policy Analysis. Sustaining Development and Biodiversity: Productivity, Efficiency, and Conservation. Cato Policy Analysis No. 175 (August 6): 18p. Accessed Online on 12/27/2005, at: http://www.cato.org/pubs/pas/pa-175.html.

Goldsmith, E. (1979). "The Steady State Economy," Ecologist, Volume 9, No. 3 (June), 3p. Accessed Online on January 2, 2006, at: http://www.edwardgoldsmith.com/page91.html.

Goudie, A. (2000). The Human Impact on the Natural Environment. Fifth Edition (Cambridge, Massachusetts, USA: Harvard University Press).

Grossman, G.M., and Krueger, A.B. (1991). Environmental Impacts of a North American Free Trade Agreement (NAFTA). National Bureau of Economic Research Working Paper 3914, NBER, Cambridge, MA.

Harf, James E., and Lombardi, Mark O. (2004). Taking Sides. Clashing Views on Controversial Global Issues. Second Edition (Guilford, Connecticut: McGraw-Hill).

Harrison, P., and F. Pearce (2000). AAAS Atlas of Population & Environment (Berkeley, USA: University of California Press).

Hawken, P. (1993). The Ecology of Commerce: A Declaration of Sustainability (New York, USA: Harper Collins).

Hubert, King M. (1987). "Exponential Growth as a Transient Phenomenon in Human History," in Margaret Strom, ed., Societal Issues, Scientific Viewpoints (New York and Washington: American Institute of Physics) pp. 76–77.

Hunter, Lori M.; John Beal; and Thomas Dickinson (2003). "Integrating Demographic and GAP Analysis Biodiversity Data: Useful Insight?" Human Dimensions of Wildlife, 8: 145–157.

Hutchinson, G.E. (1978). An Introduction to Population Ecology (New Haven, Connecticut, USA: Yale University Press).

IBRD (1992). World Development Report 1992. Development and the Environment (New York: Oxford University Press).

Institute of Biodiversity Conservation and Research (IBCR) (2001). Accessed Online on December 27, 2005, at: http://www.telecom.net.et/ibcr.

Kellert, S.R. (1984). "Assessing Wildlife and Environmental Values in Cost-Benefit Analysis," Journal of Environmental Management 18: 355–363.
Kellert, S.R. (1996). The Value of Life: Biological Diversity and Human Society (Washington, D.C.: Island Press).
Kelly, Eric D. (2004). Managing Community Growth. Second Edition (Westport, Connecticut, London: Praeger).
Kingdon, J. (1993). Self-Made Man: Human Evolution from Eden to Extinction (New York: Wiley).
Knox, P., and J. Agnew (1998). The Geography of the World Economy. Third Edition (New York, USA: John Wiley & Sons).
Krizek, K.J., Power, J. (1996). A Planner's Guide to Sustainable Development. Planning Advisory Service Report Number 467 (Chicago, IL: American Planning Association), 66p.
Kuznets, S. (1955). "Economic Growth and Income Inequality." American Economic Review, 49, 1–28.
Leopold, A. (1933). Game Management (New York, USA: Charles Scribner's Sons).
Lomborg, B. (2001). The Skeptical Environmentalist: Measuring the Real State of the World (Cambridge: Cambridge University Press).
Lorentz, A., and Shaw, K. (2000). "Are You Ready to Bet on Smart Growth?" Planning, 66 (1): 4–9.
Matthiessen, P. (1959). Wildlife in America (New York, USA: Viking Press).
McHarg, Ian L. (1994). Design With Nature, 25th Anniversary Ed. (New York: John Wiley).
Mckibben, B. (1998). May be One: A Personal and Environmental Argument for Single-Child Families (New York, USA: Simon & Schuster).
Meadows, D.H.; Meadows, D.L.; Randers, J; Behrens, W.W. (1972). The Limits to Growth (New York, NY: Universe Books) 205 p.
Meadows, D.H.; D.L. Meadows, and J. Randers, (1992). Beyond the Limits: Confronting Global Collapse, Envisioning A Sustainable Future (Post Mills, Vermont, USA: Chelsea Green) 300 p.
Myers, N. (1994). "Population and Biodiversity, "in F. Graham-Smith (ed.), Population: The Complex Reality (London: North American Press), pp. 117–136.
National Research Council (2000). Beyond Six Billion: Forecasting the World's Population. J. Bongaarts and R.A. Bulatao, editors. Panel on Population Projections, Committee on Population, Commission on Behavioral and Social Sciences and Education (Washington, D.C., USA: National Academy Press).
Nebel, B.J.and R.T. Wright (2000). Environmental Science: The Way the World Works. Seventh Edition (Upper Saddle River, New Jersey, USA: Prentice Hall).
Nelson, A.C., and Duncan, J.B. (1995). Growth Management Principles and Strategies (Chicago, IL: Planners Press).
Noss, R. (1990). "Indicators for Monitoring Biodiversity: A hierarchical Approach," Conservation Biology, 4, 355–364.
Office of Technology Assessment (OTA) (1987a). Combined Summaries: Technologies to Sustain Tropical Forest Resources and Biological Diversity (Washington, D.C.: Office of Technology Assessment).
Office of Technology Assessment (OTA) (1987b). Technologies to Maintain Biological Diversity, OTA-F-330 (Washington, D.C.: Government Printing Office).
Panayotou, T. (1993). Empirical Tests and Policy Analysis of Environmental Degradation at Different Stages of Economic Development, Working Paper WP238, Technology and Employment Programme, International Labour Office, Geneva.
Pearce, D.W. (1992). The MIT Dictionary of Modern Economics. Fourth Edition (Cambridge, Massachusetts, USA: MIT Press).
Perman, R., and Stern, D. I. (2003). "Evidence from Panel Unit Root and Cointegration Tests that the Environmental Kuznets Curve Does not Exist." Australian Journal of Agricultural and Resource Economics, 47, 325–347.

Pulliam, H.R., and N.M. Haddad (1994). "Human Population Growth and The Carrying Capacity Concept." Bulletin of the Ecological Society of America 75: 141–157.

Ricklefs, R.E., and G.L. Miller (2000). Ecology. Fourth Edition (New York, USA: W.H. Freeman).

Schor, J.B. (1997). The Overspent American: Upscaling, Downshifting, and the New Consumer (New York, USA: Basic Books).

Shafik, N., and Bandyopadhyay, S. (1992). Economic Growth and Environmental Quality: Time Series and Cross-country Evidence. Background Paper for the World Development Report 1992. The World Bank, Washington, D.C.

Simon, J.L. (1996). The Ultimate Resource 2 (Princeton, New Jersey, USA: Princeton University Press).

Smith, D. (1999). The State of the World Atlas. Sixth edition (New York, USA: Penguin Books).

Society for Conservation Biology (SCB) (2005). "Editorial: Economic Growth, Biodiversity Conservation, and SCB." Accessed Online on 12/27/2005, at: http://www.conbio.org/SCB/Publications/Newsletter/Archives/2005.

Stern, David I. (2004). "The Rise and Fall of the Environmental Kuznets Curve." World Development, Vol. 32, No. 8, pp. 1419–1439.

Suzuki, D. (1998). Earth Time: Essays (Toronto, Ontario, Canada: Stoddart).

The Wildlife Society (TWS) (2003). The Relationship of Economic Growth to Wildlife Conservation, Technical Review 03-1 (Bethesda, MD: Wildlife Society).

Trisurat, Yongyut (2006). "Community-based Wetland Management in Northern Thailand." Paper presented at The Second International Conference on Sustainable Heritage Development: Environmental, Cultural, Economic and Social Sustainability, Hanoi and Halong Bay, Vietnam 9–12, (January), 2006.

Ukanga, Okechukwu, and Afoaku, Osita G. (2005). Sustainable Development in Africa. A Multifaced Challenge (Trenton, NJ: African World Press, Inc.).

United Nations Development Program (UNDP) (1993). Global Biodiversity. UNEP/GEMS Environmental Library No. 11, Nairobi: UNEP.

United Nations Development Program (UNDP)(2004). Human Development Report 2004. Cultural Liberty in Today's Diverse World (New York: Oxford University).

UNESCO (1998). The Power of Culture. The Action Plan. See the "Preamble." The Intergovernmental Conference on Cultural Policies for Development, held at Stockholm, Sweden (30 March–2 April). Accessed Online on 10/1/1999, at http//www.unesco-sweden.org/Conference/Action-Plan.html.

U.S. Agency for International Development (2002). This is USAID. Available Online at http://www.usaid.gov/about/. Accessed 23 February 2002.

U.S. Census Bureau (2000). "Census 2000 Shows Resident Population of 281,421,906; Appointment Counts Delivered to President." U.S. Department of Commerce News, December 28, 2000. Available Online at http://www.census.gov/Press-Release/www/2000/cb00cn64.html.

U.S. Department of Commerce (2002). Mission Statement. Available Online at http://www2.osec.doc.gov/public.nsf/docs/mission-statement/. Accessed Feb. 23, 2002.

U.S. Department of the Treasury (2002). Mission, Goals and Results. Available Online at http://www.ustreas.gov/gpral/.

Wackernagel, W.M. et al (2002). "Tracking the Ecological Overshoot of the Human Economy," Proceedings of the National Academy of Sciences of the United States of America, Vol. 99, No. 14 (July) pp. 9266–9271.

Warren, D.M. (1992). "Indigenous Knowledge, Biodiversity Conservation and Development." Keynote Address. Center for Indigenous Knowledge for Agriculture and Rural Development. Iowa State University, Ames, Iowa 50011, USA. Accessed Online on December 276, 2005, at http://www.ciesin.org/docs/004-173/004-173h.

Wear, D.N., and Greis, J.G. (2002). Southern Forest Resource Assessment. Gen. Tech. Rep. SRS-53. Ashville, N.C.:U.S. Department of Agriculture, Forest Service, Southern Research Station. Available Online at: http://www.srs.fs.fed.us/sustain/.

Wilcove, D.S.; D. Rothstein; J. Dubow; A. Phillips; and E. Loses (1998). "Quantifying Threats to Imperiled Species in the United States." BioScience 48: 607–616.

Willers, B. (1994). "Sustainable Development: A New World Deception." Conservation Biology, 8: 1146–1148.

Wils, A. (1998). End-use or Extraction Efficiency in Natural Resource Utilization: Which is Better? System Dynamics Review 14: 163–188.

World Commission on Environment and Development (WCED) (1987). Our Common Future (Oxford: Oxford University Press).

World Resources Institute (WRI) (1994). World Resources, 1994–1995 (New York: Oxford University Press).

World Resources Institute (WRI) (1996). World Resources 1996–1997. A Guide to the Global Environment. The Urban Environment. A Joint Publication by: The World Resources Institute, The United Nations Environment Program, The United Nations Development Program, and The World Bank (New York, Oxford: Oxford University Press).

World Resources Institute (2000). World Resources, 2000–2001: People and Ecosystems; The Fraying Web of Life (Washington, D.C., USA: World Resources Institute).

Zovanyi, G. (1998). Growth Management for a Sustainable Future: Ecological Sustainability as the New Growth Management Focus of the 21st Century (Westport, CT: Praeger Publishers).

Zovanyi, G. (2005). "Urban Growth Management and Ecological Sustainability Confronting the 'Smart Growth Fallacy,'" in Policies for Managing Urban Growth and Landscape Change. A Key to Conservation in the 21st Century. Published by North Central Research Station, Forest Service U.S. Department of Agriculture, St. Paul, MN 55108.

Chapter Four

Gender Equity and Sustainable Socio-Economic Growth and Development

A Cross-Cultural Analysis

INTRODUCTION

Sustainable socioeconomic growth and development project has gendered aspects in favor of male, with some shades of variation from one culture to another. Gender refers to the different roles men and women play in society and the relative power they wield (Ashford, 2001). Cultural and social factors had been noted to play major role in socioeconomic development as nations progressed through different forms of social organizations (von Hayek, 1956; Hegel, 1967; Marx, 1976, 1983; Weber, 1958, 1978; Poggi, 1983). Today, there are anti-trust (anti-monopoly) laws, property right laws, contract laws, and civil and human right laws guiding human behaviors, especially in the production process. World Bank (1991, p. 135) argues that in a country's quest for sustainable socioeconomic growth and development, norms of behavior (culture) not yet adapted to the needs of a modern economy substantially increase transactions costs. The Bank also noted that political instability declines as income rises and also as education improves (p.134).

Development is a strategy of organizing social change. It is the Promethean self-conception of European ideas based on the capitalist ethos of endless accumulation of wealth as a rational economic activity. Despite the registered impressive social and economic gains by the end of the 1970s by the developing countries, most recorded negative economic growth rate (The South Center, 1993, p.3). The debt crisis of the decade punctured illusions of development and called for the reformulation of the development enterprise as a global project. Thus, there had been several demands for developing countries to adapt their productive systems as far as possible to local conditions, rather than copying the irresponsible and unsustainable models of the North (The Alternative Forum of the NGOs, 1994). This has led to the "unthinking" of development as a linear process, and a "rethinking" of social and ecological priorities to sustain human existence in the long run (McMichael, 2000, p. 298). Furthermore, socialist feminism and the feminist epistemology on development see development as androcentric (male-centered). They pointed out that the androcentric society (male-dominated society) is phallocentric with its system of hierarchical dualism. They accused Marxism of being sex-blind, based only on economy, that emphasis should be placed on the sexual division of labor (Hartman, 1984). They see women as the super exploited working class (perform unpaid labor) and call for productive and reproductive democracy. This implies collective, participatory control over family and procreative decisions, as well as control over commodity production (Jagger, 1983, pp. 148–163).

Sustainable socioeconomic growth and development is conceptualized in two ways: (a) sustained long-term growth and consumption capable of eradicating poverty. This is measured mainly in terms of material output and consumption or production and accumulation which will lead to healthy and productive life, and adequate living standard made possible through employment and income; and (b) harmonious and dynamic balance between population, resources, environment, and development. Development must be fulfilled so as to equitably meet the population, development and environment needs of present and future generations (Lassonde, 1997, see especially chapter 3). Sustainable development should ensure that actions taken today to promote development and reduce poverty do not result in environmental degradation or social exclusion tomorrow. Sustainable development movement has been concerned with the compatibility of economic development policies and environmental preservation. According to World Bank publication, responsible growth should integrate society, ecology, and the economy (World Bank, 2004a). Advocates of sustainable development have recognized the comprehensive nature of development, and human beings are at the center of concerns. Quite contrary to the previous development policies, human beings are most important and valuable resource of any nation. They are entitled to a healthy and productive life in harmony with nature (United Nations, 1997). Economic development depends on human resources or capital, defined as a well-educated, well-trained workforce that is considered essential to a successful state of economic development. For generations, "education" has been linked to economic progress for the individual and for society (Koven and Lyons, 2003, p. 50).

Gender is a key development issue. According to Joseph Stiglitz, the World Bank vice president and chief economist, gender equity is a fundamental factor in socioeconomic and environmental sustainability (Stiglitz, 1998). Women play a crucial role in the development process, although their work still remains invisible in the national accounting system. They are important builders of society and are both producers and reproducers as they carry a double workload at home and outside (Schlyter and Johal, 1990). Women are responsible for childcare and domestic work (the reproductive role) and also for earning income for the upkeep of the family (the productive role) (Muller, 1990). In the gender division of labor, women are often more aware of the environmental degradation, especially in developing countries, because they make up a significant proportion of agricultural workforce. They are heavily involved in food production from planting to cultivation, harvesting, and marketing (Ashford, 2001).

Gender inequalities may likely give rise to inefficiencies that negatively impact the development process. Reducing gender inequalities may likely bring about greater economic prosperity and help reduce poverty. Gender disadvantages intertwine with poverty, which is strongly linked to poor health, as well as the fact that women represent a disproportionate share of the poor. Reproductive health is worse among the poor in terms of highest fertility rate, poorest nutrition, limited access to skilled pregnancy and delivery care, commercial sex and sexually transmitted illness (STI). Poverty is a multidimensional phenomenon, characterized by low income and low consumption, hunger and malnutrition, poor health, lack of education and skills, lack of access to water and sanitation, and vulnerability to economic and social shocks. Low income restricts access to basic goods and services and also limits income-generating opportunities. Poverty is associated with other social factors such as race, ethnicity and gender, reflecting actual or historical patterns of discrimination against certain social groups (United Nations, 2001). Combating poverty has been a central theme of recent United Nations conferences and summits, particularly the 1992 United Nations Conference of Environment (UNCED), Brazil; Conference on Population, Cairo, 1994; Habitat, Istanbul, 1996; World Summit for Social Development, Copenhagen, 1995; Fourth World Conference on Women, Beijing, 1995; and the 2000 Millennium Summit. The Millennium Declaration included a commitment to halve in the year 2015 the proportion of world's population whose income is less than $1.00 per day. These major UN conferences have also linked together women's legal and political rights, poverty alleviation, and environmental sustainability.

In most cultures, women lack the right to equal participation in political life e.g., the proportion in parliament or legislative seats. Women's lack of political power is likely to limit their influence on laws and policies that affect their rights, as well as on priority-setting in health care and other public issues. According to World Bank, women's education is the most single influential investment that can be made in developing world as it empowers the women and provides the avenue for reducing the differences between sexes. Empowerment refers to the process by which the powerless gain greater control over the circumstances of their lives (Sen and Batliwala, 2000, p. 18).

PURPOSE OF STUDY

The purpose of this study is to investigate empirically the role of gender and culture in sustainable socioeconomic growth and environmental sustainability. The paper addresses the theoretical underpinnings of the role of gender and culture as macro-variables that affect equity, economic efficiency and sustainability across different cultures. The advantages and disadvantages of gender equity in the sustainable development process should also be explored, and some suggestions proffered on how to make gender equality have the much-needed impact on development. In other words, in proffering solutions on how to make development sustainable, the study will endeavor to tie gender issues to the five perspectives that must act together during the development process. These include, *inter alia*, financial capital, physical capital, human capital, social capital, and natural capital (World Bank, 2004b).

THEORIES AND COUNTER-MOVEMENT THEORIES OF DEVELOPMENT

Currently the world is at important crossroads regarding development as some forces have changed in the last thirty years how development is understood. It has been argued that the market is a political construct, which does not constitute a sustainable culture in itself. The question is, can social and physical world sustain current economic growth trends with current available resources? Wealth is maldistributed such that about 80 percent of the world's income is produced and consumed by 15 percent of the world population (World Bank, 2000; Borosage, 1999, p.19; Ihonvbere, 1993–1994, p.8). Acknowledging that diverse cultures inhabit the earth, there is a need to attend to global and local questions of sustainability. The environment is degraded due to urbanization, industrial development, road building, and farming. The problem of sustainability should be addressed at all levels of organization and under conditions of adequate political representation. The World Development Report (WDR) 1999/2000 urged participatory policy making and observed that institutions of good governance that embody such processes are critical and crucial for development and should encompass partnerships among all elements of civil society (World Bank, 2000, p.3). Some of the major counter-mobilizations to development (currently, globalization) project include, among others, feminism, and sustainable developmentalism. These counter-movements are reformulating development in the globalization project. The movements are united in their oppositions to resource deprivation, which means primarily the lack of material necessities and the capacity to produce them (Peet, 1999, p.205).

Feminism

Feminist movement, especially since the 1970s argue vehemently that if they can be free from drudgery they stand to be special beneficiaries of development, just as women in the developed countries. Before the advent of colonialism and the modernization of traditional societies in the less developed countries, there seemed to be some form of balance in gender roles as women led certain areas of social life, e.g., women rituals and nurturing roles (Agbola, 1990; Swantz, 1987). In most of the African countries, before external influences began to affect different sectors of societal life, women had access to productive resources such as land, which buttressed their identity. For example, in Nigeria, Zambia, and other African countries, both men and women were contributing positively to the development of traditional societies (Milimo, 1987). Both men and women, in their specialized roles, were using the commons for livestock grazing, firewood collection, game hunting, seed gathering, and herb procuring for medicinal purposes.

One of the consequences of colonialism is the introduction of new values emanating from a different cultural imposition on traditional societies. Colonial administrators reorganized existing cultures to facilitate their exploitation, which often led to cultural genocide and the marginalization of indigenous people. The result was the fragmentation of social systems built on the complementarities of male and female work. For example, colonialism brought Christianity and Islam, and while Christianity preached equality of gender, Islam preached the superiority of man over woman, that woman is to be seen and not to be heard (Agbola, 1990). In Muslim cultures, with considerable variation, women's rights remain subordinated to Islamic law, or to male interpretation of the Koran, creating what is called by some Muslim feminists "Muslim apartheid" (McMichael, 2000, p.268). Colonialism brought new concept of property, land tenure and legal systems, which were detrimental to the colonized people in general but to women in particular (Milimo, 1987; Agbola, 1990; McMichael, 2000). Private property in land emerged which typically privileged men in a patriarchy, with its attendant gender discrimination. Thus, the obvious shift of males into cash cropping disrupts the formally complementary gender roles of men and women in traditional cultures.

The women's traditional land-user rights were often displaced by the new systems of private property, which put increasing pressure on food production, normally the responsibility of women. For example, in Kenya, the gender interdependence in the Kikuyu culture was fragmented as peasant land was confiscated and men migrated to work on European farms, reducing women's control over resources and lowering their status, wealth, and authority (Wacker, 1994, pp. 132–134). Men's work became specialized and counted in the national statistics as contributing to the commercial sector, while the specialization of women's labor as "non-income earning" work remained outside the commercial sector (Hogendorn, 1992, p.20; McMichael, 2000, p.260). As such, much of the unreported products result from women's work (Rogers, 1980, p.61), despite the fact that 60 to 80 percent of the food produced in the "informal sector and 70 percent of informal entrepreneurs are women (Snyder, 1995, p.xv; Tinker, 1997). In modern national accounting systems, only "productive" work is counted or valued, leaving much of women's work invisible. Colonization brought Western-type of education in various traditional communities where parents overwhelmingly encouraged male children. Western-type education brought social, economic and political power and placed them in the hands of men at the detriment of women. Educational attainment and advancement were the prerequisites for most employment that were dominated by men.

The importance of women in the development process was brought to the fore in the 1970s by several studies, which indicated that obviously women play important role but are left out due to lack of contact. They argue that female's work should be valorized (Baserup, 1970) to eliminate the invisibility of women in the development process (Young, 1993). As mentioned earlier, feminist movement has indicated that women are the super-exploited working class (perform unpaid labor) and calls for productive and reproductive democracy, including collective, participatory control over family and procreative decisions, as well as control over commodity production (Jagger, 1983, pp.148–163). In some African countries, women do over three-fourths of agricultural work in addition to their role in human capital improvement through nurturing. For example, food crops are planted, weeded, and harvested by women, as well as storing and food preparation. Women make-up a significant portion of the agricultural workforce and are heavily involved in food production from planting to cultivation, harvesting, and marketing. Women do all kinds of work, which include, tending the children, haul firewood, carry water, etc. (Herz, 1989, p.44). Women are nurturers as they care for the environment and preserve biodiversity in market and kitchen gardens. They are environmental managers, as they are more aware of the consequences of environmental degradation. Feminist movement saw women as beasts of burden whose economic progress is seen as joining the human race (Hogendorn, 1992, p.20; Lewis, 1955, p.422). Development practices reveal a predatory relationship in which women are exploited and socially and economically marginalized, while nature is plundered.

Several studies in the late 1970s and summarized by the United Nations indicate that (Pezzullo, 1982, p.15): women are half the world's people; women perform two-thirds of the world's working hours; women receive one-tenth of the world's income; women owned only one-hundredth of the world's property; and about one-fourth of world's families are headed by women (Hogendorn, 1992, p.20). Expectedly, in the early 1980s, women, especially those from the Third World called for new theories of development that embrace feminism. It began with the movement, in the 1970s, to integrate women into the development process. They noted that production and reproduction are inseparable aspects of making of existence, which should be essential part of development theory. They reiterated that women's voices must enter the definition of development and the making of policy choices (Sen and Grown, 1987, p.82). Since then, women's conferences had called for the empowerment of women as agents, rather than seeing them as problems of development (Bunch and Carrillo, 1990). A key event was the founding of Development Alternatives with Women for a New Era (DAWN) in Bangladore, India in 1984 to link micro-level activities to macro-level perspectives on development (Peet, 1999, chapter 6). Strategies include re-establishing people's (especially women's) control over the economic decisions shaping their lives. The first United Nations world conference on women was held in Mexico City, Mexico in 1975 and concentrated on extending existing development programs to include women. The goal included involving women as decision makers concerned with empowering all women in various life situations (Jahan, 1975, p.13). The fourth conference on women was held in 1995 in Beijing, China and the "Platform for Action" concerned the human rights of women to include: right to education, food, health, greater political power and freedom from violence (Bunch et al, 1995).

The interaction between feminist theory and development has taken five main forms in the literature. The forms are: Women in Development (WID); Women and Development (WAD); Gender and Development (GAD); Women, Environment, and Development (WED); and Postmodern and Development (PAD) (Rathgeber, 1990; Young, 1992; Visvanathan et al, 1997; McMichael, 2000). Women in Development position emerged to redress the absence of gender issues in development theory and practice and argue that female's work should be visible and highly valorized. Women and Development recognizes the predatory patriarchal division of labor, which was based on a structural separation and subordination of human beings, men from women, or local people from foreigners. The perspective is focused on the relation between men and women (Rathgeber, 1990). Gender and Development's main focus is gender relations which rejects the sexual division of labor in which men have power and control over women (Young, 1993, pp. 134–135). It indicates that capitalism used gender relations to produce a reserve of labor, while women's unpaid labor in the household was a way of creating wealth for global corporations. Women, Environment, and Development focuses on the balance between economic and environmental needs and sees the environment as the living force that supports life (Shiva, 1989). It noted that the task is not simply to add women into the known (development) equation but establish a new development paradigm (Harcourt, 1994, p.5). It believes in the conservation of renewable resources and in the reduction of industrialization wastes and pollution. Thus, "sustainable development" has become a central issue in the WED perspective.

Sustainable Developmentalism

The concept of sustainable development has swept through many social movements around the world since the 1980s. Environmental movements have proposed both local and global solutions under the mantle of "sustainable development" (McMichael, 2000). According to Peet (1999), sustainable development was seen as an opportunity for challenging the "Development equals Economic Growth equation" from the perspective of a feminist methodology. The concept of sustainable development gained currency as a result of the 1987 Brundtland report, entitled, *Our Common Future.* It was given form at the 1992 Earth Summit in Rio de Janeiro, Brazil. It was presented as an agenda to simultaneously solve the global environmental problem and to facilitate the economic development of the poor, particularly those in the Third World (Newman & Kenworthy, 1999, p.2). Sustainable development is defined as: development that meets the needs of the present without compromising the ability of the future generations to meet their own needs [World Commission on Environment and Development (WCED), 1987, p.43]. Other definitions include: "development practices that meet present needs without compromising the needs of future generations through erosion of the natural resource base" (McMichael, 2000); sustainable development or sustainability means in a global context, any economic or social development should improve, not harm the environment (Newman & Kenworthy, 1999, p.10; and it also means balance between economic and environmental needs, and sees the environment as the living force that support life (Shiva 1989; Peet, 1999).

The concept of sustainability has emerged from a global political process to bring together the needs of the present generation (Newman & Kenworthy, 1999, p.4): (a) the need for economic development to overcome poverty; (b) the need for environmental protection of air, water, soil, and biodiversity, upon which we all ultimately depend; and (c) the need for social justice and cultural diversity to enable local communities to express their values in solving theses issues. Perhaps the concept that should be required is that of sustainable human development defined as: a development that not only generates economic growth but distributes its benefits equitably, that regenerates the environment rather than destroying it, that empowers people rather than marginalizing them. It is development that gives priority to the poor, enlarging their choices and opportunities and providing for the participation in decisions that affect their lives. It is development that is pro-people, pro-nature, and pro-job and pro-women (McMichael, 2000, p.268; Jahan, 1995, p.109). Nevertheless, the concept of sustainable development—and the clear knowledge of how to achieve it—remain elusive. For example, the environmental advocates stress environmental sustainability, while workers and economic development experts focus on economic sustainability, and those in human development work stress cultural and social sustainability (Hoff, 1998, p.5).

The theoretical and practical issues of sustainability include socio-cultural concerns, economic productivity and policy, planning, and governance question. The leading challenge to action includes the integration and harmonization of cultural, economic, political, and environmental factors in the process of sustainable development. There is the growing recognition that efforts to protect the environment must incorporate the economic and cultural survival of the people. Sustainable development requires socio-cultural changes in values and behavior. It also requires positive valuation of cultural continuity and social stability for the many and diverse human societies (Hoff, 1998, p.12). Sustainable development process has been proceeding at academic, laboratory, and governmental levels (Newman & Kenworthy, 1999, p.5). For example, the approach in academic discussions is on how ecological economics or "green economics" can be defined and formulated (Daly and Cobb, 1989); in laboratory" e.g., the 'clean production;" and within all levels of governments and in community process, the approach is "green planning."

There are several principles that will inform sustainable development. These are (Hoff, 1998, p.6): non-renewable resources should be used wisely and sparingly at a rate which does not restrict the options of future generations (wise use); renewable resources should be used within the limits of their capacity for regeneration (carrying capacity); the quality of the natural heritage as a whole should be maintained and improved (environmental quality); in situations of great complexity or uncertainty, society should act in a precautionary manner (precautionary approach); and there should be an equitable distribution of the costs and benefits (material and non-material) of any development (shared benefits) (Hardy and Lloyd, 1994, p.776). Sustainable development is potentially measurable in the attributes, which include, among others, non-decrease of natural stock over time, and increases or improvements in measures of human well-being, such as income, education, health, and basic freedom (Pearce et al, 1990). Therefore, the development theory emphasis has shifted from theories and concepts based on factors of production to development based on human resources (Agbola, 1990, p.171). This is an alternative conception of development of enlarging people's choices, especially in terms of increasing access to knowledge, nutrition and health services, security, leisure, and political and cultural freedom. The Human Development Index (HDI) calculated by the United Nations Development Program (UNDP) measures this. The HDI measures "development" in terms of longevity (life expectancy at birth), knowledge (adult literacy and mean years of schooling), and income sufficiency (the proportion of people with sufficient resources to live a decent life).

METHODOLOGY FOR THE STUDY

This is an explanatory study that is both longitudinal and cross-sectional in approach. Several variables and indicators, which are compared, inform the nature of and existing gender equity and sustainable socioeconomic growth and development among different cultures and regions. In this study, different cultures are synonymous with different regions of the world such as: Developing Countries, Least Developed Countries, Arab States, East Asia and the Pacific, Latin America and the Caribbean, South Asia, Sub-Saharan Africa, Central & Eastern Europe & Commonwealth of Independent States (CIS), Organization for Economic Cooperation and Development (OECD), High-Income OECD, High Human Development Countries, Medium Human Development Countries, Low Human development Countries, High Income Countries, Middle Income Countries, and Low Income Countries. These sixteen regions are as carved out and arranged by the United Nations Development Program (UNDP) (2004): *Human Development Report (HDR), 2004. Cultural Liberty in Today's Diverse World* (New York: Oxford University Press).

Several socioeconomic indicators are compared among the above regions and cultures to capture the existing gender equity, socioeconomic growth and sustainable development. The indicators for the study are derived from five major dimensions of sustainable development that include, social, cultural, economic, environmental, and institutional factors (United Nations, 2001: *Indicators of Sustainable Development*; IDRC: Development and Information Age, chapter 1). Different regions and cultures exhibit varying attitudes and approaches to support the above dimensions. Discussions on the nature of economic and social development always center on five consensus indicators such as Literacy, education, and skills; Health; Choice, democracy, and Participation; Income and economic welfare; and Technology. This study also includes other indicators or variables such as Total population; Population growth rate; Urban population; Fertility rate; and Human development indices. The above indicators are from secondary data sources from publications of the United Nations, *Human Development Reports (HDR)*; and World Bank, *World Development Report (WDR)*.

Indicators are important tools for communicating ideas, thoughts, and values. The above indicators of sustainable development can help to provide a solid basis and crucial guide for decision-making at all levels (national, regional, and global levels). They can translate physical and social science knowledge into manageable units of information that can facilitate decision-making. They help to measure and calibrate progress towards sustainable development goals and can provide early warning, sounding the alarm in time to prevent economic, social and environmental damage. Indicators are the basis for monitoring the achievement of key national goals and objectives for sustainable development. Data analyses in the study are performed by computing the totals, simple averages, ratios, percentages, and compared them among the regions or cultures and the world. Policy suggestions and recommendations for the study are based on the findings of the analyses, while the conclusions are based on the findings of the study.

REGIONAL SOCIO-DEMOGRAPHIC TREND, 1975–2015

Currently renewed discussions on development have focused on the relationships between demographic change and development factors within the framework of sustainable development. Many recent global summits (e.g., Rio de Janeiro, 1992; Cairo, 1994; Beijing, 1995) have focused on the issue of population and eradication of poverty as sine qua non for sustainable development. It has been recognized that the eradication of poverty is essential for sustainable development, and about 70 percent of all those who live in absolute poverty (less than $1.00 a day) are women (Hemmati, 2001; United Nations, 2001). Sustainable development requires the dynamic balance between population, resources, environment and development, as well as sustained long-term growth and consumption. Sustained economic growth within the framework of sustainable development is essential to eradicate poverty. According to Lassonde (1997, p.50), poverty is often accompanied by unemployment, illiteracy, low status of women, exposure to environmental risks, limited access to social and health services, inappropriate spatial distribution of population, inequitable distribution of natural resources and environmental degradation. Population growth puts strains on weak economies because of the investment necessary for meeting the needs of a growing population such as education, sanitation, water, housing, food and productive jobs. The World Bank, in its work to make development sustainable, recognized five perspectives that must act together which include, human capital, social capital, financial capital, physical capital, and natural capital. The Bank noted that responsible growth should integrate society, ecology, and the economy (World Bank, 2004a).

In the last thirty years, world population has witnessed a substantial rate of increase, which has implications for sustainable socioeconomic development and environmental sustainability. World regional demographic trends show that many regions and cultures have varied annual population growth rates; total fertility rates, as well as doubled their total populations between 1975 and 2002 (table 4.1). As their total populations are increasing, including their projections in 2015, which is the end year for the Millennium Development Goal (MDG) (UNDP, 2004), so also are their urban populations as percent of total populations, population under 15 years, and percent population age 65 and above. Apart

Table 4.1. Regional Demographic Trends, 1975–2015

Region	Total Population (millions)			Annual Population Growth Rate		Urban Population (% of total)[a]			Population Under Age 15 (% of total)		Population Age 65 and Above (% of total)		Total Fertility Rate (births per woman)	
	1975	2002[b]	2015[b]	1975–2002	2002–2015[b]	1975	2002[b]	2015[b]	2002[b]	2015[b]	2002[b]	2015[b]	1970–1975[c]	2000–2005[c]
Developing Countries	2961.2	4936.9	5868.2	1.9	1.3	26.4	41.4	48.6	32.2	28.2	5.2	6.4	5.4	2.9
Least Developed Countries	353.7	700.9	941.9	2.5	2.3	14.7	26.1	33.4	42.9	40.1	3.1	3.3	6.6	5.1
Arab States	143.4	296.6	389.7	2.7	2.1	41.7	54.2	58.8	37.1	33.5	3.7	4.3	6.7	3.8
East Asia and the Pacific	1310.5	1917.6	2124.6	1.4	0.8	20.4	40.2	51.0	25.8	21.4	6.5	8.4	5.0	2.0
Latin America and the Caribbean	317.9	530.2	622.5	1.9	1.2	61.2	76.2	80.8	31.1	26.3	5.6	7.3	5.1	2.5
South Asia	842.1	1480.3	1805.3	2.1	1.5	21.3	29.6	34.3	34.8	29.6	4.7	5.6	5.6	3.3
Sub-Saharan Africa	305.8	641.0	843.1	2.7	2.1	21.0	35.0	42.4	44.3	41.9	3.0	3.3	6.8	5.4
Central & Eastern Europe & CIS	366.6	408.9	398.4	0.4	–0.2	56.8	62.8	63.7	17.5	16.3	12.2	13.2	2.5	1.4
OECD	925.6	1148.1	1227.7	0.8	0.5	67.3	75.7	79.0	20.2	17.9	13.3	16.0	2.5	1.8
High-Income OECD	766.2	911.6	962.9	0.6	0.4	69.9	77.3	80.4	18.2	16.5	14.8	18.0	2.2	1.7
High Human Development	972.3	1201.3	1282.0	0.8	0.5	68.9	77.1	80.3	20.0	17.8	13.4	16.2	2.5	1.8
Medium Human Development	2678.4	4165.2	4759.1	1.6	1.0	28.1	42.2	49.3	29.3	24.8	6.0	7.4	4.9	2.4
Low Human Development	354.5	755.8	1021.6	2.8	2.3	19.4	32.4	39.9	44.6	42.1	3.0	3.2	6.8	5.6
High Income	782.0	941.2	997.7	0.7	0.4	70.1	77.7	80.9	18.3	16.6	14.6	17.7	2.2	1.7
Middle Income	1847.5	2720.7	3027.9	1.4	0.8	35.7	52.8	61.0	26.3	22.3	7.0	8.6	4.5	2.1
Low Income	1437.1	2560.8	3169.0	2.1	1.6	20.7	31.2	37.5	37.0	32.8	4.3	5.0	5.9	3.7
World[d]	**4068.1**	**6225.0**	**7197.2**	**1.6**	**1.1**	**37.2**	**47.8**	**53.5**	**29.4**	**26.1**	**7.1**	**8.3**	**4.5**	**2.7**

a. Because data are based on national definitions of what constitutes a city or metropolitan area, cross-country comparisons should be made with caution.
b. Data refer to medium-variant projections.
c. Data refer to estimates for the period specified.
d. Data refer to the total world population according to UN 2003. The total population of the 177 countries included in the main indicator tables was estimated to be 4,063 million in 1975, and projected to be 6,217 million in 2002 and 7,188 million in 2015.

Source: United Nations Development Program (UNDP) (2004). *Human Development Report, 2004. Cultural Liberty in Today's Diverse World* (New York, Oxford: Oxford University Press), table 5, p. 155.

from High Income countries, High Human Development countries, OECD and High-income OECD countries, and Central & Eastern Europe & CIS countries, all other regions, especially the Developing countries, Sub-Saharan Africa, Arab States, Low-income countries, and Low Human Development countries almost doubled their populations. Generally, annual population growth rate and total fertility rate in all regions slowed down since the beginning of the New Millennium. Even Central & Eastern Europe & CIS showed a negative rate. However, urban population or urbanization has been proceeding at a very rapid rate, especially in poor countries. Nevertheless, by 2002, more people lived in cities in more developed countries, as well as in Latin American and the Caribbean, and Arab States. The implication is that the population surge in cities and regions are making them unsustainable due to sprawl, coupled with environmental degradation and biodiversity destruction. Most of the current high rates of urbanization are in poor regions and cultures, which are incapable of coping with the consequent poverty and environmental stress.

REGIONAL ECONOMIC PERFORMANCE, 2002

The world economy has shown a modest increase in the last twenty-five years with much variations across regions and cultures, especially in their gross domestic product (GDP) per capita annual growth rates. Table 4.2 shows the regional economic performance in 2002, which has implications for sustainable economic development across regions and cultures. The size of the economy (GDP), purchasing power parity (PPP), and per capita income distribution show tremendous inequalities between developing countries and more developed countries. Purchasing power parity (PPP) is a rate of exchange that accounts for price differences across countries, allowing international comparisons of real output and incomes. At the PPP US $ rate, PPP US $1 has the same purchasing power in the domestic economy as $1 has in the United States (UNDP, 2004, p. 274).

Purchasing power parity (PPP) measure exposes the nature and level of poverty across regions, as well as their abilities to provide themselves with the basic needs such as food security, adequate water supply, health care, education, and shelter. The provision of basic needs itself raises productivity because productivity depends in part on health, nutrition, and education. Raising the quality of labor will thus raise output and have a negative feedback on population growth, which will likely reduce the population problem (Hogendorn, 1992, p.17). The direct provision of social services that enhance "human capital" is arguably central to both growth and development. For example, the delivery of essential social services can stimulate output, and thus growth, through higher labor productivity, and in the process reduce poverty (World Bank, *World Development*

Table 4.2. Regional Economic Performance, 2002

	Gross Domestic Product (GDP)		*GDP Per Capita*		*GDP Per Capita Annual Growth Rate*	
Region	*US$ Billions 2002*	*PPP US$ Billions 2002*	*US$ 2002*	*PPP US$ 2002*	*(Percent) 1975–2002*	*(Percent) 1990–2002*
Developing Countries	6,189.3	19,848.5	1,264	4,054	2.3	2.8
Least Developed Countries	204.7	897.7	298	1,307	0.5[a]	1.4
Arab States	712.3	1,466.3	2,462	5,069	0.1	1.0
East Asia and the Pacific	2,562.6	9,046.9	1,351	4,768	5.9	5.4
Latin America and the Caribbean	1,676.1	3,796.1	3,189	7,223	0.7	1.3
South Asia	757.1	3,898.7	516	2,658	2.4	3.2
Sub-Saharan Africa	303.5	1,157.4	469	1,790	–0.8	(.)
Central & Eastern Europe & CIS	971.1	2,914.7	2,396	7,192	–1.5[a]	–0.9
OECD	26,298.9	28,491.5	22,987	24,904	2.0	1.7
High-Income OECD	25,129.9	26368.2	27,638	29,000	2.1	1.7
High Human Development	26,924.9	29,435.4	22,690	24,806	2.0	1.7
Medium Human Development	4,659.1	17,763.5	1,120	4,269	1.7	2.1
Low Human Development	233.9	860.0	322	1,184	0.1[b]	0.3[b]
High Income	25,767.9	27,115.7	27,312	28,741	2.1	1.7
Middle Income	5,138.5	16,174.9	1,877	5,908	1.4	2.0
Low Income	1,123.9	5,359.9	451	2,149	2.2[b]	2.3[b]
World	**31,927.1**	**48,151.1**	**5,174**	**7,804**	**1.3**	**1.2**

a, b. United Nations Development Program (UNDP) (2004). *Human Development Report, 2004. Cultural Liberty in Today's Diverse World* (New York, Oxford: Oxford University Press), table 13, p. 187.

Report 1990, chapter 3). Gross domestic product (GDP), GDP per capita, and purchasing power parity (PPP) are lowest in Least developed countries, South Asia, Sub-Saharan African countries, and Low Human Development countries, while highest levels are found in OECD High-income OECD, High Human Development, and High-income countries (table 4.2).

Nonetheless, the GDP per capita annual growth rate from 1975 to 2002 is highest among East Asia and the Pacific, South Asia, Developing countries, and Low-income countries, while the Sub-Saharan Africa and Central & Eastern Europe & CIS show negative rates. Aside the 1980s that was known as the lost decade for the developing countries, GDP per capita annual growth rate shows a modest increase among the regions between 1990 and 2002. It should be mentioned that India's growth rate accounts for most of the difference in average annual growth rates of Low-income and Low Human Development countries when compared to the modest growth rates in OECD, High-income OECD, High Human Development, and High-income countries (UNDP, 2004, table 13, p.187).

HUMAN CAPITAL AND SUSTAINABLE DEVELOPMENT

As a reaction to the initial emphasis of development theories, which was on economic growth, contemporary development theories emphasize the welfare of citizens, which is based on the endowment of human resources or capital. It has been realized that a critical component of development is the development of human capital. Thus, contemporary developmentalism is conceptualized as that which puts premium on the people's ability, resourcefulness, and willingness to innovate and adopt necessary innovations for the desired structural transformation (Agbola, 1990, p.171). Human capital raises productivity, which depends in part on health, nutrition, and education. Human capital available to a country is measured by the human development index (HDI), which is a summary measure of human development. Human development index measures the average achievements in a country in three basic dimensions of human development (UNDP, 2004, p.259): a long and healthy life, as measured by life expectancy at birth; knowledge, measured by adult literacy rate; and a decent standard of living, as measured by GDP per capita (PPP US $). Human development has positive impacts on human poverty that plagues a majority of world's population; on gender related development; and on gender empowerment; all have implications for sustainable development.

Table 4.3 shows the human development indicators of different regions and cultures, which vary greatly among low-and high-GDP per capita regions. In table 4.3, Life Expectancy Index, Education Index, and Income (GDP) Index are computed by combining in each, male and female indices as each is multiplied by its population share in the nation or region (UNDP, 2004, p.261). Human development indicators and indexes show a great deal of inequalities among regions, especially in the overall human development index values. From the table, Developing countries, Least developed countries, Arab States, South Asia, Sub-Saharan Africa, and Low-income countries have below world averages than those of East Asia and the Pacific, Latin America and the Caribbean, Central& Eastern Europe & CIS, OECD, High-income OECD, and High-income countries that are above the world averages in 2002. For example, Sub-Saharan Africa, with one of the lowest GDP per capita (PPP US $1,790), has life expectancy at birth of 46.3 years, adult literacy rate (63.2 percent), combined enrolment ratio (education) as percent (44), life expectancy index (0.35), education index (0.56), GDP index (0.48), and the overall human development index value of 0.465. Comparatively, the High-income OECD, which has the highest GDP per capita (PPP US $29,000), has life expectancy at birth of 78.3 years, adult literacy rate (99.0 percent), combined enrolment ratio (93.0 percent), life expectancy index (0.89), education index (0.97), GDP index (0.95), and the overall human development index value of 0.935. The implication of the above is that, the higher the human development index (HDI) value the more developed a nation or growth is sustainable.

GENDER AND SUSTAINABLE DEVELOPMENT

Sustainable development has gendered aspects, which vary from one culture, region, or country to another. Women's development concern has been on the centrality of women's economic and social development, which is the key ingredient in population control and eradication of poverty. Concerns of the poor, and especially of women, had significant influence in generating greater awareness of the linkages between environmental issues and the global economic system (Hoff, 1998, p.8). Women are important builders of society, for example, they play critical role in primary health care and preserving hygiene; provide basic services and infrastructure in the majority of human settlements, especially among the poor; contribute a substantial portion of the labor required for shelter and infrastructure development in the rural areas and poor urban neighborhoods; provide greater portion of the energy resources consumed in settlements and in transporting rural produce; and they are constructors and maintainers of shelter (Schlyter and Johal, 1990).

Due to some cultural patterns, women carry the heaviest responsibilities for home and family, as they are the prime homemakers and maintainers of family shelter. In some cultures 60 to 80 percent of the food produced in the "informal" sector is by women, where products are not reported; and 70 percent of "informal" entrepreneurs are women (Snyder, 1995, p.xv; Tinker, 1997). This gender disproportion has been on the increase, especially in less developed countries

Table 4.3. Regional Human Development Indicators and Indexes

Region	Life Expectancy at Birth (Years) 2002	Adult Literacy (Ages 15 and above) (%) 2002[a]	Combined Gross Enrolment ratio for Primary, Secondary and Tertiary Schools (%) 2001/02[b]	GDP Per Capita (PPP US$) 2002	Life Expectancy Index	Education Index	GDP Index	Human Development Index (HDI) Value 2002
Developing Countries	64.6	76.7	60	4,054	0.66	0.71	0.62	0.663
Least Developed Countries	50.6	52.5	43	1,307	0.43	0.49	0.42	0.446
Arab States	66.3	63.3	60	5,069	0.69	0.61	0.65	0.651
East Asia and the Pacific	69.8	90.3	65	4,768	0.75	0.83	0.64	0.740
Latin America and the Caribbean	70.5	88.6	81	7,223	0.76	0.86	0.72	0.777
South Asia	63.2	57.6	54	2,658	0.64	0.57	0.55	0.584
Sub-Saharan Africa	46.3	63.2	44	1,790	0.35	0.56	0.48	0.465
Central & Eastern Europe & CIS	69.5	99.3	79	7,192	0.74	0.93	0.72	0.796
OECD	77.1	C	87	24,904	0.87	0.94	0.92	0.911
High-Income OECD	78.3	C	93	29,000	0.89	0.97	0.95	0.935
High Human Development	77.4	C	89	24,806	0.87	0.95	0.92	0.915
Medium Human Development	67.2	80.4	64	4,269	0.70	0.75	0.63	0.695
Low Human Development	49.1	54.3	40	1,184	0.40	0.50	0.41	0.438
High Income	78.3	C	92	28,741	0.89	0.97	0.94	0.933
Middle Income	70.0	89.7	71	5,908	0.75	0.84	0.68	0.756
Low Income	59.1	63.6	51	2,149	0.57	0.59	0.51	0.557
World	**66.9**	—	**64**	**7,804**	**0.70**	**0.76**	**0.73**	**0.729**

a. Data refer to estimates produced by UNESCO Institute for Statistics in July 2002, unless otherwise specified. Due to differences in methodology and timeliness of underlying data, comparisons across countries and over time should be made with caution.
b. Data refer to the 2001/02 school year, unless otherwise specified.
c. For purposes of calculating the HDI, a value of 99.0 percent was applied.

Source: United Nations Development Program (UNDP) (2004). *Human development Report, 2004. Cultural Liberty in Today's Diverse World.* (New York, Oxford: Oxford University Press), table 1, p. 142.

(Chant, 1985). Despite the fact that women are half of the world's people and head about one-fourth of world's families, 70 percent of all those who live in absolute poverty (less than $1.0 a day) are women (Pezzullo, 1982; Hogendorn, 1992; Peet, 1999). Women are the most disadvantaged group throughout the world, as they lag behind men in access to health care, nutrition, and education, while continuing to face formidable social, economic, and political barriers. Bound by stereotyping images in some societies and cultures, women's rights to land, security of tenure and housing ownership are limited, while their participation in policies and programs for development is also restricted.

Among economically active population and total population, female economic activity and participation vary greatly among regions and cultures. This has implications for female emancipation and empowerment, eradication of poverty, and sustainable development. Women's conferences have called for the empowerment of women as agents, rather than seeing as problems of development (Bunch and Carrillo, 1990). Table 4.4 shows gender inequality in economic activity among world regions and cultures in 2002. Female economic activity rate is calculated from the percent female employment in economic activities such as agriculture, industry, and services. In some developing countries where women do over three-fourths of agricultural work (Hogendorn, 1992, p.20) and engage in informal trade and services (Pezzullo, 1982; Schlyter and Johal, 1990; Tinker, 1997), female economic activity rate is higher than the world average (55.3 percent), such as in Least developed countries (64.2 percent), Sub-Saharan African (62.1 percent), and East Asia and the Pacific (68.8 percent). In Muslim cultures of the Arab States where women's rights remain subordinated to Islamic laws (McMichael, 2000, p.268), female economic activity rate is the lowest (33.0 percent). Female economic activity rate is about 50 percent in High human development, OECD, High-income OECD, and Low-income countries (which rate is modified by Latin America and the Caribbean and South Asia). In table 4.4, despite the increase in female economic activity from 1990 to 2002, its rate is still below the male rate. The highest female rate "as a percent of male rate" in 2002 was found in East Asia and the Pacific (82.0 percent), followed by Central & Eastern Europe & CIS (81.0 percent), while the lowest rate was in Arab States (42.0 percent).

Economic development cannot succeed without attention to human development. These human factors include access to healthcare, education, political voice and office, protected legal rights, and property rights. Human capital in the form of education, skills, and maintenance through health and nutrition are crucial component of development (Hogendorn, 1992, p.298). Majority of women, especially in developing countries have little access to the media and other forms of communication (Macoloo, 1990). According to Macoloo, "to educate a woman is to educate a nation," because the improvement of the conditions of women as the poorest of the poor will benefit society as a whole. Women are

Table 4.4. Regional Gender Inequality in Economic Activity[a]

	Female Economic Activity Rate (age 15 and above)[a]		
Region	*Rate (%)*[b] *2002*	*Index 1990=100*[b] *2002*	*As % of Male Rate*[b] *2002*
Developing Countries	55.8	101	67
Least Developed Countries	64.2	99	74
Arab States	33.0	118	42
East Asia and the Pacific	68.8	99	82
Latin America and the Caribbean	42.5	110	52
South Asia	43.7	107	52
Sub-Saharan Africa	62.1	99	73
Central & Eastern Europe & CIS	57.4	99	81
OECD	51.5	106	71
High-Income OECD	52.2	106	74
High Human Development	50.9	106	70
Medium Human Development	56.7	101	69
Low Human Development	56.9	102	66
High Income	52.1	106	73
Middle Income	59.1	100	73
Low Income	51.9	104	62
World	55.3	102	69

a. Female economic activity rate calculated from the percent female employment in economic activities such as agriculture, industry, and services.
b. Columns 1–3 calculated on the basis of data on the economically active population and total population in 2002.

Source: United Nations Development Program (UNDP) (2004). *Human development Report, 2004. Cultural Liberty in Today's Diverse World.* (New York, Oxford: Oxford University Press), table 27, p. 232.

central to any successful population policy, and educated women tend to have fewer and healthier children. Perhaps, the gap in national income per capita between Less-developed countries (LDCs) and more developed and advanced countries could be their differences in educational endowments.

Access to education is probably the largest single contributor to enhancing the status of women and thereby promoting development. Table 4.5 shows gender inequality in education among regions of the world in 2000. Female adult (age 15 years and above) and Youth (age 15–24 years) literacy rates are lowest among Low human development and Least developed countries, while High-income OECD, High human development, and Central & Eastern Europe & CIS have the highest rates. In adult literacy, the female rate "as a percent of male rate" is lowest among Low human development (63 percent), South Asia (66 percent), Arab States (68 percent), and Least developed countries (68 percent), while the highest rates are found in High-income OECD (99 percent), Central & Eastern Europe & CIS (99 percent), High-income countries (99 percent), and Latin American and the Caribbean countries (98 percent). In Youth literacy, the highest rate is found in Latin America and the

Table 4.5. Regional Gender Inequality in Education in 2000

	Adult Literacy		Youth Literacy	
Region	*Female Rate (Age 15 years and above) (%) 2000*	*Female Rate (As % of Male Rate) 2000*	*Female Rate (Age 15–24 years) (%) 2000*	*Female Rate (As % of Male Rate) 2000*
Developing Countries	66.0	81	80.5	91
Least Developed Countries	42.8	68	58.1	79
Arab States	50.1	68	72.5	85
East Asia and the Pacific	79.4	86	96.4	98
Latin America and the Caribbean	87.4	98	94.4	101
South Asia	43.8	66	61.2	79
Sub-Saharan Africa	53.6	77	73.0	89
Central & Eastern Europe & CIS	98.3	99	99.4	100
OECD	a	a	a	a
High-Income OECD	a	a	a	a
High Human Development	a	a	a	a
Medium Human Development	72.2	85	86.6	94
Low Human Development	38.5	63	56.7	76
High Income	a	a	a	a
Middle Income	80.9	89	94.3	98
Low Income	52.8	74	68.8	84
World	—	—	—	—

a. For the purposes of creating the ratios, a value of 99.0 percent was applied.

Source: United Nations Development Program (UNDP) (2004). *Human Development Report, 2002. Deepening Democracy in a Fragmented World.* (New York, Oxford: Oxford University Press), table 24, p. 233.

Caribbean where the female rate "as a percent of male rate" is 101 percent, followed by Central & Eastern Europe & CIS countries (100 percent), High-income OECD (99 percent), and High human development countries (99 percent). The lowest female rate "as a percent of male rate" is found in Low human development (76 percent), followed by Least-developed countries (79 percent), and South Asia countries (79 percent) (table 4.5).

In the regions and countries where female literacy are lower than those of their male counterparts, female school enrolments are always lower than those of males. The reasons have to do with state priorities, family resources that are not sufficient to educate boys and girls, female socialization, and cultural factors (Schlyter and Johal, 1990; Macoloo, 1990; Agbola, 1990; Ashford, 2001; Sharma, 2001). According to Schlyter and Johal (1990), women experience systematic discrimination in both "traditional" and "modern" legal systems, and where *de jure* discrimination has been eliminated, *de facto* discrimination and stereotyping prevail. Although gender disparities in education have been reduced in all regions (table 4.5), it is still substantially pronounced in South Asia, Sub-Saharan Africa, Arab States, and Low human development countries. Globally, about 60 percent of children not in schools are girls, and the ratio of girls to boys in primary school is 82 percent in South Asia and 84 percent in Sub-Saharan Africa, with lower ratios for secondary school (UNICEF, 2001, table 7). Furthermore, in 12 countries in Africa, over half of primary school-age children are not in school (p.4).

POLICY SUGGESTIONS AND RECOMMENDATIONS

The concept of sustainable development and the clear knowledge and understanding of how to achieve it is still intractable and elusive; especially now the world is at important crossroads regarding development. This is in part because; there exists tremendous diversity among cultures, regions, and countries of the world.

According to Hogendorn (1992, p.6), some are densely populated, others are not; some are geographically big, others are little; some rich in natural resources, some are resource poor; some were once colonies, some were not; some are honest in their public services, many are corrupt; some are democracies, some are authoritarian dictatorship; many lie somewhere in between; and some have rapid growth, while some have slow growth. The question is, what can be done to move toward socio-economic and environmental sustainability. It should be recognized that the current development project occurs within a field of power (local, national, or international), and also within a cultural field where local and national cultures are incorporated into the enterprise of endless wealth creation measured by capital accumulation or array of commodities.

The Western developmentalism or model has been irresponsible and unsustainable so far in solving key problems plaguing a majority of world populations, namely, underdevelopment, poverty, food security, and environmental balance. Essentially, the theoretical and practical issues should include socio-cultural concerns, economic productivity and the justice of economic distribution, and policy, planning, and governance question. Emphasis should be on the provision of integrating factors that allow policies simultaneously to address issues of development, sustainable resource management and poverty eradication.

It has been recognized that development is a structural change whose main development agent is the people. There are five key elements of sustainability: Cultural, Human health, Environmental, Economic, and Social, which call for broad policies that have become much of the global action. The policies are the fundamental approaches to global sustainability that must apply simultaneously and are derived from the Brundtland Report (United Nations, 1987; Newman and Kenworthy, 1999). These are: (1) the elimination of poverty, especially in the Developing countries, Low-income countries, and Low human development countries. This is necessary not just on human grounds but as an environmental issues because poverty is one factor degrading the environment. Thus, Third World economic and social developments are precursors to global sustainability (Newman and Kenworthy, 1999, p.3); (2) the developed countries (especially, OECD and High-income OECD, High-income countries, and High human development countries), must reduce their consumption of resource and production of wastes. They must be less resource-intensive in the future and such a goal cannot be achieved without economic and social change. Thus, advanced countries' economic and social developments are precursors to global sustainability (Hoff, 1998, p.11; Newman and Kenworthy, 1999, p.3; Harf and Lombardi, 2004, pp. 126–130); (3) global cooperation on environmental issues is no longer a soft option but very imperative. Nations must come to terms and agree to make necessary changes to take care of environmental problems such as hazardous wastes, greenhouse gases, and loss of biological diversity. As sustainability is a vision and a process and not an end product (Newman and Kenworthy, 1999, p.5), the spread of international best practice on the above issues is essential for the future of the world and must be adopted as a convention. Thus, a global orientation is a precursor to understanding sustainability (Newman and Kenworthy, 1999, p.3); and (4) change toward sustainability can occur only with community-based approaches that take local cultures seriously. Sustainability will come only when local communities determine how to resolve their economic and environmental conflicts in ways that create simultaneous improvement of both. At the local level, it is possible for government to more easily make the huge steps in integrating the economic, environmental, and social professions in order to make policy developments that are sustainable. Moreover, local governments are also closer to the concerned people. Therefore, an orientation to local cultures and community development is a precursor to implementing sustainability (Hoff, 1998, pp.10–11; Newman and Kenworthy, 1999, pp.3–4; Dovie, 203, pp.28–29; Hunter et al, 2003, pp.147–150; Harf and Lombardi, 2004, pp.126–130).

The above sustainable policies recommend several programs of action for socio-economic and environmental sustainability. Such programs include, among others: (1) Empowerment of women for gender equity, poverty reduction, and environmental sustainability (Buckingham-Hatfield, 2002). Empowerment is achieved through education, employment, and political representation. Women should be the focus of programs to reduce poverty and move toward environmental sustainability. Reduction of poverty is dependent on their financial independence. There should be equal access to formal education and other socioeconomic opportunities by men and women, as well as safeguards to redress the past neglects and anomalies (Agbola, 1990); (2) Emphasis on human development. Socioeconomic development and sustainability depend on human resources as human capital. There should be a universal access to education, health, and income generating activities, especially in Low human development and Low-income countries of Developing countries, as well as in the Arab States. A healthy and well-trained workforce is sine qua non for sustainable socioeconomic development, as education has been linked, for generations to economic progress for the individual and for society (Koven and Lyons, 2003, p.50); (3) Cultural policy should be one of the main components of endogenous and sustainable development policy. It has been noted that the destabilization and sheer destruction of cultures are traceable to unsustainable environmental and economic policies. Sustainable development and the flourishing of culture are interdependent (UNDP, 2004; UNESCO, 1998). According to UNESCO (1998), any policy for development must be profoundly sensitive to culture itself; cultural policies should promote creativity in all its forms, by facilitating access to cultural practices and experiences for all citizens; effective participation in the information society and the mastery by everyone of information and communications technology constitute a significant dimension of any cultural policy; and cultural policies must respect gender equality, fully recognize women's parity of right and freedom of expression and ensure their access to decision making; (4) Protection of Biodiversity/Managing Biodiversity. This is basically merging development with conservation agendas, which should be socially and culturally determined, thus, the challenges of increasing human pressure (development) and the need to provide for the livelihoods of multitudes of poor people (WRI, 1996; Dovie, 2003; Hunter et al, 2003). When conservationists work in isolation of local people, it may mean the displacement of local communities and forfeiture of their source of livelihood; and (5) Growth Management Policies and Programs. Growth management is a tool for implementing planning and land use control, which evolved in the last fifty years. Although the context for growth management varies in a democratic decision-making, growth management concerns environmental damage, especially the environmental sensitive lands. Sensitive lands such as floodplains, wetlands, steep-slopes, prime agricultural lands and forests, are vulnerable to harm from development sprawl. The philosophical emphasis of growth management is precluding development on sensitive lands, and design around them or engaging in the process McHarg (1994) called "design with nature." There are several basic types of regulatory growth management programs in effect today, which are quite different in philosophy, operation, and effect (Kelly, 2004, p.43). The programs include: adequate public facilities (concurrency) requirements; growth phasing programs, including urban growth and service boundaries; and rate-of-growth programs, which are based on the actual carrying capacity of public facilities at any given time to regulate growth rate.

CONCLUSIONS

Developmentalism and sustainable development have gendered aspects, in favor of male, with some variations from one culture to another. Development is a strategy of organizing social change, which is the Promethean self-conception of European ideas based on the capitalist ethos of endless accumulation of wealth as a rational economic activity. Thus, development is not only Eurocentric but androcentric, as seen by feminist epistemology, with its system of hierarchical dualism. Even the newly independent states also conceptualized development to mean emulating Western living stands, rationality, and scientific progress. Over time, developmentalism has turned out to be an economic nostrum, which its universal expectation in the development project has proven to be an unrealizable ideal to the Developing and Least-developed countries. Development theories and practices tend to differ according to the political positions of their adherents, their philosophical origins, and their place and time of construction and practice.

Gender is a key development issue, and gender equity is a fundamental factor in socioeconomic and environmental sustainability. Gender inequalities tend to give rise to inefficiencies that negatively impact the development process. Women play important role in the development process because they are important builders of society by being both producers and reproducers as they carry a double workload at home and outside. Women are heavily involved in food production, from planting to cultivation, harvesting, and marketing. Women are nurturers as they care for the environment and preserve biodiversity in market and kitchen gardens. Yet in modern national accounting systems, only "productive" work is counted or valued, leaving much of the women's work invisible. Gender disadvantages intertwine with poverty, and women represent a disproportionate share of the poor. Data analyses show that different regions and cultures vary greatly in terms of demographics, economic performance, environmental impact, biodiversity, human resources and capital, human development index, and gender equity. Data analyses conclude that the Developing countries, Least developed countries, Low-income countries, Arab States, Sub-Saharan African countries, East Asia, and Low human development countries fared worse than their more developed counterparts, such as the OECD, High-income OECD, High-income countries, and High human development countries. Analyses also concluded that human beings are at the center of concerns for sustainable development, as access to education seems to be the proximate variable. Thus, sustainable human development is the development that gives priority to the poor, enlarging their choices and opportunities and providing for the participation in decisions that affect their lives and livelihoods. It is the development that is pro-people, pro-nature, pro-jobs, and pro-women.

The policy suggestions and recommendations recognize the fact that the concept of sustainable development and sustainability is elusive, because of the tremendous diversity among regions and cultures. Thus, the paper suggests that sustainable development should be homegrown because it is not easily exported from one place to another (Ukanga and Afoaku, 2005, p.1). Therefore, an orientation to local cultures and community development is a precursor to implementing sustainability. It is suggested that emphasis should be on the provision of integrating factors that allow policies simultaneously to address issues of development, sustainable resource management and poverty eradication. The paper calls for several policy recommendations, which include: (a) empowerment and mainstreaming of women in the development process for gender equity, poverty reduction, and environmental sustainability; (b) emphasis on human development which is sine qua non for successful and sustainable socioeconomic development; (c) a well-articulated cultural policy which should be the main component of endogenous and sustainable development policy; (d) protection of biodiversity or managing biodiversity; and (e) growth management policies and programs through adequate public facilities (concurrency) requirements, growth phasing which includes urban service limits, urban growth boundaries which establish the ultimate limit of urbanization and where services should be available to serve urban development, and the rate-of-growth programs which are based on the actual carrying capacity of public facilities to regulate urban growth. One of the purposes of a growth management program is to prevent urban sprawl and its costs, especially in terms of environmental damage to biodiversity and fragile ecological systems.

REFERENCES

Agbola, Tunde (1990). "Women, Self-Actualization and Theories of Development," in the Special Issue: Women in Human Settlement Development and Management in Africa. *African Urban Quarterly*, Vol. Five, Nos. 3 and 4 (August and November), pp. 170–176.

Ashford, Lori S. (2001). "New Population Policies: Advancing Women's Health and Rights," *Population Bulletin*, (March), Vol. 57, No.1, pp. 21–29.

Borosage, Robert L. (1999). "The Global Turning." *The Nation*, (July, 19), pp.19–22.

Boserup, Esther (1970). *Women's Role in Economic Development* (London: George Allen & Unwin).

Buckingham-Hatfield, Susan (2002). "Gender Equality: A Prerequisite for Sustainable Development," *Geography*, (July), pp. 227–233.

Bunch, C. and R. Carillo, (1990). "Feminist Perspectives on Women in Development," in I. Tinker (ed.), *Persistent Inequalities* (Oxford, UK: Oxford University Press), pp. 70–82.

Bunch, C., M. Dutt, and S. Fried (1995). Beijing 1995: *A Global Referendum on the Human Rights of Women*. New Brunswick, N.J: Center For Women's Global Leadership.

Chant, S. (1985). "Single Parent Families: Choice or Constraint? The Formation Female-headed Households in Mexican Shanty Towns," *Development and Change*, Vol. 16, No. 4: 635–56.

Daly, H.E., and Cobb, J.B., Jr. (1989). *For the Common Good. Redirecting the Economy Toward Community, the Environment and a Sustainable Future* (Boston: Beacon Press).

Dovie, Delali B.K (2003). "Detaining Livelihood and Disputing Biodiversity: Whose Dilemma?" *Ethics, Place and Environment*, Vol. 6, No. 1, (March), pp. 27–41.

Harcourt, W. (ed.) (1994b). *Feminist Perspectives on Sustainable Development*. (London: Zed Books).

Hardy, S. and Lloyd, G. (1994). "An Impossible Dream? Sustainable Regional Economic and Environmental Development," *Regional Studies*, 28 (8), pp. 773–780.

Harf, James E. and Lombardi, Mark Owen (2004). *Taking Sides. Clashing Views on Controversial Global Issues.* Second Edition. (Guilford, Connecticut: McGraw-Hill).
Hartman, H. (1984). "The Unhappy Marriage of Marxism and Feminism: Towards a More Progressive Union," in L. Sargend (ed.), *Women and Revolution.* (Boston: SouthEnd Press).
Hegel, G. W.F. (ed.) (1967). *The Phenomenology of Mind.* Trans. J.B. Baillie (New York: Harper).
Hemmati, Minna (2001). "Women and Sustainable Development: From 2000–2002," in Dodds, F. (ed.). *Earth Summit 2002, A New Deal* (London: Earthscan), pp. 165–183.
Hoff, Marie D. (1998). *Sustainable Community Development. Studies in Economic, Environmental, and Cultural Revitalization.* (London, Washington DC: Lewis Publishers).
Hogendorn, Jan S. (1992). *Economic Development.* Second Edition. (New York: Harper Collins Publishers Inc.).
Hunter, Lori M., John Beal, and Thomas Dickinson (2003). "Integrating Demographic and GAP Analysis Biodiversity Data: Useful Insight?" *Human Dimensions of Wildlife*, 8: 145–157.
IDRC. Development and the Information Age. Chapter 1: Five Indicators of Development. http://www.idrc.ca/books/835/05chapt.1.html. Accessed online on 10/12/99.
Ihonvbere, Julius O. (1993–1994). "The Third World and the New World Order in the 1990s," in Robert J. Griffiths (ed.), *Third World 94/95: Annual Editions.* (Guilford, CT: Dushkin).
International Council on Local Environmental Initiatives (ICLEI) (1996). *The Local Agenda 21 Planning Guide. An Introduction to Sustainable Development Planning.* ICLEI, UNEP, Toronto, Canada.
Jagger, A. (1983). *Feminist Politics and Human Nature* (Totowa, NJ: Rowman & Littlefield).
Jahan, Rounaq (1995). *The Elusive Agenda: Mainstreaming Women in Development.* (London: Zed Books).
Kelly, Eric Damian (2004). *Managing Community Growth.* Second Edition. (Westport, Connecticut, London: Praeger).
Koven, Steven G. and Thomas S. Lyons (2003). *Economic Development: Strategies for State and Local Practice.* (Washington, DC: International City/County Management Association).
Lassode, Louise (1997). *Coping With Population Challenges.* (London: Earthscan Publications Ltd). See especially chapter 3: "Population and Development," pp. 49–69.
Lewis, Arthur (1955). *Theory of Economic Growth.* (Homewood, IL: Irwing).
Macoloo, G.C. (1990). "Settlement Policies and Gender Issues in Less Developed Countries: An Agenda for Action." *African Urban Quarterly*, Vol. Five, Nos. 3 and 4 (August and November), pp. 310–313.
McHarg, Ian L. (1994). *Design With Nature.* 25th Anniversary Ed. (New York: John Wiley).
McMichael, Philip (2000). *Development and Social Change: A Global Perspective (Second Edition)* (Thousand Oaks, California: Pine Forge Press).
Marx, Karl (1976). *Capital (Vol. 1).* (Harmondsworth, UK: Penguin).
Marx, Karl (1983). *Capital (Vol. 3).* (Harmondsworth, UK: Penguin).
Milimo, Mabel C. (1987). "Women, Population and Food in Africa: The Zambian Case." *Development: Seeds of Change*, Vol. 2, No. 3.
Muller, Maria S. (1990). "Women in African Cities," in Schlyter and Johal (eds.): Women in Human Settlement Development and Management. *African Urban Quarterly*, Volume Five, Nos. 3 and 4, August and November 1990, pp. 161–168.
Newman, Peter and Jeffery Kenworthy (1999). *Sustainability and Cities. Overcoming Automobile Dependency.* (Washington, D.C., Covelo, California: Island Press).
Pearce, D., Barbier, E., and Markandya (1990). *Sustainable Development: Economics and Environment in the Third World* (Aldershot: Edward Elgar).
Peet, Richard (1999). *Theories of Development.* (New York, London: The Guilford Press).
Pezzullo, C. (1982). *Women and Development: Guidelines for Program and Project Planning.* United Nations Economic Commission for Latin America and the Caribbean, Santiago, Chile.

Poggi, G. (1983). *Calvinism and Capitalist Spirit: Max Weber's Protestant Ethic*. (Amherst: University of Massachusetts Press).

Rathgeber, E.M., (1990). "WID, WAD, GAD: Trends in Research and Practice." *Journal of Developing Areas*, 24: pp. 489–502.

Rogers, B. (1980). *The Domestication of Women: Discrimination in Developing Societies* (London: Tavistock).

Schlyter, Anne and Darshan Johal (Guest Editors) (1990). "Editors Note: A Preface." Women in Human Settlement Development and Management in Africa. *African Urban Quarterly*, Volume Five, Nos. 3 and 4, August and November.

Sen, G. and C. Grown (1987). *Development Crises and Alternative Visions*. (New York: Monthly Review Press).

Sen, Gita and Srilatha Batliwala (2000). "Empowering Women for Reproductive Rights," in Harriet B. Presser and Gita Sen (eds.), *Women's Empowerment and Demographic Processes*. (Oxford, England: Oxford University Press).

Sharma, Ritu R. (2001). "Women and Development Aid." *Foreign Policy in Focus*, Vol. 6, No. 33, (September).

Shiva, V. (1989). *Staying Alive*. (London: Zed Books).

Snyder, M. (1995). *Transforming Development: Women, Poverty and Politics* (London: Intermediate Technology).

Stiglitz, Joseph (1998). "Gender and Development: The Role of the State." The World Bank Group. April 2. (Washington, DC: World Bank).

Swantz, Marja-Liisa (1987). "The Identity of Women in African Development." *Development: Seeds of Change*, Vol. 2, No. 3.

Tinker, Irene (1997). *Street Foods. Urban Food and Employment in Developing Countries*. (New York, Oxford: Oxford University Press).

Ukanga, Okechukwu and Afoaku, Osita G. (2005). *Sustainable Development in Africa. A Multifaced Challenge*. (Trenton, NJ: African World Press, Inc.).

United Nations (1997). *Report of the Secretary-General Addendum. Protecting and Promoting Human Health (Chapter 6 of Agenda 21)*, p. 3. Economic and Social Council. Commission on Sustanable Development. Fifth Session (April 7–25).

United Nations (2001). "Combating Poverty." Department of Economic and Social Affairs. Report of the Secretary-General, April 30-May 2.

United Nations (2001). *Indicators of Sustainable Development: Guidelines and Methodologies*. http://www.un.org/esa/sustdev/natlinfo/indicators. Accessed online on 10/20/2004.

United Nations Development Program (UNDP) (2004). *Human Development Report 2004. Cultural Liberty in Today's Diverse World*. (New York: Oxford University Press).

UNESCO (1998). *The Power of Culture. The Action Plan*. See the "Preamble." The Intergovernmental Conference on Cultural Policies for Development, held at Stockholm, Sweden (30 March–2 April). Accessed online on 10/1/99 at http://www.unesco-sweden.org/Conference/Action_Plan.htm.

UNICEF (2001). *The State of the World's Children 2001* (New York), table 7. See also, www.unicef.org/sowc01.

Visvanathan, N. et al (eds.) (1997). *The Women, Gender and Development Reader*. (London: Zed Books).

Von Hayek, Friedrich (1956). *The Road to Serfdom (Chicago*: University of Chicago Press).

Wacker, Corinne (1994). "Sustainable Development Through Women's Groups: A Cultural Approach to Sustainable Development," in Wendy Harcourt (ed.), *Feminist Perspectives on Sustainable Development* (London: Zed Books).

Weber, M. (ed.) (1958). *The Protestant Ethic and the Spirit of Capitalism*. Trans. T. Parsons (New York: Scribner's).

Weber, M. (1978). *Max Weber: Selections in Translation*. Ed. W.G. Runciman. Trans. E. Matthews (Cambridge, UK: Cambridge University Press).
World Bank (1990). *World Development Report 1990*. (New York, Oxford: Oxford University Press).
World Bank (1991). *World Development Report 1991. The Challenge of Development*. World Development Indicators. (New York, Oxford: Oxford University Press).
World Bank (1998). "Culture and Development Action Network Working Group Meeting Brief," (January 25–27). *Culture In Sustainable Development*. (Washington, DC: The World Bank). Accessed online on October 1, 1999, at http://www.worldbank.org/essd/kb.nsf.
World Bank (2000). *World Development Report 1999/2000: Entering the 21st Century* (New York, Oxford: Oxford University Press).
World Bank (2004a). *Responsible Growth for the New Millennium: Integrating Society, Ecology, and the Economy*. (Washington, DC: The World Bank).
World Bank (2004b). "Sustainable Development." The World Bank Group. (Washington, DC: The World Bank).
World Commission on Environment and Development (WCED) (1987). *Our Common Future*. (Oxford: Oxford University Press).
World Resources Institute (WRI) (1996). *World Resources 1996–97. A Guide to the Global Environment. The Urban Environment*. A Joint Publication by: The World Resources Institute, The United Nations Environment Program, The United Nations Development Program, and the World Bank (New York, Oxford: Oxford University Press).
Young, K. (1992). *Gender and Development Reader*. (Ottawa: Canadian Council for International Cooperation).
Young, K. (1993). *Planning Development with Women*. (New York: St Martin's Press).

Chapter Five

Socio-Economic Determinants of National Ecological Footprints in Africa

A Multiple Regression Analysis of the Factorial Impacts

INTRODUCTION

Africa is the world's second largest continent (ranks just below Asia) with an area of 11,700,000 square miles, about 906 million inhabitants, estimated in mid-2005 by *Population Reference Bureau (2005)* and 54 countries. Despite its vast natural resources endowment, Africa, especially its sub-Saharan region, is still developing with its nations ranking below other nations of the world in terms of socioeconomic and political development. The sub-Saharan region is the least urbanized (35.6 percent) among other regions of the world, with the lowest consumption level of gross domestic product (GDP) per capita (1,856 PPP US$), and according to the United Nations Human Development Index, the nations ranked below other nations, indicating lowest development (see *Human Development Report 2005*; *World Development Report 2005*). Africa that contains the poorest of the poor has not been helped by the ongoing globalization; they are largely bypassed by it. Nevertheless, as today's poor countries undergo development processes, the climate, fisheries, and forests are coming under increased strain. As their consumption of natural and economic resources increase, so also are their ecological footprints on nature.

FOCUS OF THE STUDY

Countries and Territories in Africa (see also map at the end of this chapter)

Given below are the countries of Africa from which 43 countries are selected for the study based on the availability of data and adequate population size. Several countries are excluded based on being at war (e.g., Liberia, Somalia, and Democratic Republic of Congo), non-independent (e.g., Western Sahara), not up to one million people (e.g., Djibouti), and most islands outside the continent (e.g., Cape Verde, Sao Tome and Principe, Seychelles, and Comoros).

Algeria, Angola, Benin, Botswana, Burkina Faso, Burundi, Cameroon, Cape Verde, Central African Republic, Chad, Comoros, Democratic Republic of the Congo, Congo, Republic of the, Côte d'Ivoire, Djibouti, Egypt, Equatorial Guinea, Eritrea, Ethiopia, Gabon, Gambia, The, Ghana, Guinea, Guinea-Bissau, Kenya, Lesotho, Liberia, Libyan Arab Jamahiriya, Madagascar, Malawi, Mali, Mauritania, Mauritius, Morocco, Mozambique, Namibia, Niger, Nigeria, Réunion, Rwanda, Sao Tome and Principe, Senegal, Seychelles, Sierra Leone, Somalia, South Africa, Sudan, Swaziland, Tanzania, United Republic of, Togo, Tunisia, Uganda, Zambia, Zimbabwe.

PURPOSE OF STUDY

The study analyzes and explains the national variations in total footprints and per capita footprints associated with human development indexes and ecostructural factors [socioeconomic processes within nations such as urbanization, literacy rate, gross domestic product per capita (affluence), and domestic inequality (Gini index)] in Africa. Progress towards sustainable development can be assessed using human development index (HDI) as an indicator of well-being, and the ecological footprint as a measure of demand on the biosphere. It exposes the inequitable distribution of African nations' ecological footprints as compared to those of other regions of the world. Using the tools of comparative model and regression analysis, the paper seeks to establish the key factors, among others, that highly influence and drive the footprints in different African countries. The question is what type of secure environments should meet the needs of both people and national environment in Africa? This constitutes an enormous challenge to policy makers, planners, and development experts. Thus, the study proffers some policies, recommendations, and solutions on how to reduce the footprints of nations, which also will help protect the environment and promote a more equitable and sustainable society.

THEORETICAL FRAMEWORK FOR THE STUDY

Within the biosphere, everything, including humans is interconnected, and sustainable development requires a good knowledge of human ecology. The human life-support functions of the ecosphere are maintained by nature's biocapacity which runs the risks of being depleted, the prevailing technology notwithstanding. Regardless of the humanity's mastery over the natural environment, it still remains a creature of the ecosphere and always in a state of *obligate dependency* on numerous biological goods and services (Rees, 1992, p. 123). Despite the above thesis, the prevailing economic mythology assumes a world in which carrying capacity is indefinitely expandable (Daly, 1986; Solow, 1974). The human species has continued to deplete, draw-down, and confiscate nature's biocapacity with reckless abandon. York et al (2003), indicated that population and affluence account for 95 percent of the variance in total footprints of countries. Total human impact on the ecosphere is given as: Population × Per capita impact (Ehrlich and Holdren, 1971; Holdren and Ehrlich, 1974); Hardin (1991). In other words, population size and affluence are the primary drivers of environmental impacts (Dietz et al, 2007). The footprints of nations provide compelling evidence of the impacts of consumption, thus the need for humans to change their lifestyles. According to Palmer (1998), there are in order of decreasing magnitude, three categories of consumption that contribute enormously to our ecological footprints: wood products (53 percent), food (45 percent), and degraded land (2 percent). Degraded land includes land taken out of ecological availability by buildings, roads, parking lots, etc. Palmer also indicated that about 10 percent or more of earth's forests and other ecological land should be preserved in more or less pristine condition to maintain a minimum base for global biodiversity.

The environmental impacts of urban areas should be considered because a rapidly growing proportion of world's population lives in cities, and more than one million people are added to the world's cities each week and majority of them are in developing of Africa, Latin America, and Southeast Asia (Wackernagel and Rees, 1996). The reality is that the population of all urban regions and many whole nations already exceeded their territorial carrying capacities and depend on trade for survival. Of course, such regions are running an unaccounted ecological deficit; their populations are appropriating carrying capacity from elsewhere (Pimentel, 1996; Wackernagel and Rees, 1996; Girardet, 1996, accessed Online on 12/14/07; Rees, 1992; The International Society for Ecological Economics and Island Press, 1994; Vitouset et al, 1986; Wackernagel, 1991; WRI, 1992, p. 374). Undoubtedly, the rapid urbanization and increasing ecological uncertainty have implications for world development and sustainability. Cities are densely populated areas that have high ecological footprints which leads to the perception of these populations as "parasitic," since these communities have little intrinsic biocapacity, and instead must rely upon large hinterlands.

Land consumed by urban regions is typically at least an order of magnitude greater than that contained within the usual political boundaries or the associated built-up areas (The International Society for Ecological Economics and Island Press, 1994; Rees, 1992).

According to Rees (1992), every city is an entropic black hole drawing on the concentrated material resources and low-entropy production of a vast and scattered hinterlands many times the size of the city itself. In the same vain, Vitouset et al (1986) expressed that high density settlements "appropriate" carrying capacity from all over the globe, as well as from the past and the future (see also Wackernagel, 1991). In modern cities, resources flow through the urban system without much concern either about their origin, or about the destination of their wastes, thus, inputs and outputs are considered to be unrelated. The cities' key activities such as transport, electricity supply, heating, manufacturing and the provision of services depend on a ready supply of fossil fuels, usually from far-flung hinterlands than within their usual political boundaries or their associated built-up areas. Cities are not self-contained entities, and their concentration of intense economic processes and high levels of consumption both increase and stimulate their demands on resources. Cities occupy only 2 percent of the world's land surface, but use some 75 percent of the world resources, and release a similar percentage of waste (Girardet, 1996).

Like urbanization, energy footprint, from energy use and carbon dioxide emissions, is not subject to area constraints. Energy footprint is the area of forest that would be needed to sequester the excess carbon (as carbon dioxide) that is being added to the atmosphere by the burning of fossil fuels to generate energy for travel, heating, lighting, manufacturing, etc. Actually, the demand for energy defines modern cities more than any other single factor. The natural global systems of forests and oceans for carbon sequestration are not handling the human contributions fast enough, thus the Kyoto Conference of early 1998. According to Suplee (1998), only half of the carbon we generate burning fossil fuels can be absorbed in the oceans and existing terrestrial sinks. The oceans absorb about 35 percent of the carbon in carbon dioxide (Suplee, 1998), equivalent of 1.8 giga tons of carbon every year (IPCC, 2001), while the global forests under optimum management of existing forests could absorb about 15 percent of the carbon in the CO_2 produced from the burning of fossil fuels world-wide (Brown et al, 1996). The energy footprint is caused by the unsequestrated 50 percent in the atmosphere with the potentially troubling ecological consequences, such as rapid global warming and other environmental stresses. Carbon dioxide in the atmosphere will continue to increase unless humanity finds alternative energy sources of sufficient magnitude. It is in the humanity's best interest to get off its petroleum addiction and develop sustainable consumption habits.

High literate groups concentrate in urban areas where they consume more than their fair shares of biospheric resources. Literate population generally has low domestic inequality and tends to consume more resources than their illiterate counterpart due to their higher income and urban living. Higher levels of literacy correspond with higher incomes, which allow for greater consumption (Jorgenson, 2003). This is because literate populations are subject to increased consumerist ideologies and contextual images of good life through advertising (Princen et al, 2002), what Leslie Sklair (2001) and Jennifer Clapp (2002) labeled "cultural ideology of consumerism/consumption."

METHODOLOGY FOR THE STUDY

Unit of Analysis

The unit of analysis is "country" (individual African countries).

Sample

To test for the national variations in total footprints and per capita footprints associated with human development indexes and ecostructural (socioeconomic processes) factors in Africa using comparative model, a sample of 43 countries out of 54 countries of the continent (see Global Studies, *Africa, Updated Eleventh Edition*, 2008) was analyzed in tables 5.2–5.3. Step-wise Regression Analysis was also conducted using the sample. Each table represents African countries as also represented by Redefining Progress, *Footprint of Nations 2005*, Appendix 1. Appendix 1 does not include data from countries at war such as Liberia, Somalia, Democratic Republic of Congo or Kingdom of Swaziland or non-independent Western Sahara (made up of Mauritania and Morocco zones), or countries less than 1 million population (e.g., Djibouti) and mostly the islands outside continental Africa, such as Cape Verde, Sao Tome and Principe, Seychelles and Comoros. The larger population of at least one million is chosen for the study because; population is a variable which affects the consumption rates and levels, therefore the ecological footprint.

Mode of Analysis

This is an explanatory study using Step-wise Regression as well as comparative model and descriptive statistics such as matrices, totals, averages, ratios and percentages. Regression Analysis was performed to flesh out factors that highly impacted national footprints, as well as to strengthen and buttress the comparative model analysis results. Potential technical problems are diagnosed such as the missing data (e.g., some domestic inequality or Gini index data for some countries) or the excluded countries mentioned above, which did not alter the substantive conclusions, especially regarding the ecological footprint accounts and environmental sustainability of different African countries. The national ecological footprint and ecological balance, as dependent variables, are explained recursively by the country's human development hierarchical category, population size, population density, urbanization, GDP per capita, domestic inequality (Gini index), and literacy rate, as the independent variables. The independent variables helped in explaining the varying levels of consumption among different human development indexes of the capitalist world economy. It is hypothesized that: Human Development positions and Ecostructural Factors are likely to be responsible for the variations in the Ecological Footprints and Balances. The carbon dioxide emission levels (*World Development Report*, 2005, table 1, pp. 256–257) are included to depict the consumption and environmental degradation impacts of different African countries, which reflect their national ecological footprint and ecological balance levels.

HYPOTHESIS FOR THE STUDY

Null Hypothesis (H_0)

None of the independent variables predicts the national footprint effects. The independent variables do not influence national footprints.

Alternative Hypothesis (H_1)

Some (if not all) of the independent variables predict the national footprint effects.

Variables

Dependent and independent variables are selected on the basis of the theoretical themes and underpinnings described above. The measures of the dependent and independent variables used in the present study are as follow:

Dependent Variables

The ecological footprint or consumption and ecological balance are measured in global hectares (1 hectare = 2.47 acres). A global hectare is 1 hectare of biologically productive space with world average productivity (Global Footprint Network: *Africa's Ecological Footprint-2006 Factbook*, pp. 82–90.

Independent Variables

The ecostructural mediating factors are most of the independent variables in this study that include, GDP per capita, urbanization (urban population as a percent of total population), literacy rate (%), and domestic inequality (Gini index) (see their definitions below).

Human Development Index (HDI)

The HDI is a summary measure of human development (human welfare). It measures the average achievements I a country in three basic dimensions of human development (Global Footprint Network: *Africa's Ecological Footprint-2006 Factbook*, p. 89): a long and healthy life, as measured by life expectancy at birth; knowledge, as measured by the adult literacy rate (with two-thirds weight) and the combined primary, secondary and tertiary gross enrolment ratio (with one-third weight); and a decent standard of living, as measured by GDP per capita (PPP US$). Purchasing power parity (PPP) is a rate of exchange that accounts for price difference across countries, allowing international comparisons of real output and incomes. At the PPP US$ (as used in this study), PPP US$1 has the same purchasing power in the domestic economy as $1 has in the United States of America

The high human development countries are countries with HDI of 0.800 and above, found mainly in North America, Western Europe and Australia, with Gross National Product (GNP) of $9,386 or more. The medium development countries are countries with HDI of 0.500–0.799, found in Eastern Europe, South-East Asia, South America and North Africa, with per capita GNP of $766 and 9,385 (World Bank, 2005, p. 255). The low human development countries are countries with HDI below 0.500, found mainly in Sub-Saharan Africa except the countries of South Africa, Gabon, and Ghana, which are included in the medium human development, with per capita GNP of $765 or less (World Bank, 2005, p. 255). Data for this variable in this study is taken from UNDP's *Human Development Report 2004* and *Human Development Report 2005*.

Gross Domestic Product (GDP) per capita (PPP US$)

GDP is converted to US dollars using the average official exchange rate reported by the International Monetary Fund (IMF). GDP alone does not capture the international relational characteristics as does the human development hierarchy of the world economy, which accounts for a country's relative socio-economic power and global dependence position in the modern world system. It is suggested elsewhere that GDP per capita is an inadequate measure of world-system position but a more appropriate indicator of domestic affluence or internal economic development (Burns, Kentor, and Jorgenson, 2003; Jorgenson, 2003; Dietz and Rosa, 1994). The Gross Domestic Product (GDP) per capita data for this study is taken from *Human Development Report 2004*, table 1, pp.139–142. See also *World Development Report 2005*.

Domestic Income Inequality (Gini Index)

The Gini index measures domestic income inequality of different countries, which had remained stable over a time with its impacts on other variables in the study (Bergesen and Bata, 2002; Jorgenson, 2003). Gini index measures the extent to which the distribution of income (or consumption) among individuals or households within a country deviates from a perfectly equal distribution (*Human Development Report, 2004*, p. 271)). It measures inequality over the entire distribution of income or consumption. A value of zero (0) represents perfect equality, and a value of hundred (100) represents perfect inequality. Data for domestic income inequality measured by Gini index are taken from World Bank, *World Development Report* (2005), table 2, pp. 258–259 and United Nations Development Report, *Human Development Report* (2004), 14, pp. 188–191.

Urbanization Level (Urban Population as Percent of Total Population)

Cities are not self-contained entities and their concentration of intense economic processes and high levels of consumption both increase and stimulate their demands on resources. The cities have limited intrinsic biocapacity which undoubtedly must rely upon large hinterlands. Land consumed by urban regions is typically at least an order of magnitude greater than that contained within the usual political boundaries or the associated built-up area (The International Society for Ecological Economics and Island Press, 1994; Rees, 1992). The data are taken from *Human Development Report 2004*, table 5, pp. 153–155, and *Human Development Report 2005*, table 5, pp. 232–235.

Literacy Rate

This variable refers to the percent of a nation's population over the age of fifteen (15) that can read and write in any language of their choice. Literate population generally has low domestic inequality and tends to consume more resources than their illiterate counterpart due to their higher income and urban living. High literate groups concentrate in urban areas where they consume more than their fair shares of biospheric resources. Higher levels of literacy correspond with higher incomes, which allow for greater consumption (Jorgenson, 2003). This is because literate populations are subject to increased consumerist ideologies and contextual images of good life through advertising (Princen et al, 2002), what Leslie Sklair (2001) and Jennifer Clapp (2002) labeled "cultural ideology of consumerism/consumption." The data for literacy rate is taken from the *Human Development Report 2004*, table 1, pp. 139–142; and *Human Development Report 2005*, table 12, pp. 258–261.

Population and Population Density

Apart from consumption, many have attributed to population as driving most of the sustainability problems (Palmer, 1998; Pimentel, 1996). Likewise, Dietz et al (2007) concluded that population size and affluence are the primary drivers of environmental impacts. Population growth and increases in consumption in many parts of the world have increased humanity's ecological burden on the planet. York et al (2003) indicated that population and affluence account for 95 percent of the variance in total footprints of countries. Others also see the ensuing human impact or footprint as a product of population, affluence (consumption), and technology (i.e. I = PAT (Ehrlich and Holdren, 1971; Holdren and Ehrlich, 1974; Hardin, 1991). Population as a variable in this study is taken from, *World Development Report 2005*, table 5, pp. 232–235.

COMPARATIVE MODEL ANALYSIS (RESULTS AND DISCUSSIONS)

The results and findings in this study are discussed under subtitles that include, "Total Footprint Per Capita, Biological Capacity, and Ecological Balance by World Regions," "Human Development and Affluence on Biological Capacity, Ecological Footprint and Balance in African Countries," and "Population and Ecostructural Factors on Total Footprint Per Capita in African Countries."

TOTAL FOOTPRINT PER CAPITA, BIOLOGICAL CAPACITY, AND ECOLOGICAL BALANCE BY WORLD REGIONS

Table 5.1 shows a comparison of total ecological footprint per capita, biological capacity, and ecological balance by world regions. The world regions include Africa, Middle East and Central Asia, Asia-Pacific, Latin America and Caribbean, North America, Western Europe, Central and Eastern Europe. Analysis of table 5.1 exposes the inequitable distribution of African nations' ecological footprints as compared to those of other world regions. The table shows a comparison of each region's footprint with its biocapacity depicting whether that region has an ecological reserve or is running a deficit. From the table, North America with its considerable biological capacity in 2001 (53.16 global hectares) has the largest total footprint (95.99 global hectares) and the largest ecological deficit or overshoot (–42.83 global hectares), followed by Western Europe with total footprint of 60.70 and ecological deficit of –43.86, followed by the Middle East and Central Asia, 41.61 and –28.06, and Central and Eastern Europe, 31.36 and –18.91 respectively. The above regions with overshoots must be appropriating about half of their carrying capacity from elsewhere, probably from large and vast hinterlands. Table 5.1 also shows that Africa, Asia Pacific, and Latin America and Caribbean have low ecological footprint and positive ecological balance or reserve, far better than the world averages. The world's total footprint average is 21.91 global hectares, while its average ecological balance is –6.70 global hectares. African region has the smallest ecological footprint (7.48 global hectares) and the largest ecological balance (20.03 global hectares) when compared to other regions, followed by Latin America and Caribbean, 16.90 gha and 5.31gha, and the Asia-Pacific, 19.42 gha and 10.55 gha respectively. A regional footprint that is less than the world average, and average biocapacity above that of the world, may denote sustainability at the global level (Living Planet Report, 2006).

HUMAN DEVELOPMENT AND AFFLUENCE ON BIOLOGICAL CAPACITY, ECOLOGICAL FOOTPRINT AND BALANCE IN AFRICAN COUNTRIES

Table 5.2 presents the national comparisons for both dependent (Ecological footprint and Ecological balance) and some of the independent (Human development or welfare and GDP per capita or affluence) variables in Africa. The table shows that a country's total footprint, ecological balance, and carbon dioxide emissions (waste) are a function of its human development position and affluence level in the world economy. Given the national biological capacity measured in global hectares, the higher the total ecological footprint; the lower the ecological balance and vice versa. Table 5.2 assesses the progress towards sustainability development in Africa by using human development index (HDI) as an indicator of national well-being, and the ecological footprint and balance as measures of national demand on the biosphere or biological capacity. The sustainability quotient is depicted by any country with low footprint and high human development levels. In fact, table 5.2 shows that, among African countries, improvements in the human development indexes (human welfare) have resulted in the worsening of the countries' positions on their ecological footprints and balances. Table 5.2 shows generally that African countries are not among the high human development (HDI of 0.800 and above and per capita GNP of $9,386 or more) or medium human development (HDI of 0.500–0.799 and per capita GNP of between $766 and $9,385) but generally of low human development with HDI below 0.500 and per capita GNP of $765 or less (World Bank, 2005, p. 255).

From table 5.2 and based on the existing biological capacity, it becomes apparent that the higher the HDI (human welfare), GDP per capita (affluence) i.e. consumption; the higher the carbon dioxide emissions and ecological footprint. For example, South Africa shows HDI of 0.658, GDP ($10,346), carbon dioxide emission (327.3 million tons), total footprint (40.62 global hectares), biological capacity (20.80 global hectares), and ecological balance of –19.81 global hectares. Actually, South Africa "appropriates" half of its carrying capacity from "elsewhere." Egypt shows HDI of 0.659, GDP ($3,950), carbon dioxide emissions (142.2), total footprint (12.30), biological capacity (9.25), and ecological balance of –3.05. Likewise, Algeria had HDI of 0.722, GDP ($6,107), carbon dioxide emissions (89.4), total footprint 15.21), biological capacity (20.11), and ecological balance of 4.91. The same is also true for Tunisia that shows HDI of 0.753, GDP per capita ($7,161), carbon dioxide emissions (18.4), total footprint (10.35), biological capacity (10.70), and ecological balance of 0.35. South Africa and Egypt have footprints larger than their locally available biological capacities and have ecological deficits or overshoots. Like South Africa and Egypt, Algeria and Tunisia indicate that the higher the consumption pattern; the higher the footprint, and the smaller the locally available biological capacity and ecological reserve.

Table 5.2 also depicts that the lower the HDI level and GDP per capita; the lower the carbon dioxide emissions (waste), the lower the ecological footprint, and the higher the ecological balance or reserve. For example, Sierra Leone shows the HDI of 0.298, GDP per capita ($548), carbon dioxide emission (0.60), biological capacity (14.15), total footprints (3.27), and ecological balance of 10.87. Niger has also HDI of 0.281, GDP per capita ($835), carbon dioxide emissions (1.2), biological capacity (26.84), total ecological footprints (1.92), and ecological balance of 24.91. Likewise, Burundi shows HDI of 0.378, GPD per capita ($648), carbon dioxide emissions (0.2), biological capacity (7.36), ecological footprint (1.50), and ecological balance of 5.86. In the same token, Zambia has HDI of 0.391, GDP per capita ($877), carbon dioxide emissions (1.8), biological capacity (30.40), total ecological footprint (2.89), and ecological balance of 27.51. Analysis of table 5.2 has shown that, given a nation's biocapacity, its levels of human development (welfare), GDP per capita (affluence), and carbon dioxide emissions (waste) help to explain its total ecological footprint (consumption or demand) and ecological balance (reserve).

POPULATION, ECOSTRUCTURAL FACTORS, AND TOTAL FOOTPRINT PER CAPITA IN AFRICAN COUNTRIES

Table 5.3 shows the effects of population and ecostructural factors (socioeconomic processes) on the total ecological footprints of nations. A nation's total population size and population density, as well as its urban population, literacy rate, and domestic inequality (Gini index), lead to its ecological footprint difference and variation vis-à-vis other nations. The above indicators constitute the structural causes of the ecological footprint variations among African countries. Table 5.3 shows that a nation with large population, high rates of urbanization and literacy, and low domestic inequality, has high ecological footprint, and vice versa. Large population level, high population density, high level of urbanization, high level of literacy, and low domestic inequality promote domestic consumption of biospheric resources.

Cities are not self contained entities and have little intrinsic biological capacity; their concentration of intense economic processes and high levels of consumption both increase and stimulate their demands on bioresources. Thus, cities are parasitic and in most cases must rely upon large hinterlands for their supplies which increase their footprint impacts. High density settlements appropriate carrying capacity from all over the globe, as well as from the past and the future (Vitousek et al, 1986; Wackernagel, 1991). High levels of urbanization correspond to higher levels of consumption, higher levels of energy use and carbon dioxide emissions (waste), and thus, higher energy footprints and ecological footprints. Biospheric resources are consumed at higher levels in urban areas that generally contain high literate groups with higher incomes, which allow for greater material consumption (Jorgenson, 2003; Princen et al, 2002; Sklair, 2001; Clapp, 2002).

According to table 5.3, Nigeria, Egypt, and South Africa are nations with large population, high population density, and high rates of urbanization and literacy that have high total ecological footprints. For example, Nigeria has population of 125.9 million, population density of 149 people per km^2, literacy rate of 66.8 percent, and total ecological footprint of 5.84. Likewise, South Africa has population of 46.9 million, literacy (82.4%) and total footprint of 40.62. Although Algeria and Tunisia have low population densities but they have large urban population, high literacy rates, low domestic inequalities and high total footprints. For example, Algeria has population density of 13 persons per km^2, urbanization (58.8%), literacy (69.8%), domestic inequality (35.3), and total footprint of 15.21. Nations like Burundi and Rwanda that have low urban populations, but high population densities, low domestic inequalities, and high literacy rates, have low total footprints. For example, Rwanda has the highest population density in Africa of 334 persons per km^2, but small population of 8.8 million, low urban population of 18.5 percent, high literacy rate of 64 percent, low domestic inequality of 28.9, and low footprint of 1.65. Thus, the level of consumption and not population density is a major driving force of environmental sustainability and ecological footprints.

Tables 5.2 and 5.3 depict low human development countries or the peripheral countries with low levels of urbanization and literacy, and high levels of domestic inequality as exemplified by high Gini indexes or quotients. Lower levels of literacy and urbanization correspond to lower levels of income, which allow for smaller material consumption and vice versa. The findings here parallel the theorization above, which emphasizes that the low consumption levels and the accompanying low ecological footprints and high ecological balances (reserves) are based on the nations' positions in the world capitalist economy and system. Most of the African countries were once colonized by the high human development countries such as Britain, France, Germany, Portugal, Belgium, and Spain, where the processes of underdevelopment, economic stagnation, and dependent industrialization limit their populations to consuming a greater share of unprocessed bio-productive resources. The high domestic income inequality or intra-inequality has a negative or declining effect on per capita consumption levels (Bornschier and Chase-Dunn, 1985; Kick, 2000; Kentor, 2001; and Jorgenson, 2003) that yield both low total and per capita ecological footprints.

REGRESSION ANALYSIS AND REGRESSION EQUATION

Regression analysis allows the modeling, examining, and exploring of relationships and can help explain the factors behind the observed relationships or patterns. It is also used for predictions. Regression equation, on the other hand, utilizes the Ordinary Least Square (OLS) technique, is the mathematical formula that is applied to the explanatory variables to best predict the dependent variable one is trying to model. Each independent variable or explanatory variable is associated with a regression coefficient describing the strength and the sign of that variable's relationship to the dependent variable. A regression equation might look like the one given below where Y is the dependent variable, the Xs are the explanatory variables, and the Bs are regression coefficients:

$$Y = B_0 + B_1 X_1 + B_2 X_2 + \text{-----} B_n X_n + E \text{ (Random Error Term/Residuals)}$$

Correlation or co-relation: refers to the departure of two variables from independence or they are non-independent or redundant (Antonia D'Onofrio, 2001/2002; Richard Lowry, 1999–2008).

Collinearity: Refers to the presence of exact linear relationships within a set of variables, typically a set of explanatory (predictor) variables used in a regression-type model. It means that within the set of variables, some of the variables are (nearly) totally predicted by the other variables [(Rolf Sundberg, 2002). *Encyclopedia of Environmetrics*, edited by Abdel H. El-Shaarawi and Walter W. Piegorsch (Chichester: John Wiley & Sons, Ltd), Volume 1, pp. 365–366].

Partial Correlation Coefficients (r): When large, it means that there is no mediating variable (a third variable) between two correlated variables (Antonia D'Onofrio, 2001/2002).

Pearson's Correlation Coefficient (r): This is a measure of the strength of the association between two variables. It indicates the strength and direction of a linear relationship between two random variables. Value ranges from – to +1; –1.0 to –0.7 Strong negative association; –0.7 to –0.3 Weak negative association; –03 to +0.3 Little or no association; +0.3 to +0.7 Weak positive association; +0.7 to 1.0 Strong positive correlation (Brian Luke, "Pearson's Correlation Coefficient," Learning *From The Web.net*. Accessed Online on 5/30/2008).

Multiple "R": Indicates size of the correlation between the observed outcome variable and the predicted outcome variable (based on the regression equation).

"R^2" or Coefficient of Determination: Indicates the amount of variation (%) in the dependent scores attributable to all independent variables combined, and ranges from 0 to 100 percent. It is a measure of model performance, summarizing how well the estimated Y values match the observed Y values.

"Adjusted R^2": The best estimate of R^2 for the population from which the sample was drawn. The Adjusted R-Squared is always a bit lower than the Multiple R-Squared value because it reflects model complexity (the number of variables) as it relates to the data.

R^2 and the *Adjusted R^2* are both statistics derived from the regression equation to quantify model performance (Scott and Pratt, 2009. *ArcUser*).

Standard Error of Estimate: Indicates the average of the observed scores around the predicted regression line.

Residuals: These are the unexplained portion of the dependent variable, represented in the regression equation as the random error term (E). The magnitude of the residuals from a regression equation is one measure of model fit. Large residuals indicate poor model fit. Residual = Observed – Predicted.

ANOVA: Decomposes the total sum of squares into regression (= explained) SS and residual (= unexplained) SS.

F-test in ANOVA represents the relative magnitude of explained to unexplained variation. If F-test is highly significant (p = .000), we reject the null-hypothesis that none of the independent variables predicts the effect (scores) in the population.

The "constant" represents the intercept in the equation and the coefficient in the column labeled by the independent variables.

The row of "unstandardized coefficients" or "Bs" gives us the necessary coefficient values for the multiple regression models or equations.

From table 5.2, *Human Development and Affluence on Biological Capacity, Ecological Footprint and Balance in African Countries*, the dependent variable is Total Footprint, and the independent variables are Human Development

Index (HDI), GDP Per Capita, Carbon Dioxide Emissions, Bio-Capacity, and Ecological Balance

From table 5.3, *Population, Ecostructural Factors, and Total Footprint Per Capita in African Countries*, the dependent variable is Total Footprint, and independent variables are Population, Population Density, Urban Population, Literacy Rates, and Gini Index.

REGRESSION ANALYSIS (RESULTS AND DISCUSSIONS)

Table 5.4 depicts the descriptive statistics of the data in table 5.2. The measures of variability or dispersion, from table 5.4, such as standard deviation and variance signify how homogeneous (in terms of low levels) African countries are in terms of Human Development Index (HDI) and Total Footprint, and how heterogeneous they are (in terms of high levels) in terms of GDP Per Capita and Carbon dioxide Emissions. The same is also true in table 5.10 where low variability is associated with Gini Index and high variability with Population Density, Urban Population, and Literacy Rates. Table 5.5 shows also the Partial Correlation Coefficients for the data in table 5.2. While controlling for Ecological Balance, a significant relationship or non-independence exists between Total Footprint and predictor or explanatory variables such as HDI, GDP Per Capita, and Carbon dioxide Emissions. The coefficients signify how closely these variables track or relate to each other, but nonetheless do not imply causality. Likewise, table 5.11 shows the Pearson Correlation (2-tailed) of the data from table 5.3. From table 5.11, there is strong correlation or relationship between Total Footprint and Urban Population and to some extent with Literacy Rates. Table 5.7 is the Regression model summary, which shows the Multiple R and R-Squared or Coefficient of Determination very high and significant. For example, about 91 percent of the dependent variable (Total Footprint) is accounted for by the independent (predictor, explanatory) variables (GDP Per Capita and Carbon Dioxide Emissions) entered in the stepwise regression model in table 5.6. The Multiple R is also very large (95 percent), which indicates size of the correlation between the observed outcome variables and the predicted outcome variables (based on the regression equation).

Tables 5.12, 5.13A, and 5.13B show the Multiple R and R-Squared between dependent variable (Total Footprint) and the independent variables, Urban Population and Literacy Rates. Nevertheless, their relationships as well as predictability and accountability are not as high and significant as found in table 5.7. From table 5.13A, Model Summary, about 38 percent of the variations in Total Footprint is explained or accounted for by Urban Population and Literacy Rates. Tables 5.8 and 5.14 depict the analysis of variance (ANOVA) and F-Tests. The F-tests show how strong or significant the selected independent variables in the regression models are in predicting the dependent variable. From table 5.8, the F values are high and the residual (observed – predicted) values are small. Since the F-Tests are highly significant, we reject the Null Hypothesis that none of the independent variables (predictors) predict the National footprint effects. It is worthy to note that the results in table 5.14 are not as strong as those in table 5.8 but still significant. Using the B columns of the Unstandardized Coefficients in tables 5.9 and 5.15, the regression model equations are constructed from tables 5.2 and 5.3 (see below).

(A) For table 5.2: *Human development, GDP, Carbon dioxide Emissions, Bio-Capacity Factors on Total Footprint*, the significant explanatory (predictor) variables are: GDP Per Capita and Carbon dioxide emissions. The Regression Equation is: Total Footprint = 0.852 + 1.735GDP Per Capita + 6.250 Carbon Dioxide Emissions + E (residuals).

(B) For table 5.3: *Population, Population Density, Urban Population, Literacy Rates, Domestic Inequality (Gini Index) on Total Footprint*, the significant explanatory (predictor) variables are Urban Population and Literacy Rates. The Regression Equation is: Total Footprint = –9.535 + 0.229 Urban Population + 0.131Literacy Rates + E (residuals).

From A and B above, Total National Footprint in Africa is predicted or explained by GDP Per Capita, Carbon Dioxide Emissions, Urban Population, and Literacy Rates.

POLICY RECOMMENDATIONS

The above analyses support the theorization and assertion that the global environmental stress is caused primarily by increase in resource consumption (demand) and human population size and growth. The aim of the African region and countries should be to improve on their human development to move out of poverty and at the same time prevent or reduce overshoot or deficit. Wealth (growth) can be good for the environment, only if public policy and technologies encourage sound practices and the necessary investments are made in environmental sustainability (Sachs, 2007). More affluent among them can reduce consumption and still improve their quality of life. Therefore, long-term investment will be required in many areas, including education, technology, conservation, urban and family planning, and resource certification systems, along with development of new business models and financial markets Innovative approaches to meeting human needs without environmental destruction are called for, if there should be a shift in belief that greater well-being necessarily entails more consumption. The development of low-impact energy sources will play an important role, as will a shift to sustainable agricultural and food production and distribution systems.

Factors that significantly shape African ecological footprint or demand on biocapacity include GDP Per Capita, Carbon dioxide Emissions, Urban Population, and Literacy Rates. Policies and programs should include, among others: (1) increase in population can be slowed and eventually reversed by supporting families in choosing to have fewer children. Women empowerment through their access to better education, economic opportunities, and health care is a proven approach to family, population management, and sustainable development (Aka, 2006); (2) the potential for reducing consumption depends on individual's economic situation, more affluent people can reduce consumption and still improve their quality of life; (3) in the case of footprint intensity, the amount of resources used in the production of goods and services, can be significantly reduced from energy efficiency in manufacturing and in the home, through minimizing waste and increasing recycling and reuse, to fuel-efficient cars; (4) bio-productive area can be extended, for example, degraded lands can be reclaimed through careful management that include, terracing, irrigation, and to ensure that bio-productive areas do not diminish or being lost to urbanization, salinization, or desertification; and (5) bio-productivity per hectare depends both on the type of ecosystem and the way it is managed. For most of the African nations that lag behind other world regions to gain most from their wealth-producing natural capital, they should increase their consumption levels to come out of their abject and debilitating poverty.

In order to increase their GNP/GDP per capita (affluence) and reduce domestic inequalities (Gini index or coefficient), African nations should improve their human development (welfare) indices by investing more in education (literacy), healthcare systems, and technology. Adequate and improved environmental and ecological education will reduce individual's ecological impact since it has been recognized that the environmental crisis is less an environmental and technical problem than it is a behavioral and social one (Wackernagel and Rees, 1996).

CONCLUSIONS

This is an explanatory study using comparative model and descriptive statistics such as matrices, totals, ratios, and percentages, as well as regression analysis. The results and findings show that a country's total footprint, ecological balance, and carbon dioxide emissions (waste) are a function of its human development (welfare) position and affluence level in the world economy. Given the national biological capacity, the higher the total ecological footprint, the lower the ecological balance and vice versa. Thus, as human development goes up, then environmental performance goes down.

Data analyses in the present study show that, the higher the national populations, population densities, and GDP per capita (affluence), the higher the *imbalaced* consumption total footprint, carbon dioxide emissions, and the lower the ecological balance, and vice versa. The research findings also indicate that lack of improvements in human development index (human welfare) and lack of or declining affluence result in a bettering of a nation's ecological footprint position. This is not to say that improved human development (welfare) or affluence is not desirable but should be encouraged and pursued simultaneously with adequate public policy and sound practices for environmental sustainability. It is also found out that lower levels of literacy and urbanization correspond to lower levels of income, which allow for smaller material consumption and vice versa. The higher domestic income inequality or intra-inequality in some African countries has a negative (declining) effect on per capita consumption levels that yield both low total and per capita ecological footprints.

The above analyses support the theorization and assertion that the global environmental stress is caused primarily by the increase in resource consumption (demand) and population size and growth. It is suggested also that sustainability will depend on such measures as greater emphasis on equity in international relationships, significant adjustment to prevailing terms of trade, increasing regional self-reliance, and policies to stimulate a massive increase in the material and energy efficiency of economic activities.

REFERENCES

Aka, Ebenezer (2006). "Gender Equity and Sustainable Socio-Economic Growth and Development," *The International Journal of Environmental, Cultural, Economic & Social Sustainability*, Volume 1, Number 5, pp. 53–71.

Bergesen, Albert and Michelle Bata (2002). "Global and National Inequality: Are They Connected?" *Journal of World-System Research*, 8: 130–44.

Bergesen and Bartley (2000). "World-System and Ecosystem," in *A World-Systems Reader: New Perspectives on Gender, Urbanism, Culture, Indigenous Peoples, and Ecology*, edited by Thomas Hall (Lanham, MD: Rowman and Littlefield), pp. 307–22.

Bornschier, Volker and Christopher Chase-Dunn (1985). *Transnational Corporations and Underdevelopment* (New York: Praeger).

Brown, S., Jayant, S., Cannell, M., and Kauppi, P. (1996). "Mitigation of Carbon Emissions to the Atmosphere by Forest Management," *Commonwealth Forestry Review*, 75, 79–91.

Clapp, Jennifer (2002). "The Distancing of Waste: Over-consumption in a Global Economy," in *Confronting Consumption*, edited by T. Princen, M. Maniates, and K. Conca (Cambridge, MA: MIT Press) pp. 155–76.

Daly, H.E. (1986). *Beyond Growth: The Economics of Sustainable Development* (Boston, Massachusetts, USA: Beacon Press).

Dietz, Thomas and Eugene Rosa (1994). "Rethinking the Environmental Impacts of Population, Affluence, and Technology," *Human Ecology Review*, 1: 277–200.

Dietz, Thomas, Eugene Rosa, Yoor (2007). "Driving the Human Ecological Footprint." *Frontiers in Ecology and the Environment* (February).

D'Onofrio, Antonia (2001/2002). "partial Correlation." Ed 710 Educational Statistics, Spring 2003. Accessed Online on 6/3/2008 at http://www2.widener.edu/.

Ehrlich and Holdren (1971). "Impacts of Population Growth," *Science*, 171, 1212–7.

Girardet, Herbert (1996). "Giant Footprints," Accessed Online on 12/14/07, at http:// www.gdrc.org/uem/footprints/girardet.html.

Global Footprint Network: Africa's Ecological Footprint—2006 Factbook. Global Footprint Network, 1050 Warfield Avenue, Oakland, CA 94610, USA. http://www.footprintnetwork.org/Africa.

Global Studies (2008). *Africa: Updated Eleventh Edition* (Dubuque, Iowa: McGraw-Hill).

Hardin, G. (1991). "Paramount Positions in Ecological Economics," in Constanza, R. (editor), *Ecological Economics: The Science and Management of Sustainability* (New York: Columbia University Press).

Holdren and Ehrlich (1974). "Human Population and The Global Environment," *American Scientist*, 62: 282–92.

IPCC (2001). Intergovernmental Panel on Climatic Change.

Jorgenson, Andrew K. (2003). "Consumption and Environmental Degradation: A Cross-National Analysis of Ecological Footprint," *Social Problems*, Volume 50, No. 3, pp. 374–394.

Kentor, Jeffrey (2001). "The Long Term Effects of Globalization on Income Inequality, Population Growth, and Economic Development," *Social Problems*, 48: 435–55.

Kick, Edward L. (2000). "World-System Position, National Political Characteristics and Economic Development," Journal of Political and Military Sociology, (Summer), http://www.findarticles.com/p/articles/mi-qa3719.

Lowry, Richard (1999–2008). "Subchapter 3a. Partial Correlation." Accessed Online on 6/3/08 at http://faculty.vassar.edu/lowry/cha3a.html.

Luke, Brian T. "Pearson's Correlation Coefficient." Learning From The Web.net. Accessed Online on 5/30/2008.

Palmer, A.R. (1998). "Evaluating Ecological Footprints," *Electronic Green Journal*. Special Issue 9 (December).

Pimentel, David (1996). "Impact of Population Growth on Food Supplies and Environment." American Association for the Advancement of Science (AAAS) (February, 9). See also GIGA DEATH, Accessed Online on 12/18/07, at http://dieoff.org/page13htm.

Princen, Thomas (2002). "Consumption and Its Externalities: Where Economy Meets Ecology," in *Confronting Consumption*, edited by T. Princen, M. Maniates, and K. Conca (Cambridge, MA: MIT Press) pp. 23–42.

Redefining Progress (2005). Footprints of Nations. 1904 Franklin Street, Oakland, California, 94612. See, http://www.RedefiningProgress.org.

Rees, William E. (1992). "Ecological Footprint and Appropriated Carrying Capacity: What Urban Economics Leaves Out," Environment and Urbanization, Vol. 4, No. 2, (October). Accessed Online on 12/14/07, at http://eau.sagepub.com.

Sachs, Jeffrey D. (2007). "Can Extreme Poverty Be Eliminated"? *Developing World, 07/08* (Dubuque, IA: McGraw Hill0 pp. 10–14.

Scott, Lauren and Monica Pratt (2009). "An Introduction To Using Regression Analysis With Spatial Data." *ArcUser. The Magazine for ESRI Software Users* (Spring), pp. 40–43. Lauren Scott and Monica Pratt are ESRI Geoprocessing Spatial Statistics Product Engineer and ArcUser Editor respectively.

Sklair, Leslie (2001). *The Transnational Capitalist Class* (Oxford, UK: Blackwell Press).

Solow, R. M. (1974). "The Economics of Resources or the Resources of Economics," *American Economics Review*, Vol. 64, pp. 1–14.

Sundberg, Rolf (2002). *Encyclopedia of Environmetrics*, edited by Abdel H. El-Shaarawi and Walter W. Piegorsch (Chichester: John Wiley & Sons, Ltd), Volume 1, pp. 365–366.

Suplee, D. (1998). "Unlocking the Climate Puzzle." *National Geographic*, 193 (5), 38–70.
The international Society for Ecological Economics and Island Press (1994). "Investing in Natural Capital: The Ecological Approach to Sustainability." Accessed Online on 12/18/07, at http://www.dieoff.org/page13.htm.
United Nations Development Program (2004). *Human Development Report 2004 (HDR). Cultural Liberty in Today's Diverse World* (New York, N.Y: UNDP).
United Nations Development Program (2005). *Human Development Report (HDR) 2005. International Cooperation at a Crossroads: Aid, Trade and Security in an Unequal World* (New York, N.Y: UNDP).
Vitousek, P., P. Ehrlich, A. Ehrlich and P. Matson (1986). "Human Appropriation of the Products of Photosynthesis," *Bioscience*, Vol. 36, pp. 368–374.
Wackernagel, M. (1991). "Using 'Appropriated Carrying Capacity' as an Indicator: Measuring the Sustainability of a Community." Report for the UBC Task Force on Healthy and Sustainable Communities. UBC School of Community and Regional Planning, Vancouver, Canada.
Wachernagel and Rees (1996). *Our Ecological Footprint* (New Society Publisher).
World Bank (2005). *World Development Report 2005. A Better Investment Climate for Everyone* (New York, N.Y: A Co-publication of the World Bank and Oxford University Press).
World Resources Institute (WRI) (1992). *World Resources, 1992–1993* (New York: Oxford University Press).
York, Richard, Eugene A. Rosa, and Thomas Dietz (2003). "Footprints on the Earth: The Environmental Consequences of Modernity." *American Sociological Review*, 68: 279–300.

Table 5.1. Total Footprint Per Capita, Biological Capacity, and Ecological Balance by World Regions in 2001

Region	*Total Footprint (Global Hectares) (2001)*	*Biological Capacity (Global Hectares) (2001)*	*Ecological Balance (Global Hectares) (2001)*
Africa	**7.48**	**27.51**	**20.03**
Middle East and Central Asia	41.61	13.55	–28.06
Asia-Pacific	19.42	29.97	10.55
Latin America and Caribbean	16.90	22.22	5.31
North America	95.99	53.16	–42.83
Western Europe	60.70	16.84	–43.86
Central and Eastern Europe	31.36	12.45	–18.91
World	**21.91**	**15.71**	**–6.20**

Source: Redefining Progress (2005). *Footprints of Nations*. Appendix 1. Accessed Online on 4/3/2006, at http://www.reprogress.org/education.

Table 5.2. Human Development and Affluence on Biological Capacity, Ecological Footprint and Balance in African Countries

Countries	*Human Development Index (HDI) Value (2003)*[1]	*GDP Per Capita (PPP US$) (2003)*[2]	*Carbon Dioxide Emissions Mil. Tons (2000)*[3]	*Total Footprint (Global Hectares) (2001)*[4]	*Bio-Capacity (Global Hectares) (2001)*[4]	*Ecological Balance (Global Hectares) (2001)*[4]
Algeria	0.722	6,107	89.4	15.21	20.11	4.91
Angola	0.445	2,344	6.4	6.41	44.71	38.30
Benin	0.431	1,115	1.6	3.06	10.24	7.18
Botswana	0.565	8,714	3.9	16.42	63.61	47.19
Burkina Faso	0.317	1,174	1.0	3.05	12.46	9.42
Burundi	0.378	648	0.2	1.50	7.36	5.86
Cameroon	0.497	2,118	6.5	4.43	18.66	14.23
Central African Rep.	0.355	1,089	0.3	4.19	47.40	43.21
Chad	0.341	1,210	0.1	4.21	32.96	28.75
Congo	0.512	965	1.8	3.46	55.90	52.44
Cote D'Ivoire	0.420	1,476	10.5	3.84	14.52	10.68
Egypt	0.659	3,950	142.2	12.30	9.25	–3.05
Eritrea	0.444	849	0.6	2.79	13.63	10.83
Ethiopia	0.367	711	5.6	1.56	8.53	6.97
Gabon	0.635	6,397	—	3.78	10.64	6.86
Gambia	0.470	1,859	—	3.78	10.64	6.86
Ghana	0.520	2,238	5.9	3.23	10.78	7.54
Guinea	0.466	2,097	1.3	3.51	21.10	17.59
Guinea-Bissau	0.348	711	—	3.65	28.60	24.95
Kenya	0.474	1,037	9.4	3.87	12.70	8.83
Lesotho	0.497	2,561	—	2.72	12.41	9.69
Libya	0.799	—	9.1*	39.50	71.09	31.60

Madagascar	0.499	809	2.3	3.46	23.85	20.39
Malawi	0.404	605	0.8	1.86	8.50	6.64
Mali	0.333	994	0.6	—	—	—
Mauritania	0.477	1,766	3.1	11.28	79.77	68.49
Mauritius	0.791	11,287	2.6*	22.31	58.41	36.11
Morocco	0.631	4,004	36.5	6.22	19.65	4.43
Mozambique	0.379	1,117	1.2	1.49	23.99	22.50
Namibia	0.627	6,180	1.8	9.67	106.96	97.30
Niger	0.281	835	1.2	1.92	26.84	24.91
Nigeria	0.453	1,050	36.1	5.84	9.04	3.20
Rwanda	0.450	1,268	0.6	1.65	7.13	5.48
Senegal	0.458	1,648	4.2	5.98	14.46	8.48
Sierra Leone	0.298	548	0.6	3.27	14.15	10.87
South Africa	0.658	10,346	327.3	40.62	20.80	–19.81
Sudan	0.512	1,910	0.3*	4.23	18.99	14.76
Tanzania	0.418	621	4.3	3.39	15.81	12.42
Togo	0.512	1,696	1.8	3.83	9.05	5.23
Tunisia	0.753	7,161	18.4	10.35	10.70	0.35
Uganda	0.508	1,457	1.5	3.55	10.00	6.44
Zambia	0.394	877	1.8	2.89	30.40	27.51
Zimbabwe	0.505	2,443	14.8	9.03	16.35	7.32
Sub-Saharan Africa	0.515	1,856	478.8	—	—	—
Africa	—	—	—	7.48	27.51	20.03

*Measured in per capita metric tons in 2002. *Human Development Report (HDR) 2005*, p. 290, table 22.

1. *Human Development Report (HDR) 2005.*
2. *Human Development Report (HDR) 2004*, table 1, pp. 139–142.
3. *World Development Report (WDR) 2005*. Table 1. Key Indicators of Development, pp. 256–257.
4. Wackernagel et al (2000). *Ecological Footprints and Ecological Capacities of 152 Nations: The 1996 Update*. (San Francisco, CA: Redefining Progress). See also, Redefining Progress (2005). *Footprints of Nations 2005*. Accessed Online on 4/3/2006, at http://www.reprogress.org/education. Note that ecological footprint is measured in global hectares. The enormous share of ecological overshoot attributable to carbon dioxide emissions in some African countries is made explicit in total footprint per capita (Redefining Progress, 2005; Wackernagel et al, 2005; Wackernagel and Monfreda, 2004; Dukes, 2003; Global Footprint Network: *Africa's Ecological Footprints, 2006*, p. 86).

Table 5.3. Population, Ecostructural Factors, and Total Footprint Per Capita in African Countries

Countries	*Population (Millions) (2003)*[1]	*Population Density (People Per Km^2) (2003)*[2]	*Urban Population (% of Total) (2003)*[1a]	*Literacy Rates (%) (2003)*[3]	*Domestic Inequality (Gini Index) by 2003*[4]	*Total Footprint (Global Hectares) (2001)*[5]
Algeria	39.1	13	58.8	69.8	35.3	15.21
Angola	15.0	11	35.7	66.8	—	6.41
Benin	7.9	61	44.6	33.6	—	3.06
Botswana	1.8	3	51.6	78.9	63.0	16.42
Burkina Faso	12.4	44	17.8	12.8	48.2	3.05
Burundi	7.0	281	10.0	58.9	33.3	1.50
Cameroon	15.7	35	51.4	67.9	44.6	4.43
Central African Rep.	3.9	6	42.7	48.6	61.3	4.19
Chad	9.1	7	25.0	25.5	—	4.21
Congo	3.8	11	53.5	82.8	—	3.46
Cote D'Ivoire	17.6	53	44.9	48.1	44.6	3.84
Egypt	71.3	68	42.2	55.6	34.4	12.30
Eritrea	4.1	43	20.0	—	—	2.79
Ethiopia	73.8	69	15.7	41.5	30.0	1.56
Gabon	1.3	—	83.7	—	—	24.06
Gambia	1.4	—	26.2	—	47.5	3.78
Ghana	21.2	90	45.4	54.1	40.8	3.23
Guinea	9.0	—	34.9	—	40.3	3.51
Guinea-Bissau	1.5	—	34.0	—	47.0	3.65
Kenya	32.7	56	39.3	73.6	42.5	3.87
Lesotho	1.8	59	18.0	81.4	63.2	2.27
Libya	5.6	—	86.2	81.7	—	39.50
Madagascar	17.6	29	26.6	70.6	47.5	3.46
Malawi	12.3	117	16.3	64.1	50.5	1.86

Mali	12.7	10	32.3	19.0	50.5	—
Mauritania	2.9	3	61.7	51.2	39.0	11.28
Mauritius	1.2	—	43.3	84.3	0	22.31
Morocco	30.6	67	57.4	50.7	39.5	6.22
Mozambique	19.1	24	35.6	46.5	39.6	1.49
Namibia	2.0	2	32.4	85.0	70.7	9.67
Niger	13.1	9	22.2	14.4	50.5	1.92
Nigeria	125.9	149	46.6	66.8	50.6	5.84
Rwanda	8.8	334	18.5	64.0	28.9	1.65
Senegal	11.1	52	49.6	39.3	41.3	5.98
Sierra Leone	5.1	75	38.8	29.6	62.9	3.27
South Africa	46.9	37	56.9	82.4	57.8	40.62
Sudan	34.9	—	38.9	59.0	—	4.23
Tanzania	36.9	41	35.4	69.4	38.2	3.39
Togo	5.8	89	35.2	53.0	—	3.83
Tunisia	9.9	64	63.7	74.3	39.8	10.35
Uganda	26.9	128	12.3	68.9	43.0	3.55
Zambia	11.3	14	35.9	67.9	52.6	2.89
Zimbabwe	12.9	34	19.6	90.0	56.8	9.03
Sub-Saharan Africa	674.2	30	35.6	61	—	—

1. *Human Development Report (HDR) 2004.* Table 5, pp. 153–155; also *Human Development Report (HDR) 2005.* Table 5, pp. 232–235.

1a. Because data are based on national definitions of what constitutes a city or metropolitan area, cross country comparisons should be made with caution.

2. *Human Development Report (HDR) 200.* Table 1, pp. 139–142.

3. *World Development Report (WDR) 2005.* Table 1. Key Indicators of Development, pp. 256–257.

4. Wackernagel et al (2000). *Ecological Footprints and Ecological Capacities of 152 Nations: The 1996 Update.* (San Francisco, CA: Redefining Progress). See also, Redefining Progress (2005). *Footprints of Nations 2005.* Accessed Online on 4/3/2006, at http://www.reprogress.org/education.

5. Note that ecological footprint is measured in global hectares. The enormous share of ecological overshoot attributable to carbon dioxide emissions in some African countries is made explicit in total footprint per capita (Redefining Progress, 2005; Wackernagel et al, 2005; Wackernagel and Monfreda, 2004; Dukes, 2003; Global Footprint Network: *Africa's Ecological Footprints, 2006*, p. 8).

Table 5.4. Descriptive Statistics

	N	*Minimum*	*Maximum*	*Mean*	*Std. Deviation*	*Variance*
Human Development Index	43	.28	.80	.4884	.12998	.017
GDP Per Capita	42	548.00	11287.00	2571.2381	2712.20868	7356075.9
Carbon Dioxide Emissions	43	.10	327.30	21.3837	55.32820	3061.209
Bio-Capacity	43	.00	106.96	24.7049	22.60973	511.200
Ecological Balance	43	−19.81	97.30	17.5316	20.59087	423.987
Total Footprint	43	.00	40.62	7.4323	9.00809	81.146
Valid N (listwise)	42					

Table 5.5. Partial Correlation Coefficients Controlling for Ecological Balance (ECOLBAL)

	HDI	*GDPPERCA*	*CARDIOEM*	*BIOCAP*	*TOTFTPRT*
HDI	1.0000	.8295	.4680	.6646	.6794
	(0)	(39)	(39)	(39)	(39)
	P=.	P=.000	P=.000	P=.000	P=.000
GDPPERCA	.8295	1.0000	.6154	.8386	.8764
	(39)	(0)	(39)	(39)	(39)
	P=.000	P=.	P=.000	P=.000	P=.000
CARDIOEM	.4680	.6154	1.0000	.8163	.8919
	(39)	(39)	(0)	(39)	(39)
	P=.002	P=.000	P=.	P=.000	P=.000
BIOCAP	.6646	.8386	.8163	1.0000	.8919
	(39)	(39)	(39)	(0)	(39)
	P=000	P=.000	P=.000	P=.	P=.000
TOTFTPRT	.6794	.8764	.8410	.89129	1.0000
	(39)	(39)	(39)	(39)	(0)
	P=.000	P=.000	P=.000	P=.000	P=.

(Coefficient/ (D.F.) / 2-taild Significance)
"." is printed if a coefficient cannot be computed.

Table 5.6. Regression Variables Entered/Removed[a]

Model	*Variables Entered*	*Variables Removed*	*Method*
1	GDP Per Capita		Stepwise (Criteria: Probability-of-F-to-enter <= .050, Probability-of-F-to-remove >= .100).
2	Carbon Dioxide Emission		Stepwise (Criteria: Probability –of-F-to-enter <= .050, Probability-of-F-to-remove >= .100).

a. Dependent Variable: Total Footprint.

Table 5.7. Model Summary[c]

Model	*R*	*R Square*	*Adjusted R Square*	*Std. Error of the Estimate*
1	.869[a]	.755	.749	3.79919
2	.952[b]	.906	.902	2.37728

a. Predictors: (Constant), GDP Per Capita
b. Predictors: (Constant), GDP Per Capita, Carbon Dioxide Emissions
c. dependent Variable: Total Footprint

Table 5.8. ANOVA[c]

Model		*Sum of Squares*	*df*	*Mean Square*	*F*	*Significance*
1	Regression	1777.945	1	1777.945	123.179	.000[a]
	Residual	577.353	40	14.434		
	Total	2355.298	41			
2	Regression	2134.890	2	1067.445	188.879	.000[b]
	Residual	220.407	39	5.651		
	Total	2355.298	41			

a. Predictors: (Constant), GDP Per Capita
b. Predictors: (Constant), GDP Per Capita, Carbon Dioxide Emissions
c. Dependent Variable: Total Footprint

Table 5.9. Coefficients[a]

Model		*Unstandardized Coefficients*		*Standardized Coefficients*		
		B	*Std. Error*	*Beta*	*t*	*Sig.*
1	(Constant)	.426	.812		.524	.603
	GDP Per Capita	2.428E-03	.000	.869	11.099	.000
2	(Constant)	.852	.511		1.667	.104
	GDP Per Capita	1.735E-03	.000	.621	10.693	.000
	Carbon Dioxide Emissions	6.250E-02	.008	.462	7.947	.000

a. Dependent Variable: Total Footprint

Table 5.10. Descriptive Statistics

	N	*Minimum*	*Maximum*	*Mean*	*Std. Deviation*	*Variance*
Population	43	1.2	125.90	18.7186	23.86795	569.679
Population Density	36	2.00	334.00	60.7778	71.07425	5051.549
Urban Population	43	10.00	86.20	38.6233	17.51494	306.773
Literacy Rates	38	12.80	90.00	58.7368	20.58042	423.554
Gini Index	33	28.90	70.70	46.5364	10.29845	106.058
Total Footprint	42	1.49	40.62	7.6093	9.04132	81.745
Valid N (listwise)	29					

Table 5.11. Correlations

	Population	*Population Density*	*Urban Population*	*Literacy Rates*	*Gini Index*	*Total Footprint*
Population Pearson Correlation	1	.152	.013	.025	–.275	.028
Sig. (2-tailed)		.377	.932	.881	.121	.858
N	.43	36	43	38	33	42
Population Density Pearson Correlation	.152	1	–.383*	.065	–.448*	–.236
Sig. (2-tailed)	.377	.	.021	.711	.013	.172
N	36	36	36	35	30	35
Urban Population Pearson Correlation	.013	–.383*	1	.245	–.077	.664**
Sig. (2-tailed)	.932	.021	.	.138	.971	.000
N	43	36	43	38	33	42
Literacy Rates Pearson Correlation	.025	.065	.245	1	.183	.437**
Sig. (2-tailed)	.881	.711	.138	.	.344	.007
N	38	35	38	38	30	37
Gini Index Pearson Correlation	–.275	–.448*	–.007	.183	1	.234
Sig. (2-tailed)	.121	013	.971	.334	.	.196
N	33	30	33	30	33	42
Total Footprint Pearson Correlation	.028	–.236	.664**	.437**	.234	1
Sig. (2-tailed)	.858	.172	.000	.007	.196	.
N	42	35	42	37	32	42

* Correlation is significant at the 0.05 level (2-tailed).
** Correlation is significant at the 0.01 level (2-tailed).

Table 5.12. Regression Variables Entered/Removed[a]

Model	*Variables Entered*	*Variables Removed*	*Method*
1	Urban Population		Stepwise (Criteria: Probability-of-F-to-enter <= .050, Probability-of-F-to-remove >= .100).
2	Literacy Rates		Stepwise (Criteria: Probability-of-F-to-enter <= .050, Probability-of-F-remove >= .100).

a. Dependent Variable: Total Footprint

Table 5.13A. Model Summary[c]

Model	*R*	*R Square*	*Adjusted R Square*	*Std. Error of the Estimate*
1	.520[a]	.271	.244	6.73110
2	.614[b]	.377	.329	6.33863

a. Predictors: (Constant), urban population
b. Predictors: (Constant), urban population, literacy rates
c. Dependent Variable: total footprint

Table 5.13B. Model Summary[c] Change Statistics

Model	*R Square Change[a]*	*F Change[b]*	*df1*	*df2*	*Sig. F Change*	*Durbin-Watson*
1	.271	10.030	1	27	.004	
2	.106	4.447	1	26	.045	2.200

a. Predictors: (Constant), urban population
b. Predictors: (Constant), urban population, literacy rates
c. Dependent Variable: total footprint

Table 5.14. ANOVA[c]

Model		*Sum of Squares*	*df*	*Mean Square*	*F*	*Significance*
1	Regression	454.427	1	454.427	10.030	.004[a]
	Residual	1223.310	27	45.308		
	Total	1677.736	28			
2	Regression	633.101	2	316.551	7.879	.002[b]
	Residual	1044.635	26	40.178		
	Total	1677.736	28			

a. Predictors: (Constant), urban population
b. Predictors: (Constant), urban population, literacy rates
c. Dependent Variable: total footprint

Table 5.15. Coefficients[a]

Model		Unstandardized Coefficients		Standardized Coefficients		
		B	*Std. Error*	*Beta*	*t*	*Sig.*
1	(Constant)	–2.384	3.134		–.761	.453
	Urban Population	.247	.078	.520	3.167	.004
2	(Constant)	–9.535	4.495		–2.121	.044
	Urban Population	.229	.074	.482	3.092	.005
	Literacy Rates	.131	.062	.329	2.109	.045

a. Dependent Variable: Total Footprint

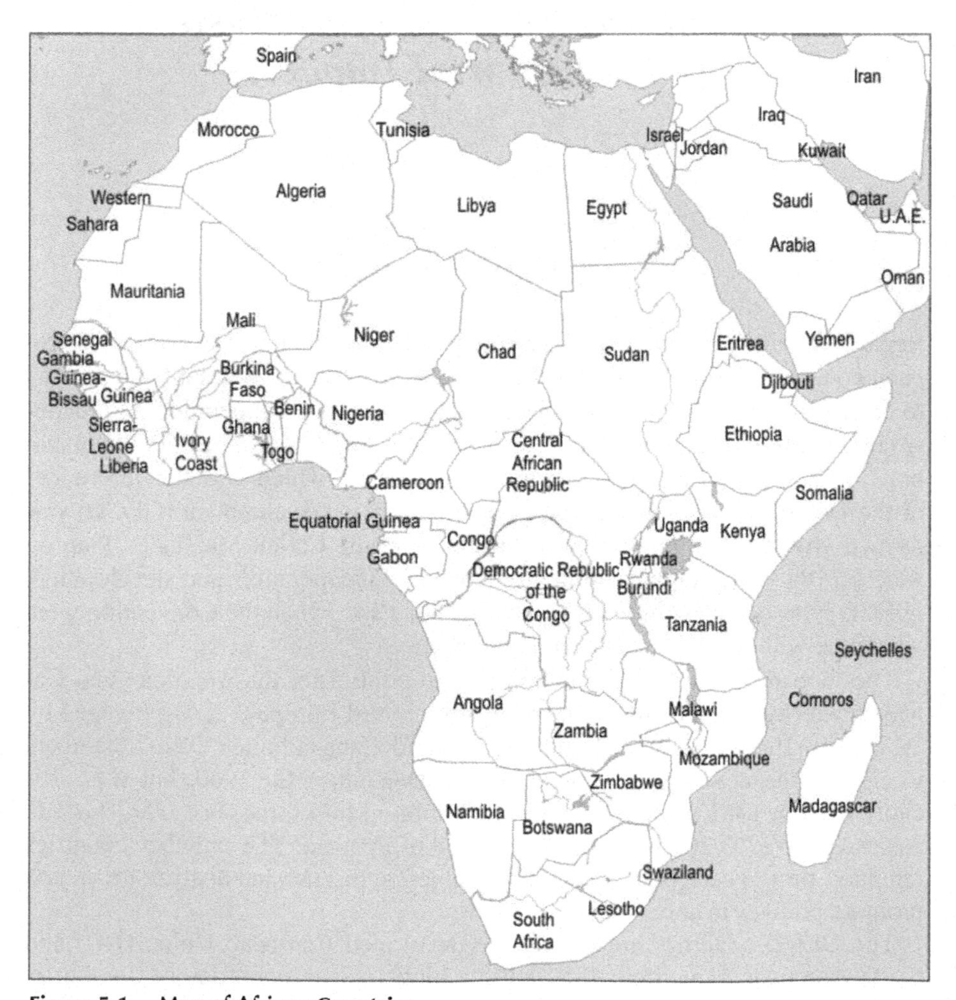

Figure 5.1. Map of African Countries

Chapter Six

Socio-Economic Determinants of National Ecological Footprints in Organization for Economic Cooperation and Development (OECD) Countries

A Multiple Regression Analysis of the Factorial Impacts

INTRODUCTION

Regardless of the bitterness and acrimony that prevailed among warring European countries during World War II, their leaders recognized that the best way to ensure lasting peace, after the war, was to encourage cooperation and reconstruction, rather than punish the defeated. According to *OECD at 50*, Organization for European Economic Cooperation (OEEC) which was the forerunner of the Organization for Economic Cooperation and Development (OECD) was formed after the famous speech of General George Catlett Marshall's Plan on June 5, 1947. The organization for Economic Cooperation and development (OECD) was officially born on September 30, 1961, when the Convention went into force with its headquarters in Paris, France.

The forerunner (OEEC) formation was to administer the American and Canadian aid (Marshall Plan) for the reconstruction of Europe that was ravaged by World War II. With its global outlook, U.S and Canada joined OEEC members in signing the new OECD Convention on December 14, 1960, but was officially born in 1961 when the Convention entered into force (See *OECD at 50*. Accessed on 9/23/2011 at http://www.oecd.org/pages). The member countries regularly turn to each other to identify problems, discuss and analyze them, and promote policies to solve them.

The OECD countries are mostly the developed European Union (EU) and North American Countries. They are the high income countries of the world, with highest gross domestic product (GDP) per capita of $9,306 or more in 2005 when compared to other world region countries (*World development Re-*

port 2005). In 2005 the region also had GDP PPPUS$29197.0 billion (about 50 percent of world's total), 6833 million gha in total footprints (about 40 percent of world's total), biological capacity of 3968 global gha (about 30 percent of world's total), and negative ecological balance or deficit of –2864.1 or about 70 percent of world's total (see table 6.1). They also included the most livable countries of the world according to the *United Nations Human Development Index, 2005*, for example, in descending order of livability, Norway, Iceland, Australia, Luxembourg, Canada, Sweden, Switzerland, Ireland, Belgium, United States, Japan, Netherlands, Finland, Denmark, United Kingdom, France, Austria, Italy, New Zealand, and Germany. They are also countries with the highest per capita government social spending in healthcare, education, basic infrastructure, pension, defense, welfare, interest, and other social services, which have positive effects on economic growth, human capital formation, and capital accumulation in the private sector (Diamond, 1989). For example, the region, in 2010, had the highest expenditure on health per capita 4,222PPP$; followed by the developed Non-OECD (1,807), Latin America and Caribbean (732), Europe and Central Asia (623), Arab States (287), East Asia and the Pacific (207), Sub-Saharan Africa (127), and South Asia (123) (*Human Development Report 2010*, p. 201).

Among world regions in 2010, the developed OECD region was third largest with population of 1,026.3 million and average annual growth rate of 0.4 percent, after East Asia and the Pacific (1,974.3 million and 0.8%) and South Asia (1,719.1 and 1.4%), then followed by Sub-Saharan Africa (808.8 million and 2.4%), Latin America and the Caribbean (582.7 million and 1.0%), Europe and Central Asia (410.3 million and 0.2%), Arab States (348.2 million and 1.9%), and Non-OECD (29.7 million and 1.2%). The developed OECD was equally, in 2010, the third highest urbanized region (77.1%), after the developed Non-OECD (91.7%) and Latin America and the Caribbean (79.5%), then followed by Europe and Central Asia (64.4%), Arab States (55.3%), East Asia and the Pacific (45.3%), Sub-Saharan Africa (37.0%), and South Asia (31.7%) (*Human Development Report 2010*, table 11, pp. 184–187).

The OECD region is a high consumption region (highest among all regions) of gross national income per capita of $11,906 or more in 2008 (*World Development Report 2010*, p. 377). At the same time the region had the highest GDP per capita of $40,976.0, and distantly followed by Europe and Central Asia ($8,361.0), Latin America and the Caribbean ($7,567.0), Arab States ($4,774.0o), East Asia and the Pacific ($3,032.0), Sub-Saharan Africa ($1,233.0), and South Asia ($954.0) (*Human Development Report 2010*, p. 210). In 2010, the OECD region had the highest Human Development Index (HDI) of 0.879, followed by the developed Non-OECD (0.844), Latin America and the Caribbean (0.704), Europe and Central Asia (0.702), East Asia and the Pacific (0.643), Arab States (0.588), South Asia (0.516), and Sub-Saharan Africa (0.389) (*Human Development Report 2010*, p. 155).

Gender inequality, a major factor in overall human and national development, is lowest in this region. For example, in 2008, the developed OECD region showed the lowest Gender Inequality Index of 0.317, followed by the developed Non-OECD (0.376), East Asia and Pacific (0.467), Europe and Central Asia (0.498), Latin America and the Caribbean (0.609), Arab States (0.699, Sub-Saharan Africa (0.735), and South Asia (0.739).

FOCUS OF THE STUDY

The study area comprised of member countries of Organization for Economic Co-operation and Development as listed in table 6.2. See also Countries in the Major World Aggregates. *Human Development Report (HDR) 2005*, p. 365. United Nations Development Program (UNDP), New York N.Y, 10017, USA. Listed below are the 28 countries that were selected for the study out of 34 OECD countries. Their selection was based on the availability of data and adequate population size (see Global Footprint Network. *The Ecological Footprint Atlas 2008*). Some countries were excluded from the list based on the fact that their populations were less than one million people (e.g. Iceland and Luxembourg), or the dates of their ratification as members of OECD were after 2005 (e.g. Chile, Estonia, Israel, and Slovenia). Most country data (ecological footprints and bio-capacity) were collected and made available by 2005. These data had not been available since 2007. Even Luxembourg had no separate footprint data in 2005. In fact, some data sources combined Belgium and Luxembourg, for example, Redefining Progress, *Footprint of Nations, 2005*, and Global Footprint Network, *Living Planet Report, 2006*. The selected countries for the study are: Australia, Austria, Belgium, Canada, Czech Republic, Denmark, Finland, France, Germany, Greece, Hungary, Ireland, Italy, Japan, Korea, Mexico, Netherlands, New Zealand, Norway, Poland, Portugal, Slovak Republic, Spain, Sweden, Switzerland, Turkey, United Kingdom, and United States of America.

PURPOSE OF STUDY

The study analyzes and explains the national variations in total footprints and per capita footprints associated with human development indexes and ecostructural factors [socioeconomic processes within nations such as urbanization, literacy rate, gross domestic product per capita (affluence), and domestic inequality (Gini index)] in Organization for Economic Co-operation and Development (OECD) countries. Progress towards sustainable development can be assessed using human development index (HDI) as an indicator of well-being, and the ecological footprint as a measure of demand on the biosphere. Table 6.3 exposes the inequitable distribution of Organization for Economic Co-operation and Development nations' ecological footprints as compared to those of other regions of the world. From this table, OECD region had about 18 percent of world population in 2005 but with about 30 percent or almost one-third of world's total biological capacity. The region also has about 39 percent or two-fifth of world's total ecological footprint, as well as the world's highest total negative ecological balance or deficit (–2864.1 million global hectares). Table 6.4 also shows that OECD had in 2005 highest per capita ecological footprint of 5.83 (higher than the world's 2.69); biological capacity of 3.39 per capita (about one and one-half the world's 2.06); and per capita negative ecological balance of –2.44 global hectares, second highest in the world after North America (–2.71). The OECD's data for total footprints, bio-capacity, and ecological balance in 2005 included only those countries represented in this study. A regional footprint per capita that is less than the world average and per capita ecological balance more than the world average may denote sustainability at the global level (Living Planet Report, 2006). The less than world's average in footprint per capita and more than world's average in per capita ecological balance were found in developing countries in 2005, especially in Africa, Middle East and Central Asia, and Latin America and Caribbean regions.

Using the tools of comparative model and regression analysis, the paper seeks to establish the key factors, among others, that highly influence and drive the ecological footprints in different Organization for Economic Co-operation and Development countries. The question is what type of secure and sustainable environments should meet the needs of both people and natural environment in OECD? This constitutes an enormous challenge to policy makers, planners, and development experts. Thus, the study proffers some policies, recommendations, and solutions on how to reduce the ecological footprints of nations, which also will help protect the natural environment and promote a more equitable and sustainable society.

THEORETICAL FRAMEWORK FOR THE STUDY

Within the biosphere, everything is interconnected, including humans, thus sustainable development requires a good knowledge of human ecology. The human life-support functions of the ecosphere are maintained by nature's biocapacity which runs the risks of being depleted, the prevailing technology notwithstanding. Regardless of the humanity's mastery over the natural environment, it still remains a creature of the ecosphere and always in a state of *obligate dependency* on numerous biological goods and services (Rees, 1992, p. 123). Despite the above *dependency* thesis, the prevailing economic mythology assumes a world in which carrying capacity is indefinitely expandable (Daly, 1986; Solow, 1974).

The human species has continued to deplete, draw-down, and confiscate nature's biocapacity with reckless abandon. York et al (2003), indicated that population and affluence account for 95 percent of the variance in total footprints of countries. Thus, large human population all over the world and their excessive consumption of the scarce natural resources are responsible for the national footprint of nations. Thus, total human impact on the ecosphere is given as: population × per capita impact (Ehrlich and Holdren, 1971; Holdren and Ehrlich, 1974); Hardin (1991). In other words, population size and affluence are the primary drivers of environmental impacts (Dietz et al, 2007). The footprints of nations provide compelling evidence of the impacts of consumption, thus the need for humans to change their lifestyles and conserve scarce natural capital. According to Palmer (1998), there are in order of decreasing magnitude, three categories of consumption that contribute enormously to our ecological footprints: wood products (53 percent), food (45 percent), and degraded land (2 percent). Degraded land includes land taken out of ecological availability by buildings, roads, parking lots, recreation, businesses, and industries. Palmer also indicated that about 10 percent or more of earth's forests and other ecological land should be preserved in more or less pristine condition to maintain a minimum base.

The environmental impacts of urban areas should be considered because a rapidly growing proportion of world's population lives in cities, and more than one million people are added to the world's cities each week and majority of them are in developing countries of Africa, Latin America, and Southeast Asia (Wackernagel and Rees, 1996). The reality is that the populations of urban regions of many nations had already exceeded their territorial carrying capacities and depend on trade for survival. Of course, such regions are running an unaccounted ecological deficit; their populations are appropriating and meeting their carrying capacity from elsewhere (Pimentel, 1996; Wackernagel and Rees, 1996; Girardet, 1996, accessed Online on 12/14/07; Rees, 1992; The International Society for Ecological Economics and Island Press, 1994; Vitouset et al, 1986; Wackernagel, 1991; WRI, 1992, p. 374). Undoubtedly, the rapid urbanization occurring in many regions and the increasing ecological uncertainty have implications for world development and sustainability. Cities are densely populated areas that have high ecological footprints which leads to the perception of these populations as "parasitic," since these communities have little intrinsic biocapacity, and instead, must rely upon large hinterlands. Land consumed by urban regions is typically at least an order of magnitude greater than that contained within the usual political boundaries or the associated built-up areas (The International Society for Ecological Economics and Island Press, 1994; Rees, 1992).

According to Rees (1992), every city is an entropic black hole drawing on the concentrated material resources and low-entropy production of a vast and scattered hinterlands many times the size of the city itself. In the same vein,

Vitouset et al (1986) asserted that high density settlements "appropriate" or augment their carrying capacity from all over the globe, as well as from the past and the future (see also Wackernagel, 1991). In modern industrial cities, resources flow through the urban system without much concern either about their origins, or about the destination of their wastes, thus, inputs and outputs are considered to be unrelated. The cities' key activities such as transportation, provision of electricity supply, heating, manufacturing and the provision of socio-economic services depend on a ready supply of fossil fuels, usually from far-flung hinterlands than within their usual political boundaries or their associated built-up areas. Cities are not self-contained entities, and their concentration of intense economic processes and high levels of consumption both increase and stimulate their demands on resources. Cities occupy only 2 percent of the world's land surface, but use some 75 percent of the world resources, and release a similar percentage of waste (Girardet, 1996).

Like urbanization, energy footprint, created from energy use and carbon dioxide emissions, is not subject to area constraints. Energy footprint is the area of forest that would be needed to sequester the excess carbon (as carbon dioxide) that is being added to the atmosphere by the burning of fossil fuels to generate energy for travel, heating, lighting, manufacturing, recreation, among other uses. Actually, the demand for energy defines modern cities more than any other single factor. Cities contain enormous concentration of economic activities that consume enormous quantities of energy. The natural global systems of forests and oceans for carbon sequestration are not handling the human carbon contributions fast enough, thus the Kyoto Conference of early 1998. According to Suplee (1998), only half of the carbon humans generate burning fossil fuels can be absorbed in the oceans and existing terrestrial sinks. The oceans absorb about 35 percent of the carbon in carbon dioxide (Suplee, 1998), equivalent of 1.8 giga tons of carbon every year (IPCC, 2001), while the global forests under optimum management of existing forests could absorb about 15 percent of the carbon in the CO_2 produced from the burning of fossil fuels world-wide (Brown et al, 1996). The energy footprint is caused by the un-sequestrated 50 percent in the atmosphere with the potentially troubling ecological consequences, such as rapid global warming and other environmental stresses, including climate change. Carbon dioxide in the atmosphere will continue to increase unless humanity finds alternative energy sources of sufficient magnitude. It is in the humanity's best interest to get off its petroleum addiction (control and minimize fossil fuel use) and develop sustainable consumption habits.

Literacy affects the consumption of natural capital resources. Highly literate groups concentrate in urban areas where they consume more than their fair shares of biospheric resources. Literate populations generally have lower rates of domestic inequality and tend to consume more resources than their illiterate counterparts due to their higher incomes and higher standards of urban living.

Furthermore, higher levels of literacy correspond with higher incomes, which allow for greater consumption (Jorgenson, 2003). This is because literate populations are subject to increased consumerist ideologies and contextual images of good life through advertising (Princen et al, 2002); what Leslie Sklair (2001) and Jennifer Clapp (2002) labeled "cultural ideology of consumerism/consumption."

METHODOLOGY FOR THE STUDY

Unit of Analysis

The unit of analysis is "country" (individual Organization for Economic Cooperation and Development countries).

Sample

To test for the national variations in total footprints and per capita footprints associated with human development index, Comparative Model Analysis and Stepwise Regression Analysis were used as the tools of analysis. Using comparative model, a sample of 28 countries out of 34 countries of the OECD region (see Global Footprint Network. *The Ecological Footprint Atlas 2008*) was analyzed in tables 6.7–6.8. Step-wise Regression Analysis was also conducted using the sample. Each table represents OECD countries as also represented by Global Footprint Network, *The Ecological Footprint Atlas 2008*, tables 2,3, and 4. These tables do not include countries that were less than one million people in 2005 (e.g. Iceland and Luxembourg), or those countries with dates of ratification as members of OECD were after 2005 when footprints data for this study were collected for the region (e.g. Chile, Estonia, Israel, and Slovenia). The larger population of at least one million is chosen for the study because; population is a variable which affects the consumption rates and levels, therefore the ecological footprint.

Mode of Analysis

This is an explanatory study using Step-wise Regression as well as comparative model and descriptive statistics such as matrices, totals, averages, ratios and percentages. Regression Analysis was performed to flesh out factors that highly impacted national footprints, as well as to strengthen the comparative model analysis results. Potential technical problems are diagnosed that might affect the validity of the result of the analysis such as the missing data (e.g. public expenditure data for Turkey, 2000–2004 and Portugal in 2005) or the excluded countries mentioned above, which did not alter the substantive conclusions, especially regarding the ecological footprint accounts and environmental sustainability of different OECD countries. The national ecological footprint and ecological balance, as dependent variables, are explained recursively by the country's human development hierarchical category, population size, population density, urbanization, GDP per capita, domestic inequality (Gini index), and literacy rate, as the independent variables. The independent variables mediated in explaining the varying levels of consumption among different human development indexes of the capitalist world economy. It is hypothesized that Human Development positions and Ecostructural Factors are likely to be responsible for the variations in the Ecological Footprints and Balances. The carbon dioxide emission levels [*Human Development Report*, 2007/2008, table 24, pp. 310–313; IEA (2008), CO_2 Emissions from Fuel Combustion: 2008 Edition, IEA, Paris] are included to depict the consumption and environmental degradation impacts of different OECD countries that reflect their national ecological footprint and ecological balance levels.

HYPOTHESIS FOR THE STUDY

Null Hypothesis (H_0)

None of the independent variables predicts the national footprint effects. The independent variables do not influence national footprints.

Alternative Hypothesis (H_1)

Some (if not all) of the independent variables predict the national footprint effects.

Variables

Dependent and independent variables are selected on the basis of the theoretical themes and underpinnings described above. The measures of the dependent and independent variables used in the present study are as follow:

Dependent Variables

The ecological footprint or consumption and ecological balance are measured in global hectares (1 hectare = 2.47 acres). A global hectare is 1 hectare of biologically productive space with world average productivity (Global Footprint Network: *Africa's Ecological Footprint-2006 Factbook*, pp. 82–90).

Independent Variables

The ecostructural mediating factors are most of the independent variables in this study that include, GDP per capita, urbanization (urban population as a percent of total population), literacy rate (%), public expenditure, carbon dioxide emission, and domestic inequality (Gini index) (See Appendix A for Variable definitions).

COMPARATIVE MODEL ANALYSIS (RESULTS AND DISCUSSIONS)

The results and findings in this study are discussed under subtitles that include, "Carbon Dioxide Emissions from Fuel Combustion," "Total Public (Government) Social Expenditure per Head," "Human Development and Affluence on Biological Capacity, Ecological Footprint and Balance in OECD Countries," and "Population and Ecostructural Factors on Total Footprint Per Capita in OECD Countries."

CARBON DIOXIDE (CO_2) EMISSIONS FROM FUEL COMBUSTION (IN MILLION TONS) IN ORGANIZATION FOR ECONOMIC COOPERATION AND DEVELOPMENT (OECD) COUNTRIES: 2000–2005

Table 6.5 shows carbon dioxide emissions from fuel combustion in OECD countries from 2000 to 2005. The OECD countries are high income countries of the world that are highly developed and industrialized. They are nations with more voracious appetites for consumer goods, fossil fuels, and urban sprawl. They are most reliant on fossil fuels; do not live within their biological capacities which they import from outside their defined political boundaries, and carry the largest responsibility for global environmental stress [see Dr. John Talberth, Director of Sustainability Indicators Program, Redefining Progress (2005). *Footprints of Nations*].

In 2005, the OECD region was responsible for over 70 percent of the world ecological overshoots, which could be attributable to carbon dioxide emissions (see table 6.3). Table 6.5 shows that OECD region had, since 2000, consistently been responsible for more than one-half of the world's total carbon dioxide emissions. The table also shows that the United States had consistently produced about half of the total OECD region's carbon dioxide emissions, followed by 27 more developed European Union countries, followed by Japan, then Germany and United Kingdom. During the same periods, Iceland and Luxembourg that were not included in this study had respectively the lowest carbon dioxide emissions. Excessive carbon dioxide emissions are responsible for greenhouse effects, climate modifications and change, as well as environmental stress in most of the OECD countries.

TOTAL PUBLIC (GOVERNMENT) SOCIAL EXPENDITURE PER HEAD (AT CURRENT PRICES AND PPPS IN US DOLLARS) IN ORGANIZATION FOR ECONOMIC COOPERATION AND DEVELOPMENT (OECD) COUNTRIES: 2000–2005

Table 6.6 shows total public (government) social expenditure per head in OECD countries. From 2000 to 2005, most countries showed consistent increases in government social spending. Most countries public expenditure policy objective is to sustain equitable economic growth. Public expenditure plays an important role in physical and human capital formation over time, especially when it is directed towards infrastructure and skilled manpower development (Diamond, 1989; Barro, 1991; Niloy Bose et al, 2007; Easterly and Rebelo, 1993; World Bank, 1993a). No wonder nations with highest human development index are found among OECD countries which consistently spend more than the average in public spending, e.g. Norway, Austria, Sweden, Denmark, Belgium, France, and Germany, among others (see, UNDP *Human Development Reports 2007/2008; 2010*). The question is whether high levels of public spending affect national footprint levels.

HUMAN DEVELOPMENT AND AFFLUENCE ON BIOLOGICAL CAPACITY, ECOLOGICAL FOOTPRINT AND BALANCE IN ORGANIZATION FOR ECONOMIC COOPERATION AND DEVELOPMENT (OECD) COUNTRIES

Table 6.7 presents the national comparisons for both dependent (Ecological footprint and Ecological balance) and some of the independent (Human development or welfare, GDP per capita or affluence, and Carbon dioxide emissions) variables in Organization for Economic Cooperation and Development Countries. The table shows that a country's total footprint, ecological balance, and carbon dioxide emissions (waste) are a function of its human development position and affluence level in the world economy. Given the national biological capacity measured in global hectares, the higher the total ecological footprint; the lower the ecological balance and vice versa. Table 6.7 assesses the progress towards sustainability development in OECD countries by using human development index (HDI) as an indicator of national well-being, and the ecological footprint and balance as measures of national demand on the biosphere and biological capacity respectively. The sustainability quotient is depicted by any country with low footprint and high human development levels (e.g. Finland, New Zealand, Sweden, and Australia). In fact, table 6.7 shows that, among OECD countries, improvements in the human development indexes (human welfare) have resulted in the worsening of the countries' positions on their ecological footprints and balances e.g. Germany, Italy, Japan, Mexico, Spain, UK, and United States). Table 6.7 shows that OECD countries are among the high human development (HDI of 0.800 and above and per capita GNP of $9,386 or more (World Bank, 2005, p. 255).

From table 6.7 and based on the existing biological capacity, it becomes apparent that the higher the HDI (human welfare), GDP per capita (affluence) i.e. consumption; the higher the carbon dioxide emissions and ecological footprint. For example, United States shows HDI of 0.951, GDP per capita ($41,890), carbon dioxide emission (5985 million tons), total footprint (2809.8 global hectares), biological capacity (1496.4 global hectares), and ecological balance of –1313.3 global hectares. Actually, United States "appropriates" half of its carrying capacity from "elsewhere." Japan shows HDI of 0.953, GDP per capita ($31,267), carbon dioxide emission (1228), and total footprint (626.6), biological capacity (77.2 global hectares), and ecological balance of –549.4 global hectares. In fact, Japan appropriates more than eight times of its carrying capacity from elsewhere. UK shows HDI of 0.946, GDP per capita ($33,328), carbon dioxide emission (535), and total footprint (319.2), biological capacity of 98.6 global hectares, and ecological balance of –220.6. Thus, UK appropriates more than three times its carrying capacity beyond its borders. Likewise, Italy shows HDI of 0.941, GDP per capita ($28,529), carbon dioxide emission of (454), and total footprint of 276.5, biological capacity of 71.2 global hectares, and ecological balance of –205.3 global hectares. Once more, Italy appropriates about four times its carrying capacity from elsewhere. On the other hand, Turkey shows the lowest HDI of 0.775, GDP per capita($8,407), carbon dioxide emission (216), total footprint (198.6), biological capacity (120.9), and ecological balance (–77.7). Likewise, Slovakia shows HDI of 0.863, GDP per capita ($15,871), carbon dioxide emission (38), total footprint (17.8), biological capacity (15.2), and ecological balance (–2.6). United States, Japan, and UK have footprints larger than their biological capacity, thus, a negative ecological balance, while Finland, New Zealand, Sweden, Australia, and Canada have their footprints less than their biological capacities, thus, a positive ecological balance. Subsequently, analysis of table 6.7 shows that, given a nation's biological capacity, its levels of human development (welfare), GDP per capita (affluence), and carbon dioxide emissions (waste), they help to explain its total ecological footprint (consumption or demand) and ecological balance (reserve or deficit).

POPULATION, ECOSTRUCTURAL FACTORS, AND TOTAL FOOTPRINT PER CAPITA IN ORGANIZATION FOR ECONOMIC COOPERATION AND DEVELOPMENT COUNTRIES

Table 6.8 shows the effects of population and ecostructural factors (socioeconomic processes) on the total ecological footprints of nations. A nation's total population size and population density, as well as its urban population, literacy rate, and domestic inequality (Gini index), lead to its ecological footprint difference and variation vis-à-vis other nations. The above indicators constitute the structural causes of the ecological footprint variations among OECD countries. Table 6.8 shows that a nation with large population, high rates of urbanization and literacy, and low domestic inequality, has high ecological footprint, and vice versa. Large population level, high population density, high level of urbanization, high level of literacy, and low domestic inequality promote domestic consumption of biospheric resources (e.g. US, UK. And Australia).

Cities are not self-contained entities and have little intrinsic biological capacity; therefore their concentration of intense economic processes and high levels of consumption both increase and stimulate their demands on bioresources. Thus, cities are parasitic and in most cases must rely upon large hinterlands for their supplies which increase their footprint impacts. Correspondingly, high density settlements appropriate carrying capacity from all over the globe, as well as from the past and the future (Vitousek et al, 1986; Wackernagel, 1991). High levels of urbanization correspond to higher levels of consumption, higher levels of energy use and carbon dioxide emissions (waste), and thus, higher energy footprints and ecological footprints. As mentioned earlier, biospheric resources are consumed at higher levels in urban areas that generally contain high literate groups with higher incomes, which allow for greater material consumption (Jorgenson, 2003; Princen et al, 2002; Sklair, 2001; Clapp, 2002).

According to table 6.8, United States (New York, Los Angeles, Chicago, and Houston), Japan (Tokyo and Osaka), UK (London), Canada (Lima), Turkey (Istanbul), and Denmark (Copenhagen) are nations with large population and large cities. United States (80.8), UK (89.7), New Zealand (86.2), Denmark (85.6), Belgium (97.2), and Australia (88.2) have high urban population as percent of their total populations. All the OECD countries have high adult literacy rates. Generally, low domestic inequality in OECD countries does not show high per capita footprints but mixed footprint levels. What may be also deduced from table 6.8 is that, high population density does not depict high footprint per capita but the opposite. For example, Republic of Korea (485), Japan (349), Belgium (342), Netherlands (479), and Germany (237), except Denmark (127) and UK (246), all showed low footprints per capita that are less than 4.5, which is less than the region's average (5.3). Thus, the level of consumption, and not population density, spurred by large urban population and high levels of adult literacy is the major driving force of environmental sustainability and ecological footprints.

Tables 6.7 and 6.8 depict high human development countries or more developed countries with high levels of urbanization, literacy, and domestic inequality in 2005. These correspond to medium-high levels of income, which allow for high levels of material consumption and footprint. The findings here parallel the theorization above, which emphasizes that the high consumption levels and the accompanying high ecological footprints and low ecological balances (reserves) and vice versa, are based on the nations' positions in the world capitalist economy and system. In fact, some of the medieval OECD sovereign nations (e.g. UK, Spain, Portugal, France, and Belgium) sometime colonized other developing countries of the world and had been constantly consuming a greater share of unprocessed bio-productive resources as well as industrial processed finished products. The low domestic income inequality or intra-inequality in this region has a positive effect on per capita consumption levels (Bornschier and Chase-Dunn, 1985; Kick, 2000; Kentor, 2001; and Jorgenson, 2003), which yields both high total and per capita ecological footprints.

REGRESSION ANALYSIS AND REGRESSION EQUATION

On one hand regression analysis allows the modeling, examining, and exploring of relationships and can help explain the factors behind the observed relationships or patterns. It is also used for predictions. Regression equation, on the other hand, utilizes the Ordinary Least Square (OLS) technique. This mathematical formula when applied to the explanatory variables is best used to predict the dependent variable that one is attempting to model. Each independent variable or explanatory variable is associated with a regression coefficient describing the strength and the sign of that variable's relationship to the dependent variable. A regression equation might look like the one given below where Y is the dependent variable, the Xs are the explanatory variables, and the Bs are regression coefficients:

$$Y = B_0 + B_1 X_1 + B_2 X_2 + \text{-----} B_n X_n + E \text{ (Random Error Term/Residuals)}$$

See Appendix B for Regression Analysis Terms.

The row of "unstandardized coefficients" or "Bs" gives us the necessary coefficient values for the multiple regression models or equations.

From table 6.7, *Human Development and Affluence on Biological Capacity, Ecological Footprint and Balance in Organization for Economic Cooperation and Development Countries*, the dependent variable is Total Footprint, and the independent variables are Human Development Index (HDI), GDP Per Capita,

Carbon Dioxide Emissions, Bio-Capacity, and Ecological Balance. Also, from table 6.8, *Population, Ecostructural Factors, and Total Footprint Per Capita in Organization for Economic Cooperation and Development Countries*, the dependent variable is Total Footprint, and independent variables are Population, Population Density, Urban Population, Literacy Rates, and Gini Index.

REGRESSION ANALYSIS (RESULTS AND DISCUSSIONS)

Table 6.9 depicts the descriptive statistics of the data in table 6.7. The measures of variability or dispersion, from table 6.9, expressed using standard deviation and variance, signify how homogeneous (in terms of low levels) Organization for Economic Cooperation and Development countries are in terms of Human Development Index (HDI) and how heterogeneous they are (in terms of high levels) in terms of GDP Per Capita, Carbon dioxide Emissions, Public Expenditure Total Footprint, Bio-capacity, and Ecological Balance. The same is also true in table 6.15 where descriptive statistics of the data in table 6.8 show low levels of variability associated with Total Footprint Per Capita, Adult Literacy Rates, Gini Index, and Urban Population; and high variability with Public Expenditure, Population Density, and Population. Table 6.10 shows the Pearson Correlation (1-tailed test) of the data from table 6.7. From table 6.10, there is a strong correlation or relationship (at the 0.01 level) between Human Development Index and GDP Per Capita; Total Footprint and Carbon Dioxide Emission; Bio-Capacity and Carbon Dioxide Emissions; Total Footprint and Bio-Capacity; Total Footprint and Ecological Balance; and Ecological Balance and Bio-Capacity. Likewise, table 6.16 shows correlation at 0.01 levels between Population and Gini Index, as well as with Adult Literacy and Gini Index; and Adult Literacy and Public Expenditure. There is also fairly strong correlation between Total Footprint Per Capita and Adult Literacy; Total Footprint Per Capita and Urban Population; and Gini Index and Urban Population.

Table 6.12 is the Regression model summary, which shows the Multiple R and R-Squared or Coefficient of Determination are very high and significant. For example, about 99.2 percent of the dependent variable (Total Footprint) is accounted for by the independent variable, Carbon Dioxide Emissions; about 99.3 percent of Total Footprint is explained by Carbon Dioxide Emissions and Ecological Balance; and 100 percent of Total Footprint is explained by Carbon Dioxide, Ecological Balance, and Bio-Capacity. These are entered in Stepwise Regression Model in table 6.11. The Multiple R is very large (100 percent), which indicates size of the correlation between the observed outcome variables and the predicted outcome variables (based on the regression equation).

Table 6.18 shows the Multiple R and R-Squared between dependent variable (Total Footprint Per Capita) and the independent variables, Public Expenditure,

Population Density, and Urban Population. Nevertheless, their relationships as well as predictability and accountability are not as high and significant as found in table 6.12. These are entered in Stepwise Regression Model as shown in table 6.17. From table 6.18, the Model Summary, about 47 percent of the variations in Total Footprint Per Capita is explained or accounted for by Public Expenditure, Population Density, and Urban Population. Tables 6.13 and 6.19 depict the analysis of variance (ANOVA) and F-Tests. The F-tests show how strong or significant the selected independent variables in the regression models are in predicting the dependent variable. From table 6.13, the F values and degree of freedom (24 or 86 percent) are high and the residual (observed – predicted) values are small. Since the F-Tests are high and significant, the Null Hypothesis may be rejected, which says that none of the independent variables (predictors) predict the National footprint and per capita footprint effects. It is worthy to note that the results in table 6.19 are not as strong as those in table 6.13 but still significant. In fact, the results are weak indicating that the national footprints in OECD region could be said to depend primarily on their positions on the world capitalist economy and system. These countries appropriate more of their natural capital resources beyond their prescribed political boundaries, far into their vast hinterlands.

The "Coefficients[a]" table of the step-wise regression process is used to construct the regression model equation. Using the B columns of the Unstandardized Coefficients in tables 6.14 and 6.20, the regression model equations are constructed from tables 6.7 and 6.8; including Public Expenditure data from table 6.6.

(A) For table 6.7: *Human Development, GDP per Capita, Carbon Dioxide Emissions, Bio-Capacity Factors on Total Footprint*, the significant explanatory (predictor) variables are: Carbon Dioxide Emissions, Ecological Balance and Bio-Capacity. The Regression Equation is: Total Footprint = –131 + 0.445Carbon Dioxide Emissions –Ecological Balance + Bio-Capacity + E (residuals). (B) Also for table 6.8: *Population, Population Density, Urban Population, Literacy Rates, Domestic Inequality (Gini Index) on Total Footprint Per Capita*, the significant explanatory (predictor) variables are Public Expenditure, Population Density, and Urban Population. The Regression Equation is: Total Footprint Per Capita = 0.677 + 0.000Public Expenditure – 0.005Population Density + .058Urban Population + E (residuals). Therefore, given the factors A and B above, Total National Footprint in Organization for Economic Cooperation and Development (OECD) region is predicted or explained by Carbon Dioxide Emissions, Ecological Balance, Bio-Capacity, Public Expenditure, Population Density, and Urban Population.

POLICY RECOMMENDATIONS

The above analyses support the theorization and assertion that the global environmental stress is caused primarily by increase in resource consumption (demand) and human population size and growth. The analyses indicate that large human population all over the world and their excessive consumption of the scarce natural resources are responsible for the national footprint of nations. The large absolute population of Organization for Economic Cooperation and Development region (about 18 percent of world population) and the large national and city populations in the region are exerting tremendous pressures on natural resources. This region in 2005 had footprint approximately one and half the size of its biological capacity, and an overshoot (ecological balance) of about two and half times the size of its biological capacity (see tables 6.1 and 6.3). Its net demand on the planet is far more than its available capacity. Thus, the region appropriates about two and half times its natural capital from outside its political boundary. However, as a high income with per capita income $9,305 or more in 2005 and high human development region of the world capitalist economy, resource consumption is high and increasing, as well as taking its toll on bio-capacity's natural resources. Wealth (growth) can be good for the environment, only if public policy and technologies encourage sound practices and the necessary investments are made in environmental sustainability (Sachs, 2007). The affluent countries of the region, especially US, Japan, UK, France, Germany, and Spain can reduce consumption and still improve their quality of life, if adequate population management policies are in place. Therefore, long-term investment will be required in many areas, including education, technology, resource conservation, and urban and family planning. Innovative approaches to meeting human needs without environmental destruction are called for as well as a shift in belief that greater well-being necessarily entails more consumption. The way additional population or material growth can be sustained, without ravaging biodiversity and ultimately destroying the ecological basis of human life is through cutbacks in resource consumption by the above high consumption countries. The above action is necessary to vacate the ecological space necessary to justify growth in low income and developing countries of the world (Rees, 2012).

Factors that significantly shape Organization for Economic Cooperation and Development (OECD) ecological footprint or demand on bio-capacity include Carbon dioxide Emissions, Urban Population, Population Density, Public Expenditure, enormous pressure on bio-capacity (especially, forest and mineral resources), and ecological balance. Policies and programs should include: (1) the potential for reducing consumption depends on individual's economic situation, more affluent people can reduce consumption and still improve their quality of life. It is a matter of conscience and equity; (2) in the case of footprint intensity, the amount of resources used in the production of goods and services, can be significantly reduced, from energy efficiency in manufacturing and in the home, through minimizing waste and increasing recycling and reuse, to fuel-efficient cars. Innovative approaches to meeting human needs without environmental destruction are called for; (3) bio-productive area can be extended, for example, degraded lands can be reclaimed through careful management that include, terracing, irrigation, and to ensure that bio-productive areas do not diminish or being lost to urbanization, salinization, or desertification; (4) bio-productivity per hectare depends both on the type of ecosystem and the way it is managed; (5) Carbon sequestration can be improved by (i) by reducing the rate of deforestation in the region necessary for carbon sinking, and (ii) more efficient manufacturing and use of adequate technology. There should be the development of non-wood products and adequate forest management and services; and (6) concern over deforestation and its ecological, social, and economic repercussions should call for increased efforts in conservation and improved forest management. There should be a balance between national economic needs and increase in natural resources forest harvesting activities. Adequate and improved environmental and ecological education will reduce an individual's ecological impact since it has been recognized that the environmental crisis is less an environmental and technical problem than it is a behavioral and social one (Wackernagel and Rees, 1996).

CONCLUSIONS

This is an explanatory study using comparative model and descriptive statistics such as matrices, totals, ratios, and percentages, as well as regression analysis. The results and findings show that a country's total footprint, ecological balance, and carbon dioxide emissions (waste) are a function of its human development (welfare) position and affluence level in the world economy. Given the national biological capacity, it goes that the higher the total ecological footprint, then, the lower the ecological balance. Thus and generally, as the human development and income go up, then the environmental performance goes down.

Data analyses in the present study show that the higher the national populations, urban populations, public expenditures and GDP per capita (affluence), the higher the *imbalaced* consumption, total footprint, carbon dioxide emissions, and the lower the ecological balance. The research findings also indicate that improvements in human development index (human welfare) result, in some cases such as Norway, New Zealand, Switzerland, Ireland, Sweden, Australia, and Denmark, in lowering the nation's ecological footprint position. Thus, improved human development (welfare) is likely to have been pursued simultaneously with adequate public policy, awareness, education, and sound practices for environmental sustainability. It is also found out that higher levels of literacy and urbanization correspond to higher levels of income, which allow for larger material consumption. The low domestic income inequality or intra-inequality in some of the Organization for Economic Cooperation and Development countries has positive or increasing effect on per capita consumption levels that yield both high total and per capita ecological footprints. Hence, OECD countries should be conscious of their large and voracious appetites for fossil fuels and enormous consumption patterns for natural resources; and cutback and conserve accordingly for socio-economic and environmental sustainability.

The above analyses support the theorization and assertion that the global environmental stress is caused primarily by the increase in resource consumption (demand) and population size and growth. It is suggested also that sustainability will depend on such measures as greater emphasis on equity in international relationships and consumption patterns; significant adjustment to prevailing terms of trade; increasing regional self-reliance; in some cases, adequate population management initiatives; and policies to stimulate a massive increase in the material and energy efficiency of economic activities.

APPENDIX A: VARIABLE DEFINITIONS

Human Development Index (HDI): The HDI is a summary measure of human development (human welfare). It measures the average achievements of a country in three basic dimensions of human development (Global Footprint Network: *Africa's Ecological Footprint-2006 Factbook*, p. 89): a long and healthy life, as measured by life expectancy at birth; knowledge, as measured by the adult literacy rate (with two-thirds weight) and the combined primary, secondary and tertiary gross enrolment ratio (with one-third weight); and a decent standard of living, as measured by GDP per capita (PPP US$). Purchasing power parity (PPP) is a rate of exchange that accounts for price difference across countries, allowing international comparisons of real output and incomes. At the PPP US$ (as used in this study), PPP US$1 has the same purchasing power in the domestic economy as $1 has in the United States of America.

The high human development countries are countries with HDI of 0.800 and above, found mainly in North America, Western Europe and Australia, with Gross National Product (GNP) of $9,386 or more. The medium development countries are countries with HDI of 0.500–0.799, found in Eastern Europe, South-East Asia, Latin America and Caribbean, and North Africa, with per capita GNP of $766 and 9,385 (World Bank, 2005, p. 255). The low human development countries are countries with HDI below 0.500, found mainly in Sub-Saharan Africa except the countries of South Africa, Gabon, and Ghana, which are included in the medium human development, with per capita GNP of $765 or less (World Bank, 2005, p. 255). Data for this variable in this study is taken from UNDP's *Human Development Report 2004* and *Human Development Report 2005*.

Gross Domestic Product (GDP) per capita (PPP US$): GDP is converted to US dollars using the average official exchange rate reported by the International

Monetary Fund (IMF). GDP alone does not capture the international relational characteristics as does the human development hierarchy of the world economy, which accounts for a country's relative socio-economic power and global dependence position in the modern world system. It is suggested elsewhere that GDP per capita is an inadequate measure of world-system position but a more appropriate indicator of domestic affluence or internal economic development (Burns, Kentor, and Jorgenson, 2003; Jorgenson, 2003; Dietz and Rosa, 1994). The Gross Domestic Product (GDP) per capita data for this study is taken from *Human Development Report 2007/2008*, table 1, and pp. 229–232. See also *World Development Report 2005*.

Domestic Income Inequality (Gini Index): The Gini index measures domestic income inequality of different countries, which had remained stable over a time with its impacts on other variables in the study (Bergesen and Bata, 2002; Jorgenson, 2003). Gini index measures the extent to which the distribution of income (or consumption) among individuals or households within a country deviates from a perfectly equal distribution (*Human Development Report, 2004*, p. 271)). It measures inequality over the entire distribution of income or consumption. A value of zero (0) represents perfect equality, and a value of hundred (100) represents perfect inequality. Data for domestic income inequality measured by Gini index are taken from World Bank, *World Development Report* (2005), table 2, pp. 258–259 and United Nations Development Report, *Human Development Report* 2007/2008, 15, pp. 281–284.

Urbanization Level (Urban Population as percent of Total Population): Cities are not self-contained entities and their concentration of intense economic processes and high levels of consumption both increase and stimulate their demands on resources. The cities have limited intrinsic biocapacity which undoubtedly must rely upon large hinterlands. Land consumed by urban regions is typically at least an order of magnitude greater than that contained within the usual political boundaries or the associated built-up area (The International Society for Ecological Economics and Island Press, 1994; Rees, 1992). The data are taken from *Human Development Report 2007/2008*, table 5, and pp. 243–246.

Literacy Rate: This variable refers to the percent of a nation's population over the age of fifteen (15) that can read and write in any language of their choice. Literate population generally has low domestic inequality and tends to consume more resources than their illiterate counterpart due to their higher income and urban living. High literate groups concentrate in urban areas where they consume more than their fair shares of biospheric resources. Higher levels of literacy correspond with higher incomes, which allow for greater consumption (Jorgenson, 2003). This is because literate populations are subject to increased consumerist ideologies and contextual images of good life through advertising (Princen et al, 2002), what Leslie Sklair (2001) and Jennifer Clapp (2002) labeled "cultural ideology of consumerism/consumption." The data for literacy rate is taken from the *Human Development Report 2007/2008*, table 1, pp. 229–232.

Population and Population Density: Apart from consumption, many have attributed to population as driving most of the sustainability problems (Palmer, 1998; Pimentel, 1996). Likewise, Dietz et al (2007) concluded that population size and affluence are the primary drivers of environmental impacts. Population growth and increases in consumption in many parts of the world have increased humanity's ecological burden on the planet. York et al (2003) indicated that population and affluence account for 95 percent of the variance in total footprints of countries. Others also see the ensuing human impact or footprint as a product of population, affluence (consumption), and technology (i.e. I = PAT (Ehrlich and Holdren, 1971; Holdren and Ehrlich, 1974; Hardin, 1991). Population as a variable in this study is taken from, *World Development Report 2005*, table 1, and pp. 256–259.

Government (Public)Social Spending Per Capita: Spending by a government (federal, state, and local), municipality, or local authority, which covers such things as spending on healthcare, education, pensions, defense, welfare, interest, and other social services, and is funded by tax revenue, seigniorage, or government borrowing. Public expenditure exerts an effect on economic growth rate through the positive externality in the productivity of the capital stock. Investment in education and health is also investment in people and in the future (See, Education: Crisis Reinforces Importance of a Good Education, says OECD. Accessed Online on 9/23/2011 at http://www.oecd.org/document/21/0,3746.

APPENDIX B: REGRESSION ANALYSIS TERMS

Correlation or co-relation: refers to the departure of two variables from independence or they are non-independent or redundant (Antonia D'Onofrio, 2001/2002; Richard Lowry, 1999–2008).

Collinearity: Refers to the presence of exact linear relationships within a set of variables, typically a set of explanatory (predictor) variables used in a regression-type model. It means that within the set of variables, some of the variables are (nearly) totally predicted by the other variables [(Rolf Sundberg, 2002). *Encyclopedia of Environmetrics*, edited by Abdel H. El-Shaarawi and Walter W. Piegorsch (Chichester: John Wiley & Sons, Ltd), Volume 1, pp. 365–366].

Partial Correlation Coefficients (r): When large, it means that there is no mediating variable (a third variable) between two correlated variables (Antonia D'Onofrio, 2001/2002).

Pearson's Correlation Coefficient (r): This is a measure of the strength of the association between two variables. It indicates the strength and direction of a linear relationship between two random variables. Value ranges from - to +1; –1.0 to –0.7 Strong negative association; –0.7 to –0.3 Weak negative association; –03 to +0.3 Little or no association, +0.3 to +0.7 Weak positive association; +0.7 to 1.0 Strong positive correlation (Brian Luke, "Pearson's Correlation Coefficient," Learning *From The Web.net*. Accessed Online on 5/30/2008).

Multiple "R": Indicates size of the correlation between the observed outcome variable and the predicted outcome variable (based on the regression equation).

"R^2" or Coefficient of Determination: Indicates the amount of variation (%) in the dependent scores attributable to all independent variables combined, and ranges from 0 to 100 percent. It is a measure of model performance, summarizing how well the estimated Y values match the observed Y values.

"Adjusted R^2": The best estimate of R^2 for the population from which the sample was drawn. The Adjusted R-Squared is always a bit lower than the Multiple R-Squared value because it reflects model complexity (the number of variables) as it relates to the data.

R^2 and the *Adjusted R^2* are both statistics derived from the regression equation to quantify model performance (Scott and Pratt, 2009. *ArcUser*).

Standard Error of Estimate: Indicates the average of the observed scores around the predicted regression line.

Residuals: These are the unexplained portion of the dependent variable, represented in the regression equation as the random error term (E). The magnitude of the residuals from a regression equation is one measure of model fit. Large residuals indicate poor model fit. Residual = Observed – Predicted.

ANOVA: Decomposes the total sum of squares into regression (= explained) SS and residual (= unexplained) SS.

F-test in ANOVA represents the relative magnitude of explained to unexplained variation. If F-test is highly significant (p = .000), we reject the null-hypothesis that none of the independent variables predicts the effect (scores) in the population.

The "constant" represents the intercept in the equation and the coefficient in the column labeled by the independent variables.

REFERENCES

Aka, Ebenezer (2006). "Gender Equity and Sustainable Socio-Economic Growth and Development," *The International Journal of Environmental, Cultural, Economic & Social Sustainability*, Volume 1, Number 5, pp. 53–71.

Bergesen, Albert and Michelle Bata (2002). "Global and National Inequality: Are They Connected?" *Journal of World-System Research*, 8: 130–44.

Bergesen and Bartley (2000). "World-System and Ecosystem," in *A World-Systems Reader: New Perspectives on Gender, Urbanism, Culture, Indigenous Peoples, and Ecology*, edited by Thomas Hall (Lanham, MD: Rowman and Littlefield), pp. 307–22.

Bornschier, Volker and Christopher Chase-Dunn (1985). *Transnational Corporations and Underdevelopment* (New York: Praeger).

Brown, S., Jayant, S., Cannell, M., and Kauppi, P. (1996). "Mitigation of Carbon Emissions to the Atmosphere by Forest Management," *Commonwealth Forestry Review*, 75, 79–91.

Centro Internacional de Agricultura Tropical CIAT), United Nation Environment Program (UNEP), Center for International Earth Science Information Network (CIESIN), Columbia University, and the World Bank (2005). Latin America and Caribbean Population Database, Version 3. http://www.na.unep.net/datasets/datalist.php3.

City Mayors (2006). *The World's Largest Cities and Urban Areas*. http://www.citymayors.com/statistics/urban_2006.

Clapp, Jennifer (2002). "The Distancing of Waste: Over-consumption in a Global Economy," in *Confronting Consumption*, edited by T. Princen, M. Maniates, and K. Conca (Cambridge, MA: MIT Press) pp. 155–76.

Daly, H.E. (1986). *Beyond Growth: The Economics of Sustainable Development* (Boston, Massachusetts, USA: Beacon Press).

Dietz, Thomas and Eugene Rosa (1994). "Rethinking the Environmental Impacts of Population, Affluence, and Technology," *Human Ecology Review*, 1: 277–200.

Dietz, Thomas, Eugene Rosa, Yoor (2007). "Driving the Human Ecological Footprint." *Frontiers in Ecology and the Environment* (February).

D'Onofrio, Antonia (2001/2002). "Partial Correlation." Ed 710 Educational Statistics, Spring 2003. Accessed Online on 6/3/2008 at http://www2.widener.edu/.

Economic Commission for Latin America and Caribbean (CEPAC in Spanish) 2009–2010). *Economic Survey of Latin America and the Caribbean 2009–2010.*

Ehrlich and Holdren (1971). "Impacts of Population Growth," *Science*, 171, 1212–7.

Ferreira Francisco H and David de Ferranti et al (2004). "Inequality in Latin America: Breaking the History?" The World Bank. Washington, DC, USA.

Frank, Andre Gunder. See him for an in-depth discussion of Dependency Theory and Underdevelopment in Latin America. Some of his works include: *The Development of Underdevelopment*, 1966, MRP; *Capitalism and Underdevelopment in Latin America*, 1967; *Latin America: Underdevelopment or Revolution*, 1972; and *Theoretical Introduction to Five Thousand Years of World System History*, 1990, Review.

Girardet, Herbert (1996). "Giant Footprints," Accessed Online on 12/14/07, at http:// www.gdrc.org/uem/footprints/girardet.html.

Global Footprint Network: Africa's Ecological Footprint—2006 Fact book. Global Footprint Network, 1050 Warfield Avenue, Oakland, CA 94610, USA. http://www.footprintnetwork.org/Africa.

Global Footprint Network (2008). *The Ecological Footprint Atlas 2008*, October 28.

Goodwin, Paul B. (2007). Global Studies. *Latin America, Twelfth Edition* (Dubuque, Iowa: McGraw-Hill Contemporary Learning Series).

Hardin, G. (1991). "Paramount Positions in Ecological Economics," in Constanza, R. (editor), *Ecological Economics: The Science and Management of Sustainability* (New York: Columbia University Press).

Holdren and Ehrlich (1974). "Human Population and The Global Environment," *American Scientist*, 62: 282–92.

IPCC (2001). Intergovernmental Panel on Climatic Change.

Jorgenson, Andrew K. (2003). "Consumption and Environmental Degradation: A Cross-National Analysis of Ecological Footprint," *Social Problems*, Volume 50, No. 3, pp. 374–394.

Kentor, Jeffrey (2001). "The Long Term Effects of Globalization on Income Inequality, Population Growth, and Economic Development," *Social Problems*, 48: 435–55.

Kick, Edward L. (2000). "World-System Position, National Political Characteristics and Economic Development," Journal of Political and Military Sociology, (Summer), http://www.findarticles.com/p/articles/mi qa3710

Lowry, Richard (1999–2008). "Subchapter 3a. Partial Correlation." Accessed Online on 6/3/08 at http://faculty.vassar.edu/lowry/cha3a.html.

Luke, Brian T. "Pearson's Correlation Coefficient." Learning From The Web.net. Accessed Online on 5/30/2008.

New Internationalist (2003). Latin America and the Caribbean the Facts (May).

Palmer, A.R. (1998). "Evaluating Ecological Footprints," *Electronic Green Journal*. Special Issue 9 (December).

Pimentel, David (1996). "Impact of Population Growth on Food Supplies and Environment." American Association for the Advancement of Science (AAAS) (February, 9). See also GIGA DEATH, Accessed Online on 12/18/07, at http://dieoff.org/page13htm.

Princen, Thomas (2002). "Consumption and Its Externalities: Where Economy Meets Ecology," in *Confronting Consumption*, edited by T. Princen, M. Maniates, and K. Conca (Cambridge, MA: MIT Press) pp. 23–42.

Redefining Progress (2005). Footprints of Nations. 1904 Franklin Street, Oakland, California, 94612. See http://www.RedefiningProgress.org.

Rees, William E. (1992). "Ecological Footprint and Appropriated Carrying Capacity: What Urban Economics Leaves Out," Environment and Urbanization, Vol. 4, No. 2, (October). Accessed Online on 12/14/07, at http://eau.sagepub.com.

Sachs, Jeffrey D. (2007). "Can Extreme Poverty Be Eliminated"? *Developing World, 07/08* (Dubuque, IA: McGraw Hill0 pp. 10–14.

Scott, Lauren and Monica Pratt (2009). "An Introduction To Using Regression Analysis With Spatial Data." *ArcUser. The Magazine for ESRI Software Users* (Spring), pp. 40–43. Lauren Scott and Monica Pratt are ESRI Geo-processing Spatial Statistics Product Engineer and ArcUser Editor respectively.

Sklair, Leslie (2001). *The Transnational Capitalist Class* (Oxford, UK: Blackwell Press).

Solow, R. M. (1974). "The Economics of Resources or the Resources of Economics," *American Economics Review*, Vol. 64, pp. 1–14.

Sundberg, Rolf (2002). *Encyclopedia of Environmetrics*, edited by Abdel H. El-Shaarawi and Walter W. Piegorsch (Chichester: John Wiley & Sons, Ltd), Volume 1, pp. 365–366.

Suplee, D. (1998). "Unlocking the Climate Puzzle." *National Geographic*, 193 (5), 38–70.

The international Society for Ecological Economics and Island Press (1994). "Investing in Natural Capital: The Ecological Approach to Sustainability." Accessed Online on 12/18/07, at http://www.dieoff.org/page13.htm.

United Nations (2009). *World Population Prospects: The 2008 Revision. The Highlights*. Population Division of the Department of Economic and Social Affairs of the United Nations Secretariat. United Nations, New York. See table 1.1.

United Nations Development Program (2004). *Human Development Report 2004 (HDR). Cultural Liberty in Today's Diverse World* (New York, N.Y: UNDP).

United Nations Development Program (2005). *Human Development Report (HDR) 2005. International Cooperation at a Crossroads: Aid, Trade and Security in an Unequal World* (New York, N.Y: UNDP).

United Nations Development Program (2007/2008). *Human Development Report 2007/2008. Fighting Climate Change: Human Solidarity in a Divided World.*

United Nations Development Program (2010). *Human Development Report 2010. The Real Wealth of Nations: Pathway to Human Development*, p. 187.

United Nations Educational, Scientific and Cultural Organization (UNESCO) (2006–2008). *International Hydrological Program (IHP) in Latin America and the Caribbean Report.*

United Nations, *World Population Prospects: The 2004 Revision*, March 2005.
Vitousek, P., P. Ehrlich, A. Ehrlich and P. Matson (1986). "Human Appropriation of the Products of Photosynthesis," *Bioscience*, Vol. 36, pp. 368–374.
Wackernagel, M. (1991). "Using 'Appropriated Carrying Capacity' as an Indicator: Measuring the Sustainability of a Community." Report for the UBC Task Force on Healthy and Sustainable Communities. UBC School of Community and Regional Planning, Vancouver, Canada.
Wackernagel and Rees (1996). *Our Ecological Footprint* (New Society Publisher).
World Bank (2005). *World Development Report 2005. A Better Investment Climate for Everyone* (New York, N.Y: A Co-publication of the World Bank and Oxford University Press).
World Bank (2010). *World Development Report 2010. Development and Climate Change*. Selected World Development Indicators, table 1, p. 379.
World Resources Institute (WRI) (1992). *World Resources, 1992–1993* (New York: Oxford University Press).
York, Richard, Eugene A. Rosa, and Thomas Dietz (2003). "Footprints on the Earth: The Environmental Consequences of Modernity." *American Sociological Review*, 68: 279–300.

Table 6.1. Regional Population, Gross Domestic Product (GDP), Total Footprint, Biological Capacity, and Ecological Balance by World Income Regions in 2005

World Income Regions	*Regional Population (Millions) (2005)*	*GDP PPP US$ Billions (2005)**	*Total Footprint Million gha (2005)*	*Biological Capacity Global gha (2005)*	*Ecological Balance Million gha (2005)*
Low Income	2370.6	5879.1	2377.2	2089.7	–287.5
% World	36.6	9.7	13.6	15.6	7.0
Middle Income	3097.9	22586.3	6787.0	6684.8	–102.2
% World	47.8	37.3	38.9	50.0	2.5
High Income[1]	971.8	32680.7	6196.0	3561.5	–2634.5
% World	15.0	53.9	35.5	26.7	64.5
OECD[2]	**1171.6**	**29197.0**	**6832.9**	**3968.8**	**–2864.1**
% World	18.1	48.2	39.5	29.7	70.2
World Total	**6475.6**	**60597.3**	**17443.6**	**13361.0**	**–4082.7**

*Human Development Report (HDR) 2007/2008. *Fighting Climate Change: Human Solidarity in a Divided World*, table 14, p. 280.

1. High Income World includes all the 24 High Income OECD countries (out of 30 OECD countries in 2005). See HDR 2005, p. 365.
2. OECD countries include most of the High Income countries.

Source: Global Footprint Network (2008). *Ecological Footprint Atlas 2008*, Appendix F, table 2, pp. 46–50.

Table 6.2. Organization for Economic Cooperation and Development (OECD) Member Countries

OECD Member Countries	*Ratification Dates*
Australia	June 7, 1971
Austria	September 29, 1961
Belgium	September 13, 1961
Canada	April 10, 1961
Chile*	May 7, 2010
Czech Republic	December 21, 1995
Denmark	May 30, 1961
Estonia*	December 9, 2010
Finland	January 28, 1969
France	August 7, 1961
Germany	September 27, 1961
Greece	September 27, 1961
Hungary	May 7, 1996
Iceland*	June 5, 1961
Ireland	August 17, 1961
Israel*	September 7, 2010
Italy	March 29, 1962
Japan	April 28, 1964
Korea	December 12, 1996
Luxembourg*	December 7, 1961
Mexico	May 18, 1994
Netherlands	November 13, 1961
New Zealand	May 29, 1973
Norway	July 4, 1961
Poland	November 22, 1996
Portugal	August 4, 1961
Slovak Republic	December 14, 2000
Slovenia*	July 21, 2010
Spain	August 3, 1961
Sweden	September 28, 1961
Switzerland	September 28, 1961
Turkey	August 2, 1961
United Kingdom	May 2, 1961
United States	April 12, 1961

*Not included in the study either because the population is less than 1 million people (e.g. Iceland and Luxembourg) or the date of ratification was after 2005 (e.g. Chile, Estonia, Israel, and Slovenia). Most country data were collected and made available by 2005.

Source: OECD Home Page. Accessed on October 7, 2011.

Table 6.3. Total Population, Ecological Footprint, Biological Capacity, and Ecological Balance by World Regions in 2005

Region	*Total Population (Million) 2005*	*Total Ecological Footprint (Million gha) 2005*	*Total Biological Capacity (Million gha) 2005*	*Total Ecological Balance (Million gha) 2005*
OECD*	**1171.6**	**6832.9**	**3968.8**	**–2864.1**
% World	**18.1**	**39.2**	**29.7**	**70.2**
Africa	902.0	1237.5	1627.1	389.6
% World	13.9	7.1	12.2	
Middle East and Central Asia	365.7	846.8	466.9	–379
% World	5.6	4.9	3.5	9.3
Asia-Pacific	3562.1	5758.6	2923.3	–2835.3
% World	55.0	33.0	21.9	69.5
Latin America and Caribbean	553.0	1350.8	2655.7	1304.9
% World	8.5	7.7	19.9	
North America	330.5	3037.8	2143.3	–894.5
% World	5.1	17.4	16.0	21.9
European EU	487.3	2291.8	1128.2	–1163.6
% World	7.5	13.1	8.4	28.5
European Non-EU	239.6	842.4	1391.6	549.2
% World	3.7	4.8	10.4	
World	**6475.6**	**17443.6**	**13361.0**	**–4082.7**

*OECD countries include EU and North American countries

EU: European Union

Percentages may not add up to 100 percent due: (i) to rounding up; (ii) some countries that were not up to 1 million people in 2005 were not included in the study; and (iii) date of ratification of some OECD countries was after 2005, thus not included in the study.

Source: Global Footprint Network (2008). *The Ecological Footprint Atlas 2008,* Appendix F, table 2, pp. 46–50.

Table 6.4. Per Capita Ecological Footprint, Biological Capacity, and Ecological Balance by World Regions in 2005

Region	*Ecological Footprint (gha per person) 2005*	*Biological Capacity (gha per person) 2005*	*Ecological Balance (gha per person) 2005*
OECD[1]	**5.83**	**3.39**	**–2.44**
Africa	1.37	1.8	0.43
Middle East and Central Asia	2.32	1.28	–1.04
Asia-Pacific	1.62	0.82	–0.80
Latin America and Caribbean	2.44	4.80	2.36
North America	9.19	6.49	–2.71
Europe EU	4.69	2.32	–2.38
Europe Non-EU	3.52	5.81	2.29
World	**2.69**	**2.06**	**–0.63**

Note: EU: European Union

1. OECD countries include most of the European Union (EU) countries

Source: Global Footprint Network (2008). *The Ecological Footprint Atlas 2008,* Appendix F, table 1, pp. 41–45.

Table 6.3. Carbon Dioxide (CO_2) Emissions from Fuel Combustion (in Million Tons) in Organization for Economic Cooperation and Development (OECD) Countries: 2000–2005

Country	*2000*	*2001*	*2002*	*2003*	*2004*	*2005*
Australia	339	351	360	361	370	387
Austria	62	66	68	74	74	75
Belgium	127	128	119	127	124	120
Canada	533	526	534	554	550	556
Czech Republic	122	122	117	121	122	120
Denmark	50	51	51	56	51	47
Finland	54	59	62	72	67	55
France	376	384	376	384	384	387
Germany	827	845	833	842	843	811
Greece	87	90	90	94	93	95
Hungary	55	56	55	58	57	57
Iceland	2	2	2	2	2	2
Ireland	41	43	43	42	42	43
Italy	425	427	434	452	450	454
Japan	1192	1178	1214	1223	1222	1228
Korea	431	449	457	459	479	469
Luxembourg	8	8	9	10	11	11
Mexico	357	356	362	371	374	402
Netherlands	173	179	179	184	185	183
New Zealand	32	34	35	36	36	36
Norway	34	33	33	35	36	37
Poland	292	291	280	291	295	294
Portugal	59	59	63	58	60	63
Slovak Republic	37	38	38	38	37	38
Spain	284	285	302	310	327	339
Sweden	53	52	54	55	54	50
Switzerland	42	43	41	43	44	44
Turkey	201	182	192	202	207	216
United Kingdom	526	539	524	536	536	535
United States	5693	5673	5614	5689	5772	5785
European Union (27)	3842	3916	3886	4005	4010	3979
OECD-Total	**12514**	**12549**	**12541**	**12779**	**12904**	**12942**
Brazil	303	312	311	304	321	327
China	3038	3084	3309	3830	4547	5060
India	977	986	1017	1043	1114	1161
Indonesia	265	282	291	299	316	331
Russian Republic	1514	1514	1505	1540	1524	1531
South Africa	299	284	295	321	338	330
World	**23509**	**23666**	**24065**	**25108**	**26332**	**27146**

Source: IEA (2008), CO_2 Emissions from Fuel Combustion: 2008 Edition, IEA, Paris

Table 6.6. Total Public (Government) Social Expenditure per Head (at Current Prices and PPPs in US Dollars) in Organization for Economic Cooperation and Development (OECD) Countries: 2000–2005

Countries	*2000*	*2001*	*2002*	*2003*	*2004*	*2005*
Australia	4856	4894	5174	5528	5742	5807
Austria	7683	7793	8236	8666	9039	9277
Belgium	6963	7328	7849	7994	8270	8469
Canada	4728	5036	5172	5439	5526	5858
Czech Republic	2964	3200	3473	3734	3814	3954
Denmark	7431	7679	8232	8483	8946	9084
Finland	6237	6450	6886	7160	7772	7953
France	7030	7432	7932	7950	8241	8648
Germany	6785	7063	7438	7813	7982	8157
Greece	3523	4096	4340	4491	4788	5234
Hungary	2459	2735	3149	3452	3525	3835
Ireland	3886	4406	5064	5434	5895	6366
Italy	5947	6383	6427	6623	6771	6957
Japan	4221	4548	4774	4991	5298	5651
Korea	814	944	992	1068	1294	1465
Mexico	532	603	648	726	761	840
Netherlands	5813	6079	6534	6719	7005	7249
New Zealand	4017	4056	4221	4233	4386	4621
Norway	7688	8242	8745	9405	9830	10307
Poland	2164	2404	2580	2675	2788	2854
Portugal	3343	3542	3922	4313	4433	—
Slovak Republic	1964	2123	2301	2305	2408	2638
Spain	4326	4527	4904	5201	5496	5792
Sweden	7913	8083	8568	9135	9585	9645
Switzerland	5648	5906	6407	6757	7015	7193
Turkey	—	—	—	—	—	1065
United Kingdom	4963	5508	5766	6102	6672	6816
United States	4943	5288	5699	5985	6254	6531
OECD-Total	**4820**	**5103**	**5475**	**5771**	**6071**	**6298**

Source: Social Expenditure: Aggregate Data, OECD Social Expenditure Statistics (database), Published 07 August 2009. Key tables from OECD-ISSN 2074–3904- © OECD 2009

Table 6.7. Human Development and Affluence on Biological Capacity, Ecological Footprint and Balance in Organization for Economic C•rporation and Development (OECD) Countries

Countries	*Human Development Index (HDI) Value 2005*[1]	*GDP Per Capita PPP US($) 2005*[2]	*Carbon Dioxide Emissions Mil. Tons 2005*[3]	*Total Footprint Million gha 2005*[4]	*Bio-Capacity Million gha 2005*[4]	*Ecological Balance Million gha 2005*[4]
Australia	0.962	31794	387	157.4	310.9	153.5
Austria	0.948	33700	75	40.8	23.4	–17.4
Belgium	0.946	32118	120	53.4	11.7	–41.6
Canada	0.961	33375	556	228.1	646.9	418.8
Czech Rep.	0.891	20538	120	54.8	28.0	–26.7
Denmark	0.949	33973	47	43.6	31.0	–12.7
Finland	0.952	32153	55	27.5	61.6	34.0
France	0.952	30386	387	298.1	184.4	–113.7
Germany	0.935	29461	811	349.5	160.5	–189.0
Greece	0.926	23381	95	65.2	18.8	–46.4
Hungary	0.874	17887	57	35.8	28.5	–7.4
Ireland	0.959	38505	43	26.0	17.6	–8.3
Italy	0.941	28529	454	276.5	71.2	–205.3
Japan	0.953	31267	1228	626.6	77.2	–549.4

Korea Rep. of	0.921	22029	469	178.9	33.4	–145.5
Mexico	0.829	10751	402	361.9	178.4	–183.5
Netherlands	0.953	32684	183	71.5	18.4	–53.1
New Zealand	0.943	24996	36	31.0	56.6	25.6
Norway	0.968	41420	37	32.0	28.3	–3.7
Poland	0.870	13847	294	152.6	81.0	–71.6
Portugal	0.897	20410	63	46.6	12.9	–33.6
Slovakia	0.863	15871	38	17.8	15.2	–2.6
Spain	0.949	27167	339	247.2	57.6	–189.6
Sweden	0.956	32525	50	46.1	90.2	44.1
Switzerland	0.955	35633	44	36.3	9.2	–27.1
Turkey	0.775	8407	216	198.6	120.9	–77.7
UK	0.946	33238	535	319.2	98.6	–220.6
United States	0.951	41890	5785	2809.8	1496.4	–1313.3
OECD	**0.916**	**29197**	**12942**	**6832.9**	**3968.8**	**–2864.1**

1 and 2. Human Development Report (HDR) 2007/2008. *Fighting Climate Change: Human Solidarity in a Divided World*, pp. 229–230.

3. IEA (2008), CO_2 Emissions from Fuel Combustion: 2008 Edition, IEA Parts. http://www.oecd-ilibrary.org/environment/co_2-emissions-from-fuel-combustion-2009_207. Accessed Online on 9/21/2011.

4. For OECD countries represented.

Table 6.8. Population, Ecostructural Factors, and Total Footprint Per Capita in Organization for Economic Cooperation and Development (OECD) Countries

Countries	*Population (Millions) 2005*[1]	*Population Density People per km*^{2} *2003*[2]	*Urban Population (% of Total) 2005*[3]	*Adult Literacy Rates (%) by 2005*[4]	*Domestic Inequality (Gini Index) Late 2000s*[5]	*Total Footprint gha per Capita 2005*[6]
Australia	20.3	3	88.2	99.0	34.0	7.8
Austria	8.3	97	66.0	99.0	26.0	5.0
Belgium	10.4	342	97.2	99.0	27.0	5.1
Canada	32.2	3	80.1	99.0	32.0	7.1
Czech Rep.	10.2	123	73.5	99.0	26.0	5.4
Denmark	5.4	127	85.6	99.0	25.0	8.0
Finland	5.2	17	61.1	99.0	26.0	5.3
France	61.0	109	76.7	99.0	29.0	4.9
Germany	82.7	237	75.2	99.0	30.0	4.2
Greece	11.1	83	59.0	96.0	32.0	5.9
Hungary	10.1	110	66.3	99.4	27.0	3.6
Ireland	4.1	57	60.5	99.0	30.0	6.3
Italy	58.6	196	67.6	98.4	34.0	4.8
Japan	127.9	349	65.8	99.0	33.0	4.9
Korea Rep. of	47.9	485	80.8	99.0	32.0	3.7

Mexico	104.3	54	76.0	91.6	48.0	3.4
Netherlands	16.3	479	80.2	99.0	29.0	4.4
New Zealand	4.1	15	86.2	99.0	33.0	7.7
Norway	4.6	15	77.4	99.0	25.0	6.9
Poland	38.2	125	62.1	99.8	31.0	4.0
Portugal	10.5	111	57.6	93.8	36.0	4.4
Slovakia	5.4	110	56.2	99.0	24.0	3.3
Spain	43.4	82	76.7	99.0	31.0	5.7
Sweden	9.0	22	84.2	99.0	26.0	5.1
Switzerland	7.4	186	75.2	99.0	28.0	5.0
Turkey	73.0	92	67.3	87.4	41.0	2.7
UK	60.2	246	89.7	99.0	34.0	5.3
United States	299.8	32	80.8	99.0	38.0	9.4
OECD	**1171.6**	**34***	**75.6**	**99.0**	**31.0**	**5.3**

1, 2, 5. OECD figures are for the OECD countries represented in the study.
3, 4. OECD figures are for all the OECD countries.
5. Source: Provisional data from OECD Income Distribution and Poverty Database (www.oecd.org/els/social/Inequality).
6. Total Footprint per capita is for the number of OECD countries represented in the study and not for all the OECD countries.
*For OECD Population Density. See: *Wikipedia, The Free Encyclopedia*. Organization for Economic Cooperation and Development.

Table 6.9. Descriptive Statistics

	N	*Minimum*	*Maximum*	*Mean*	*Std. Deviation*
Human Development Index	28	.775	.968	.92582	.046135
GDP Per Capita	28	8407.000	41890.000	27783.46429	8789.138594
Carbon Dioxide Emissions	28	36.000	5785.000	461.64286	1080.039650
Public Expenditure	28	840.000	10307.000	5953.53571	2620.201568
Total Footprint	28	17.800	2809.800	243.92143	523.650753
Bio-Capacity	28	9.200	1496.400	141.74286	295.046479
Ecological Balance	28	–1313.300	418.800	–102.27857	284.650984
Valid N (listwise)	28				

Table 6.10. Correlations

		Total Footprint	*Human Development Index*	*GDP Per Capita*	*Carbon Dioxide Emissions*	*Public Expenditure*	*Bio-Capacity*	*Ecological Balance*
Pearson Correlation	Total Footprint	1.000	.091	.277	.996	–.003	.907	–.899
	Human Development Index	.091	1.000	.907	.128	.791	.136	–.026
	GDP Per Capita	.277	.907	1.000	.309	.840	.309	–.190
	Carbon Dioxide Emissions	.996	.128	.309	1.000	.016	.915	–.884
	Public Expenditure	–.003	.791	.840	.016	1.000	.018	.024
	Bio-Capacity	.907	.136	.309	.915	.018	1.000	–.632
	Ecological Balance	–.899	–.026	–.190	–.884	.024	–.632	1.000
Sig. (1-tailed)	Total Footprint	.	.323	.076	.000	.494	.000	.000
	Human Development Index	.323	.	.000	.258	.000	.246	.447
	GDP Per Capita	.076	.000	.	.055	.000	.055	.167
	Carbon Dioxide Emissions	.000	.258	.055	.	.469	.000	.000
	Public Expenditure	.494	.000	.000	.469	.	.464	.452
	Bio-Capacity	.000	.246	.055	.000	.464	.	.000
	Ecological Balance	.000	.447	.167	.000	.452	.000	.
N	Total Footprint	28	28	28	28	28	28	28
	Human Development Index	28	28	28	28	28	28	28
	GDP Per Capita	28	28	28	28	28	28	28
	Carbon Dioxide Emissions	28	28	28	28	28	28	28
	Public Expenditure	28	28	28	28	28	28	28
	Bio-Capacity	28	28	28	28	28	28	28
	Ecological Balance	28	28	28	28	28	28	28

Table 6.11. Variables Entered/Removed[a]

Model	*Variables Entered*	*Variables Removed*	*Method*
1	Carbon Dioxide Emissions	.	Stepwise (Criteria: Probability-of-F-to-enter <= .050, Probability-of-F-to-remove >= .100).
2	Ecological Balance	.	Stepwise (Criteria: Probability-of-F-to-enter <= .050, Probability-of-F-to-remove >= .100).
3	Bio-Capacity	.	Stepwise (Criteria: Probability-of-F-to-enter <= .050, Probability-of-F-to-remove >= .100).
4	.	Carbon Dioxide Emissions	Stepwise (Criteria: Probability-of-F-to-enter <= .050, Probability-of-F-to-remove >= .100).

a. Dependent Variable: Total Footprint

Table 6.12. Model Summary[e]

Model	R	R Square	Adjusted R Square	Std. Error of the Estimate	Change Statistics					Durbin-Watson
					R Square Change	F Change	df1	df2	Sig. F Change	
1	.996[a]	.992	.991	48.658096	.992	3101.066	1	26	.000	
2	.997[b]	.993	.993	44.073237	.002	6.691	1	25	.016	
3	1.000[c]	1.000	1.000	.602060	.007	133947.029	1	24	.000	
4	1.000[d]	1.000	1.000	.589912	.000	.001	1	24	.971	2.012

a. Predictors: (Constant), Carbon Dioxide Emissions
b. Predictors: (Constant), Carbon Dioxide Emissions, Ecological Balance
c. Predictors: (Constant), Carbon Dioxide Emissions, Ecological Balance, Bio-Capacity
d. Predictors: (Constant), Ecological Balance, Bio-Capacity
e. Dependent Variable: Total Footprint

Table 6.13. ANOVA[e]

Model		Sum of Squares	df	Mean Square	F	Sig.
1	Regression	7342115.139	1	7342115.139	3101.066	.000[a]
	Residual	61557.868	26	2367.610		
	Total	7403673.007	27			
2	Regression	7355111.752	2	3677555.876	1893.256	.000[b]
	Residual	48561.255	25	1942.450		
	Total	7403673.007	27			
3	Regression	7403664.308	3	2467888.103	6808421.788	.000[c]
	Residual	8.699	24	.362		
	Total	7403673.007	27			
4	Regression	7403664.307	2	3701832.154	10637578.836	.000[d]
	Residual	8.700	25	.348		
	Total	7403673.007	27			

a. Predictors: (Constant), Carbon Dioxide Emissions
b. Predictors: (Constant), Carbon Dioxide Emissions, Ecological Balance
c. Predictors: (Constant), Carbon Dioxide Emissions, Ecological Balance, Bio-Capacity
d. Predictors: (Constant), Ecological Balance, Bio-Capacity
e. Dependent Variable: Total Footprint

Table 6.14. Coefficients[a]

		Unstandardized Coefficients		Standardized Coefficients		
Model		B	Std. Error	Beta	t	Sig.
1	(Constant)	21.029	10.029		2.097	.046
	Carbon Dioxide Emissions	.483	.009	.996	55.687	.000
2	(Constant)	21.888	9.090		2.408	.024
	Carbon Dioxide Emissions	.445	.017	.917	26.510	.000
	Ecological Balance	–.165	.064	–.089	–2.587	.016
3	(Constant)	–.131	.138		–.951	.351
	Carbon Dioxide Emissions	–4.472E-5	.001	.000	–.036	.971
	Ecological Balance	–1.000	.002	–.544	–409.409	.000
	Bio-Capacity	1.000	.003	.564	365.988	.000
4	(Constant)	–.129	.124		–1.039	.309
	Ecological Balance	–1.000	.001	–.544	–1944.095	.000
	Bio-Capacity	1.000	.000	.564	2015.521	.000

a. Dependent Variable: Total Footprint

Table 6.15. Descriptive Statistics

	N	*Minimum*	*Maximum*	*Mean*	*Std. Deviation*
Population	28	4.10	299.80	41.8464	60.63965
Population Density	28	3.00	485.00	143.1071	140.72546
Urban Population	28	56.20	97.20	74.0429	10.80571
Adult Literacy	28	87.40	99.80	98.0500	2.73015
Gini Index	28	24.00	48.00	30.9643	5.35054
Public Expenditure	28	840.00	10307.00	5953.5357	2620.20157
Total Footprint	28	2.70	9.40	5.3321	1.59375
Valid N (listwise)	28				

Table 6.16. Correlations

		Total Footprint	*Population*	*Population Density*	*Urban Population*	*Adult Literacy*	*Gini Index*	*Public Expenditure*
Pearson Correlation	Total Footprint	1.000	.226	–.394	.418	.396	–.095	.455
	Population	.226	1.000	.069	.104	–.133	.555	–.132
	Population Density	–.394	.069	1.000	.158	.148	–.057	–.066
	Urban Population	.418	.104	.158	1.000	.233	.018	.319
	Adult Literacy	.396	–.133	.148	.233	1.000	–.698	.521
	Gini Index	–.095	.555	–.057	.018	–.698	1.000	–.569
	Public Expenditure	.455	–.132	–.066	.319	.521	–.569	1.000
Sig. (1-tailed)	Total Footprint	.	.123	.019	.013	.018	.315	.007
	Population	.123	.	.364	.300	.250	.001	.251
	Population Density	.019	.364	.	.211	.225	.387	.370
	Urban Population	.013	.300	.211	.	.116	.463	.049
	Adult Literacy	.018	.250	.225	.116	.	.000	.002
	Gini Index	.315	.001	.387	.463	.000	.	.001
	Public Expenditure	.007	.251	.370	.049	.002	.001	.
N	Total Footprint	28	28	28	28	28	28	28
	Population	28	28	28	28	28	28	28
	Population Density	28	28	28	28	28	28	28
	Urban Population	28	28	28	28	28	28	28
	Adult Literacy	28	28	28	28	28	28	28
	Gini Index	28	28	28	28	28	28	28
	Public Expenditure	28	28	28	28	28	28	28

Table 6.17. Variables Entered/Removed[a]

Model	*Variables Entered*	*Variables Removed*	*Method*
1	Public Expenditure	.	Stepwise (Criteria: Probability-of-F-to-enter <= .050, Probability-of-F-to-remove >= .100).
2	Population Density	.	Stepwise (Criteria: Probability-of-F-to-enter <= .050, Probability-of-F-to-remove >= .100).
3	Urban Population	.	Stepwise (Criteria: Probability-of-F-to-enter <= .050, Probability-of-F-to-remove >= .100).

a. Dependent Variable: Total Footprint

Table 6.18. Model Summary[d]

Model	*R*	*R Square*	*Adjusted R Square*	*Std. Error of the Estimate*	*Change Statistics*					*Durbin-Watson*
					R Square Change	*F Change*	*df1*	*df2*	*Sig. F Change*	
1	.455[a]	.207	.177	1.44594	.207	6.802	1	26	.015	
2	.584[b]	.341	.288	1.34480	.133	5.058	1	25	.034	
3	.688[c]	.473	.407	1.22712	.132	6.025	1	24	.022	1.844

a. Predictors: (Constant), Public Expenditure
b. Predictors: (Constant), Public Expenditure, Population Density
c. Predictors: (Constant), Public Expenditure, Population Density, Urban Population
d. Dependent Variable: Total Footprint

Table 6.19. ANOVA[d]

Model		Sum of Squares	df	Mean Square	F	Sig.
1	Regression	14.222	1	14.222	6.802	.015[a]
	Residual	54.359	26	2.091		
	Total	68.581	27			
2	Regression	23.369	2	11.684	6.461	.005[b]
	Residual	45.212	25	1.808		
	Total	68.581	27			
3	Regression	32.441	3	10.814	7.181	.001[c]
	Residual	36.140	24	1.506		
	Total	68.581	27			

a. Predictors: (Constant), Public Expenditure
b. Predictors: (Constant), Public Expenditure, Population Density
c. Predictors: (Constant), Public Expenditure, Population Density, Urban Population
d. Dependent Variable: Total Footprint

Table 6.20. Coefficients[a]

		Unstandardized Coefficients		Standardized Coefficients		
Model		B	Std. Error	Beta	t	Sig.
1	(Constant)	3.683	.689		5.347	.000
	Public Expenditure	.000	.000	.455	2.608	.015
2	(Constant)	4.363	.708		6.159	.000
	Public Expenditure	.000	.000	.431	2.651	.014
	Population Density	–.004	.002	–.366	–2.249	.034
3	(Constant)	.677	1.635		.414	.682
	Public Expenditure	.000	.000	.302	1.918	.067
	Population Density	–.005	.002	–.436	–2.885	.008
	Urban Population	.058	.023	.391	2.455	.022

a. Dependent Variable: Total Footprint

Chapter Seven

Socio-Economic Determinants of National Ecological Footprints in Latin America and Caribbean

A Multiple Regression Analysis of the Factorial Impacts

INTRODUCTION

The United Nations defines the region of Latin America and the Caribbean as comprising of Central America that includes Mexico (Land in Turmoil), South America (An Imperfect Prism), and the Caribbean (Sea of Diversity) (Paul B. Goodwin, *Latin America*, 12th Edition, 2007). It is a region of the Americas where Romance languages, particularly Spanish, Portuguese, and French are primarily spoken. In 2010, the region is the third largest in the world with a population of 582.7 million (8.5 percent of world population), after Asia and Africa respectively (*Human Development Report 2010*, p.187). By 2050, the population is expected to increase to over 800 million people [United Nations, *World Population Prospects: The 2008 Revision*, table 1.1; CIAT (2006): Latin America and Caribbean Population].

The total area of the region in 2002 was 20,469,646 km^2 which comprised of South America, 17,755,725 km^2; Central America, 517,692 km^2; and Caribbean, 2,196,229km^2, while their respective populations were 349.4 million, 36 million, and 375 million (United Nations, *World Population Prospects: The 2004 Revision*, March 2005; CIAT and UNDP, 2005; UNESCO, *2006–2008 Report*). Currently, Latin America and Caribbean region is one of the highest urbanized regions of the world (79.5%), even higher than that of the Organization of Economic Cooperation and Development (OECD) region (77.1%) (*Human Development Report 2010*, table 11: Demographic Trend, p. 187). The region contains the world's largest and fastest growing metropolitan areas, such as Mexico City, Mexico with population of 19.2 million; Sao Paulo, Brazil (18.6

million); Buenos Aires, Argentina (13.5 million); Rio de Janeiro, Brazil (11.6 million); Lima, Peru (8.4 million); Bogota, Colombia (7.8 million); Santiago, Chile (5.7 million); Guadalajara, Mexico (4.0 million); Monterrey, Mexico (3.6 million); Brasilia, Brazil (3.5 million) (City Mayors, 2006).

Among developing countries of the world, Latin America and the Caribbean (also expressed as LAC region), has medium consumption level of gross domestic product (GDP) per capita ($7,567 PPP US$), second to Europe and Central Asia ($8,361 PPP US$) (*HDR 2010*, table 16, p. 210). In 2005, apart from Haiti and Nicaragua which belong to low income countries (LIC) with per capita income of $765.00 or less, all other Latin America and the Caribbean countries were either lower middle income (LMC: $766.00-$3,035.00) or upper middle income (UMC: $3,036.00-$9,385.00) (*World Development Report 2005*, p. 255). In 2010, all LAC countries belonged to either lower middle income (LMC) or upper middle income (UMC), except Haiti that still remained LIC (*World Development Report 2010*, p. 377). The richest countries in LAC are Mexico, Brazil, Argentina, while the poorest are Haiti, Nicaragua, Bolivia, and Honduras (Economic Commission for Latin America and Caribbean, 2009–2010). Apart from the Organization for Economic Cooperation and Development (OECD) countries (0.879) and Non-OECD countries (0.844), the LAC region is the third highest (0.704) among developed and developing regions in terms of human development index (HDI), indicating a medium-high development (*HDR 2010*, table 3, p. 155).

Socio-economic inequality is a major feature and a serious problem in the LAC region. High inequality is an institution that had been perpetuated ever since colonial times, which had survived different political and economic regimes. It is undermining the region's economic potential and well-being of its population. Socio-economic inequality has been reproduced and transmitted through generations because Latin American political systems allow a differentiated access on the influence that social groups have in the decision making process. Differences in socio-economic opportunities and endowments tend to be based on race, ethnicity, rurality, and gender (Ferreira and Ferranti et al, 2004). The deep historical roots of inequality in Latin America had over time constrained the poor's social mobility. According to the United Nations Development Program, *Human Development Reports, 2010*, table 3, pp. 152–155, the countries with the highest inequality (measured with Gini index) in the region were Haiti (59.5), Colombia (58.5), Bolivia (58.2), Honduras (55.3), Brazil (55.0), and Panama (54.9), while the countries with the lowest inequality in the region were Venezuela (43.4), Uruguay (46.4), and Costa Rica (47.2).

On one hand, table 7.1 shows that Middle Income Region has about 48 percent of the world population. It has also 37 percent of the world GDP, about 39 percent of the total footprint, 50 percent of world biological capacity, and about 2.5 percent of world ecological balance overshoots. On the other hand, the High Income Region has only 15 percent of the world population, with the highest GDP (53.9 percent), and highest ecological balance overshoot (64.5 percent). Since LAC belongs to middle income region where human development is relatively high, though with the lowest ecological balance overshoot when compared to Low Income Region and High Income Region (See table 7.1), the question is: what factors are responsible for or are driving the total and per capita footprints.

FOCUS OF THE STUDY

Countries and Territories in Latin America and the Caribbean (LAC) (see also the map at the end of this chapter).

Listed below are the 22 countries that were selected for the study out of 46 Latin America and the Caribbean countries. Their selection was based on the availability of data and adequate population size (see Global Footprint Network. *The Ecological Footprint Atlas 2008*). Several countries were excluded from the list based on the fact that their populations were less than one million people (e.g. Barbados, Belize, Guyana, Suriname, and The Bahamas), or that they were dependent territories of other large countries of France, United Kingdom, United States, and Netherlands (e.g. Antigua and Barbuda, Dominica, Grenada, Guadeloupe, Martinique, Montserrat, Navassa Island, Puerto Rico, Saba, Saint Barthelemy, Saint Kitts and Nevis, Saint Lucia, Saint Martin, Saint Vincent and the Grenadines, Saint Eustatius, Saint Maarten, Turks and Caicos Islands, and United States Virgin Islands). They are colonies and military outposts of the large countries. The selected countries for the study are: Argentina, Bolivia, Brazil, Chile, Colombia, Costa Rica, Cuba, Dominican Republic, Ecuador, El Salvador, Guatemala, Haiti, Honduras, Jamaica, Mexico, Nicaragua, Panama, Paraguay, Peru, Trinidad & Tobago, Uruguay, and Venezuela.

PURPOSE OF STUDY

The study analyzes and explains the national variations in total footprints and per capita footprints associated with human development indexes and ecostructural factors [socioeconomic processes within nations such as urbanization, literacy rate, gross domestic product per capita (affluence), and domestic inequality (Gini index)] in Latin America and the Caribbean (LAC) countries. Progress towards sustainable development can be assessed using human development index (HDI) as an indicator of well-being, and the ecological footprint as a measure of demand on the biosphere. Table 7.2 exposes the inequitable distribution of Latin

America and Caribbean nations' ecological footprints as compared to those of other regions of the world. From this table, LAC region had only 8.5 percent of world population in 2005 but with about 20 percent or one-fifth of world's total biological capacity. The region also has 7.7 percent of world's total ecological footprint, as well as the world's highest total ecological balance (1304.9 million global hectares). Table 7.3 also shows that LAC had in 2005 per capita ecological footprint of 2.44 (less than the world's 2.69); biological capacity of 4.80 per capita (double the world's 2.06); and per capita ecological balance of 2.36 global hectares, highest in the world. Using the tools of comparative model and regression analysis, the paper seeks to establish the key factors, among others, that highly influence and drive the ecological footprints in different Latin America and the Caribbean countries. The question is what type of secure and sustainable environments should meet the needs of both people and natural environment in Latin America and Caribbean? This constitutes an enormous challenge to policy makers, planners, and development experts. Thus, the study proffers some policies, recommendations, and solutions on how to reduce the ecological footprints of nations, which also will help protect the natural environment and promote a more equitable and sustainable society.

THEORETICAL FRAMEWORK FOR THE STUDY

Within the biosphere, everything is interconnected, including humans, thus sustainable development requires a good knowledge of human ecology. The human life-support functions of the ecosphere are maintained by nature's biocapacity which runs the risks of being depleted, the prevailing technology notwithstanding. Regardless of the humanity's mastery over the natural environment, it still remains a creature of the ecosphere and always in a state of *obligate dependency* on numerous biological goods and services (Rees, 1992, p. 123). Despite the above *dependency* thesis, the prevailing economic mythology assumes a world in which carrying capacity is indefinitely expandable (Daly, 1986; Solow, 1974). The human species has continued to deplete, draw-down, and confiscate nature's biocapacity with reckless abandon. York et al (2003), indicated that population and affluence account for 95 percent of the variance in total footprints of countries. Thus, large human population all over the world and their excessive consumption of the scarce natural resources are responsible for the national footprint of nations. Thus, total human impact on the ecosphere is given as: population × per capita impact (Ehrlich and Holdren, 1971; Holdren and Ehrlich, 1974); Hardin (1991). In other words, population size and affluence are the primary drivers of environmental impacts (Dietz et al, 2007). The footprints of nations provide compelling evidence of the impacts of consumption, thus the need for humans to change their lifestyles and conserve scarce natural capital. According to Palmer

(1998), there are in order of decreasing magnitude, three categories of consumption that contribute enormously to our ecological footprints: wood products (53 percent), food (45 percent), and degraded land (2 percent). Degraded land includes land taken out of ecological availability by buildings, roads, parking lots, recreation, businesses, and industries. Palmer also indicated that about 10 percent or more of earth's forests and other ecological land should be preserved in more or less pristine condition to maintain a minimum base

The environmental impacts of urban areas should be considered because a rapidly growing proportion of world's population lives in cities, and more than one million people are added to the world's cities each week and majority of them are in developing countries of Africa, Latin America, and Southeast Asia (Wackernagel and Rees, 1996). The reality is that the populations of urban regions of many nations had already exceeded their territorial carrying capacities and depend on trade for survival. Of course, such regions are running an unaccounted ecological deficit; their populations are appropriating and meeting their carrying capacity from elsewhere (Pimentel, 1996; Wackernagel and Rees, 1996; Girardet, 1996, accessed Online on 12/14/07; Rees, 1992; The International Society for Ecological Economics and Island Press, 1994; Vitouset et al, 1986; Wackernagel, 1991; WRI, 1992, p. 374). Undoubtedly, the rapid urbanization occurring in many regions and the increasing ecological uncertainty have implications for world development and sustainability. Cities are densely populated areas that have high ecological footprints which leads to the perception of these populations as "parasitic," since these communities have little intrinsic biocapacity, and instead, must rely upon large hinterlands. Land consumed by urban regions is typically at least an order of magnitude greater than that contained within the usual political boundaries or the associated built-up areas (The International Society for Ecological Economics and Island Press, 1994; Rees, 1992).

According to Rees (1992), every city is an entropic black hole drawing on the concentrated material resources and low-entropy production of a vast and scattered hinterlands many times the size of the city itself. In the same vein, Vitouset et al (1986) asserted that high density settlements "appropriate" or augment their carrying capacity from all over the globe, as well as from the past and the future (see also Wackernagel, 1991). In modern industrial cities, resources flow through the urban system without much concern either about their origins, or about the destination of their wastes, thus, inputs and outputs are considered to be unrelated. The cities' key activities such as transportation, provision of electricity supply, heating, manufacturing and the provision of socio-economic services depend on a ready supply of fossil fuels, usually from far-flung hinterlands than within their usual political boundaries or their associated built-up areas. Cities are not self-contained entities, and their concentration of intense economic processes and high levels of consumption both increase and stimulate their demands on resources. Cities occupy only 2 percent of the world's land surface, but use some 75 percent of the world resources, and release a similar percentage of waste (Girardet, 1996).

Like urbanization, energy footprint, created from energy use and carbon dioxide emissions, is not subject to area constraints. Energy footprint is the area of forest that would be needed to sequester the excess carbon (as carbon dioxide) that is being added to the atmosphere by the burning of fossil fuels to generate energy for travel, heating, lighting, manufacturing, recreation, among other uses. Actually, the demand for energy defines modern cities more than any other single factor. Cities contain enormous concentration of economic activities that consume enormous quantities of energy. The natural global systems of forests and oceans for carbon sequestration are not handling the human carbon contributions fast enough, thus the Kyoto Conference of early 1998. According to Suplee (1998), only half of the carbon humans generate burning fossil fuels can be absorbed in the oceans and existing terrestrial sinks. The oceans absorb about 35 percent of the carbon in carbon dioxide (Suplee, 1998), equivalent of 1.8 giga tons of carbon every year (IPCC, 2001), while the global forests under optimum management of existing forests could absorb about 15 percent of the carbon in the CO_2 produced from the burning of fossil fuels world-wide (Brown et al, 1996). The energy footprint is caused by the un-sequestrated 50 percent in the atmosphere with the potentially troubling ecological consequences, such as rapid global warming and other environmental stresses, including climate change. Carbon dioxide in the atmosphere will continue to increase unless humanity finds alternative energy sources of sufficient magnitude. It is in the humanity's best interest to get off its petroleum addiction (control and minimize fossil fuel use) and develop sustainable consumption habits.

Literacy affects the consumption of natural capital resources. Highly literate groups concentrate in urban areas where they consume more than their fair shares of biospheric resources. Literate populations generally have lower rates of domestic inequality and tend to consume more resources than their illiterate counterparts due to their higher incomes and higher standards of urban living. Furthermore, higher levels of literacy correspond with higher incomes, which allow for greater consumption (Jorgenson, 2003). This is because literate populations are subject to increased consumerist ideologies and contextual images of good life through advertising (Princen et al, 2002); what Leslie Sklair (2001) and Jennifer Clapp (2002) labeled "cultural ideology of consumerism/consumption."

METHODOLOGY FOR THE STUDY

Unit of Analysis

The unit of analysis is "country" (individual Latin America and Caribbean countries).

Sample

To test for the national variations in total footprints and per capita footprints associated with human development index, Comparative Model Analysis and Stepwise Regression Analysis were used as the tools of analysis. Using comparative model, a sample of 22 countries out of many countries and dependent territories of the region was analyzed in tables 4–5 (see Global Footprint Network. *The Ecological Footprint Atlas 2008*; Goodwin, Paul (2007), *Global Studies. Latin America. Twelfth Edition*). Step-wise Regression Analysis was also conducted using the sample. Each table represents LAC countries as also represented by Global Footprint Network, *The Ecological Footprint Atlas 2008*, tables 2,3, and 4. These tables do not include countries that were less than one million people in 2005, as well as the dependent territories of France, United Kingdom, United States, and Netherlands. The larger population of at least one million is chosen for the study because; population is a variable which affects the consumption rates and levels, therefore the ecological footprint.

Mode of Analysis

This is an explanatory study using Step-wise Regression as well as comparative model and descriptive statistics such as matrices, totals, averages, ratios and percentages. Regression Analysis was performed to flesh out factors that highly impacted national footprints, as well as to strengthen the comparative model analysis results. Potential technical problems are diagnosed that might affect the validity of the result of the analysis such as the missing data (e.g., domestic inequality or Gini index data for Cuba) or the excluded countries mentioned above, which did not alter the substantive conclusions, especially regarding the ecological footprint accounts and environmental sustainability of different LAC countries. The national ecological footprint and ecological balance, as dependent variables, are explained recursively by the country's human development hierarchical category, population size, population density, urbanization, GDP per capita, domestic inequality (Gini index), and literacy rate, as the independent variables. The independent variables mediated in explaining the varying levels of consumption among different human development indexes of the capitalist world economy. It is hypothesized that Human Development positions and Eco-structural Factors are likely to be responsible for the variations in the Ecological Footprints and Balances. The carbon dioxide emission levels (*Human Development Report*, 2007/2008, table 24, pp. 310–313) are included to depict the consumption and environmental degradation impacts of different LAC countries that reflect their national ecological footprint and ecological balance levels.

HYPOTHESIS FOR THE STUDY

Null Hypothesis (H_0)

None of the independent variables predicts the national footprint effects. The independent variables do not influence national footprints.

Alternative Hypothesis (H_1)

Some (if not all) of the independent variables predict the national footprint effects.

Variables

Dependent and independent variables are selected on the basis of the theoretical themes and underpinnings described above. The measures of the dependent and independent variables used in the present study are as follow:

Dependent Variables

The ecological footprint or consumption and ecological balance are measured in global hectares (1 hectare = 2.47 acres). A global hectare is 1 hectare of biologically productive space with world average productivity (Global Footprint Network: *Africa's Ecological Footprint-2006 Factbook*, pp. 82–90).

Independent Variables

The ecostructural mediating factors are most of the independent variables in this study that include, GDP per capita, urbanization (urban population as a percent of total population), literacy rate (%), and domestic inequality (Gini index) (See *Appendix A* for Variable definitions).

COMPARATIVE MODEL ANALYSIS (RESULTS AND DISCUSSIONS)

The results and findings in this study are discussed under subtitles that include, "Per Capita Ecological Footprint, Biological Capacity, and Ecological Balance by World Regions," "Human Development and Affluence on Biological Capacity, Ecological Footprint and Balance in LAC Countries," and "Population and Ecostructural Factors on Total Footprint Per Capita in LAC Countries."

PER CAPITA ECOLOGICAL FOOTPRINT, BIOLOGICAL CAPACITY, AND ECOLOGICAL BALANCE BY WORLD REGIONS

Different regions of the world have different per capita ecological footprints, biological capacities, and ecological balances. Table 7.3 shows a comparison of per capita ecological footprint, biological capacity, and ecological balance by world regions. The world regions include Latin America and Caribbean, Africa, Middle East and Central Asia, Asia-Pacific, North America, Europe EU, Europe Non-EU. Analysis of table 7.3 exposes the inequitable distribution of LAC nations' ecological footprints as compared to those of other world regions. Furthermore, this table shows a comparison of each region's footprint with its biocapacity depicting whether that region has an ecological reserve or is running a deficit. Also from the table, one may find that North America, with its considerable per capita biological capacity in 2005 (6.49 gha) had the largest per capita footprint (9.19 gha) and the largest per capita ecological deficit or overshoot (–2.71 gha); followed by Europe EU (Western Europe) with per capita footprint of 4.69 and per capita ecological deficit of –2.38; followed by Europe Non-EU (Central and Eastern Europe), 3.52 and ecological reserve of 2.29; Latin America and Caribbean, 2.44 and ecological balance or reserve of 2.36; and Middle East and Central Asia, 2.32 and –1.04 respectively. The lowest per capita ecological footprint of 1.37 was in Africa with per capita ecological reserve of 0.43; followed by Asia-Pacific, 1.62 and ecological deficit of –1.04. The highest per capita ecological reserve is found in LAC (2.36), followed by Europe Non-EU (2.29), and Africa (0.43). LAC per capita ecological footprint (2.44), biological capacity (4.80), and ecological balance (2.36) were more than the world average in 2005 of 2.69, 2.06, and –0.63 respectively. A regional footprint per capita that is less than the world average and per capita ecological balance more than the world average may denote sustainability at the global level (Living Planet Report, 2006). The less than world's average in footprint per capita and more than world's average in per capita ecological balance were found in developing countries in 2005, especially in Africa, Middle East and Central Asia, and Latin America and Caribbean regions.

HUMAN DEVELOPMENT AND AFFLUENCE ON BIOLOGICAL CAPACITY, ECOLOGICAL FOOTPRINT AND BALANCE IN LATIN AMERICA AND CARIBBEAN COUNTRIES

Table 7.4 presents the national comparisons for both dependent (Ecological footprint and Ecological balance) and some of the independent (Human development or welfare, GDP per capita or affluence, and Carbon dioxide emissions) variables in Latin America and Caribbean. The table shows that a country's total footprint, ecological balance, and carbon dioxide emissions (waste) are a function of its human development position and affluence level in the world economy. Given the national biological capacity measured in global hectares, the higher the total ecological footprint; the lower the ecological balance and vice versa. Table 7.4 assesses the progress towards sustainability development in LAC by using human development index (HDI) as an indicator of national well-being, and the ecological footprint and balance as measures of national demand on the biosphere and biological capacity respectively. The sustainability quotient is depicted by any country with low footprint and high human development levels. In fact, table 7.4 shows that, among LAC countries, improvements in the human development indexes (human welfare) have resulted in the worsening of the countries' positions on their ecological footprints and balances. Table 7.4 shows that LAC countries are among the high human development (HDI of 0.800 and above and per capita GNP of $9,386 or more) or medium human development (HDI of 0.500–0.799 and per capita GNP of between $766 and $9,385), even Haiti that is one of the poorest countries of the world had HDI of more than 0.500 (World Bank, 2005, p. 255).

From table 7.4 below and based on the existing biological capacity, it becomes apparent that the higher the HDI (human welfare), GDP per capita (affluence) i.e. consumption; the higher the carbon dioxide emissions and ecological footprint. For example, Mexico shows HDI of 0.829, GDP per capita ($10,751), carbon dioxide emission (437.8 million tons), total footprint (361.9 global hectares), biological capacity (178.4 global hectares), and ecological balance of –183.5 global hectares. Actually, Mexico "appropriates" half of its carrying capacity from "elsewhere." Brazil shows HDI of 0.800, GDP per capita ($8,402), carbon dioxide emission (331.6), and total footprint (439.2). Brazil's biological capacity is highest (1353.8 global hectares) among the LAC countries, as well as its ecological balance (914.6 global hectares). Chile shows HDI of 0.867, GDP per capita ($12,027), carbon dioxide emission (62.4), and total footprint (48.9). Likewise, Argentina shows HDI of 0.869, GDP per capita ($14,280), carbon dioxide emission of (141.7), and total footprint of 95.2. On one hand, Haiti shows the lowest HDI of 0.529, GDP per capita($1,663), carbon dioxide emission (1.8), total footprint (4.6), biological capacity (2.3), and ecological balance (–2.3). Likewise, Jamaica shows HDI of 0.736, GDP per capita ($4,291), carbon dioxide emission (10.6), total footprint (2.9), biological capacity (1.7), and ecological balance (–1.20). Like Mexico, Haiti, and Jamaica have footprints larger than their biological capacity, thus, a negative ecological balance, while Argentina and Brazil have their footprints less than their biological capacities, thus, a positive ecological balance. Subsequently, analysis of table 4 shows that, given a nation's biological capacity, its levels of human development (welfare), GDP per capita (affluence), and carbon dioxide emissions (waste), they help to explain its total ecological footprint (consumption or demand) and ecological balance (reserve or deficit).

POPULATION, ECOSTRUCTURAL FACTORS, AND TOTAL FOOTPRINT PER CAPITA IN LATIN AMERICA AND CARIBBEAN COUNTRIES

Table 7.5 shows the effects of population and ecostructural factors (socioeconomic processes) on the total ecological footprints of nations. A nation's total population size and population density, as well as its urban population, literacy rate, and domestic inequality (Gini index), lead to its ecological footprint difference and variation vis-à-vis other nations. The above indicators constitute the structural causes of the ecological footprint variations among LAC countries. Table 7.5 shows that a nation with large population, high rates of urbanization and literacy, and low domestic inequality, has high ecological footprint, and vice versa. Large population level, high population density, high level of urbanization, high level of literacy, and low domestic inequality promote domestic consumption of biospheric resources.

Cities are not self contained entities and have little intrinsic biological capacity; therefore their concentration of intense economic processes and high levels of consumption both increase and stimulate their demands on bioresources. Thus, cities are parasitic and in most cases must rely upon large hinterlands for their supplies which increase their footprint impacts. Correspondingly, high density settlements appropriate carrying capacity from all over the globe, as well as from the past and the future (Vitousek et al, 1986; Wackernagel, 1991). High levels of urbanization correspond to higher levels of consumption, higher levels of energy use and carbon dioxide emissions (waste), and thus, higher energy footprints and ecological footprints. As mentioned earlier, biospheric resources are consumed at higher levels in urban areas that generally contain high literate groups with higher incomes, which allow for greater material consumption (Jorgenson, 2003; Princen et al, 2002; Sklair, 2001; Clapp, 2002).

According to table 7.5, Mexico (Mexico City, Guadalajara, and Monterrey), Brazil (Sao Paulo, Rio de Janeiro, Brasilia), Argentina (Buenos Aires), Peru (Lima), Colombia (Bogota), and Chile (Santiago) are nations with large population and large cities. Argentina (90.1), Brazil (84.2), Chile (87.6), Mexico (76.0), Uruguay (92.0), and Venezuela (93.4) have high urban population as percent of their total populations. Argentina (97.2), Brazil (88.6), Chile (95.7), Costa Rica (94.9), Mexico (91.6), Panama (91.9), Paraguay (93.5), Uruguay (96.8), and Venezuela (93) have high adult literacy rates. Nonetheless, Mexico (46.1), Uruguay (44.9), and Venezuela (48.2) with low domestic inequality, have high per capita footprints. What may be deduced from table 5 is that, high population density does not depict high footprint per capita but the opposite. For example, Cuba (99), Dominican Republic (181), El Salvador (315), Guatemala (114), Haiti (306), and Jamaica (244), except Trinidad & Tobago, all showed low footprints per capita that are less than 2.0, which is less than the region's average (2.44). Thus, the level of consumption, and not population density, spurred by large urban population, high levels of adult literacy, and low levels of inequality are the major driving forces of environmental sustainability and ecological footprints.

Tables 7.4 and 7.5 depict medium-high human development countries or higher level developing countries with medium-high levels of urbanization, literacy, and domestic inequality in 2005. These correspond to medium-high levels of income, which allow for medium levels of material consumption and footprint. The findings here parallel the theorization above, which emphasizes that the low consumption levels and the accompanying low ecological footprints and high ecological balances (reserves) and vice versa, are based on the nations' positions in the world capitalist economy and system. Most of the Latin America and Caribbean countries were once colonized by the medieval sovereign nation states of Spain, Portugal, and France, thus the spoken languages in the region. The region is where the process of underdevelopment, economic stagnation, and dependent industrialization limited their populations to consuming a greater share of unprocessed bio-productive resources, rather than industrial processed finished products (see Frank, Gunder for more discussion on dependency theory in Latin America). The high domestic income inequality or intra-inequality has a negative or declining effect on per capita consumption levels (Bornschier and Chase-Dunn, 1985; Kick, 2000; Kentor, 2001; and Jorgenson, 2003) that yield both low total and per capita ecological footprints.

REGRESSION ANALYSIS AND REGRESSION EQUATION

On one hand regression analysis allows the modeling, examining, and exploring of relationships and can help explain the factors behind the observed relationships or patterns. It is also used for predictions. Regression equation, on the other hand, utilizes the Ordinary Least Square (OLS) technique. This mathematical formula when applied to the explanatory variables is best used to predict the dependent variable that one is attempting to model. Each independent variable or explanatory variable is associated with a regression coefficient describing the strength and the sign of that variable's relationship to the dependent variable. A regression equation might look like the one given below where Y is the dependent variable, the Xs are the explanatory variables, and the Bs are regression coefficients:

$$Y = B_0 + B_1 X_1 + B_2 X_2 + \text{-----} B_n X_n + E \text{ (Random Error Term/Residuals)}$$

See Appendix B for Regression Analysis Terms

The row of "unstandardized coefficients" or "Bs" gives us the necessary coefficient values for the multiple regression models or equations.

From table 7.4, *Human Development and Affluence on Biological Capacity, Ecological Footprint and Balance in Latin America and Caribbean Countries*, the dependent variable is Total Footprint, and the independent variables are Human Development Index (HDI), GDP Per Capita, Carbon Dioxide Emissions, Bio-Capacity, and Ecological Balance. Also, from table 7.5, *Population, Ecostructural Factors, and Total Footprint Per Capita in Latin America and Caribbean Countries*, the dependent variable is Total Footprint, and independent variables are Population, Population Density, Urban Population, Literacy Rates, and Gini Index.

REGRESSION ANALYSIS (RESULTS AND DISCUSSIONS)

Table 7.6 depicts the descriptive statistics of the data in table 7.4. The measures of variability or dispersion, from table 7.6, expressed using standard deviation and variance, signify how homogeneous (in terms of low levels) Latin America and Caribbean countries are in terms of Human Development Index (HDI) and how heterogeneous they are (in terms of high levels) in terms of GDP Per Capita, Carbon dioxide Emissions, Total Footprint, Biocapacity, and Ecological Balance. The same is also true in table 7.12 where descriptive statistics of the data in table 7.5 show low levels of variability associated with Gini Index, Total Footprint Per Capita, Adult Literacy Rates, and Urban Population; and high variability with Population and Population Density. Table 7.7 shows the Pearson Correlation (2-tailed test) of the data from table 7.4. From table 7.7, there is a strong correlation or relationship (at the 0.01 level) between Human Development Index and GDP Per Capita; Total Footprint and Carbon Dioxide Emission; Bio-Capacity and Carbon Dioxide Emissions; Total Footprint and Bio-Capacity; Total Footprint and Ecological Balance; and Ecological Balance and Bio-Capacity. Likewise, table 7.13 shows a strong correlation (at 0.05 levels, 2-tailed) between Population Density and Urban Population, and at 0.01 level between Population Density and Adult Literacy Rates, as well as with Total Footprint Per Capita. There is also strong correlation between Urban Population and Adult Literacy; Total Footprint Per Capita and Urban Population; and Total Footprint Per Capita and Adult Literacy Rates.

Table 7.9 is the Regression model summary, which shows the Multiple R and R-Squared or Coefficient of Determination are very high and significant. For example, about 88 percent of the dependent variable (Total Footprint) is accounted for by the independent variable, Carbon Dioxide Emissions; about 97 percent of Total Footprint is explained by Carbon Dioxide Emissions and Bio-Capacity; and 100 percent of Total Footprint is explained by Carbon Dioxide, Bio-Capacity, and Ecological Balance. These are entered in Stepwise Regression Model in table 7.8. The Multiple R is very large (99 percent), which indicates size of the correlation between the observed outcome variables and the predicted outcome variables (based on the regression equation).

Tables 7.14 and 7.15 show the Multiple R and R-Squared between dependent variable (Total Footprint Per Capita) and the independent variables, Urban Population. Nevertheless, their relationships as well as predictability and accountability are not as high and significant as found in table 7.9. From table 7.15, the Model Summary, about 44 percent of the variations in Total Footprint Per Capita is explained or accounted for by Urban Population. Tables 7.10 and 7.16 depict the analysis of variance (ANOVA) and F-Tests. The F-tests show how strong or significant the selected independent variables in the regression models are in predicting the dependent variable. From table 7.10, the F values are high and the residual (observed – predicted) values are small. Since the F-Tests are high and significant, the Null Hypothesis may be rejected by saying that none of the independent variables (predictors) predict the National footprint and per capita footprint effects. It is worthy to note that the results in table 7.16 are not as strong as those in table 10 but still significant.

The "Coefficients[a]" table of the step-wise regression process is used to construct the regression model equation. Using the B columns of the Unstandardized Coefficients in tables 7.11 and 7.17, the regression model equations are constructed from tables 7.4 and 7.5 (see below).

(A) For table 7.4: *Human Development, GDP Per Capita, Carbon Dioxide Emissions, Bio-Capacity Factors on Total Footprint*, the significant explanatory (predictor) variables are: Carbon Dioxide Emissions, Bio-Capacity, and Ecological Balance. The Regression Equation is: Total Footprint = 0.225–0.006 Carbon Dioxide Emissions + 1.007 Bio-Capacity–1.008Ecological Balance + E (residuals). Also for table 7.5: *Population, Population Density, Urban Population, Literacy Rates, Domestic Inequality (Gini Index) on Total Footprint Per Capita*, the significant explanatory (predictor) variable is Urban Population. The Regression Equation is: Total Footprint Per Capita = –0.744 + 0.45 Urban Population + E (residuals). Therefore, given the factors A and B above, Total National Footprint in Latin America and Caribbean is predicted or explained by Carbon Dioxide Emissions, Bio-Capacity, Ecological Balance and Urban Population.

POLICY RECOMMENDATIONS

The above analyses support the theorization and assertion that the global environmental stress is caused primarily by increase in resource consumption (demand) and human population size and growth. The analyses indicate that large human population all over the world and their excessive consumption of the scarce natural resources are responsible for the national footprint of nations. Absolute population growth in Latin America and the Caribbean countries and their rapid growing world largest cities are exerting tremendous pressures on natural resources. This region in 2005 had bio-capacity approximately twice the size of its footprint (see tables 7.2 and 7.3). Its net demand on the planet is less than its available capacity. However, as a middle-high income (per capita income between $3,036 and $9,385 in 2005) and relatively high human development region of the world capitalist economy, resource consumption is high and increasing, as well as taking its toll on bio-capacity's forest resources. Wealth (growth) can be good for the environment, only if public policy and technologies encourage sound practices and the necessary investments are made in environmental sustainability (Sachs, 2007). More affluent countries among them (e.g. Argentina, Chile, Mexico, and Trinidad & Tobago) can reduce consumption and still improve their quality of life, if adequate population management policies are in place. Therefore, long-term investment will be required in many areas, including education, technology, conservation, urban and family planning, and resource certification systems. Innovative approaches to meeting human needs without environmental destruction are called for as well as a shift in belief that greater well-being necessarily entails more consumption.

The high level of domestic inequality in the region calls for countries in the region to embark on public policies geared at protecting the most vulnerable (e.g. women, the landless, and the very poor) as part of a broader strategy that encompasses not only social areas but also macroeconomic and production policies. Some aspects of indigenous cultures that perpetuate inequality should be dismantled for more progressive socio-economic equity policies.

Factors that significantly shape Latin America and the Caribbean ecological footprint or demand on bio-capacity include Carbon dioxide Emissions, Urban Population, enormous pressure on bio-capacity (especially, forest resources), and ecological balance. Policies and programs should include: (1) increase in population can be slowed and eventually reversed by supporting families in choosing to have fewer children. Women empowerment through their access to better education, economic opportunities, and health care is a proven approach to family, population management, and sustainable development (Aka, 2006); (2) the potential for reducing consumption depends on individual's economic situation, more affluent people can reduce consumption and still improve their quality of life; (3) in the case of footprint intensity, the amount of resources used in the production of goods and services, can be significantly reduced, from energy efficiency in manufacturing and in the home, through minimizing waste and increasing recycling and reuse, to fuel-efficient cars; (4) bio-productive area can be extended, for example, degraded lands can be reclaimed through careful management that include, terracing, irrigation, and to ensure that bio-productive areas do not diminish or being lost to urbanization, salinization, or desertification; (5) bio-productivity per hectare depends both on the type of ecosystem and the way it is managed; (6) Carbon sequestration can be improved by (i) by reducing the rate of deforestation in the region necessary for carbon sinking, and (ii) more efficient manufacturing and use of adequate technology. The rate of deforestation (biodiversity destruction) in LAC is the highest in the world. For example, out of the 418 million hectares of national forests lost during the past 30 years, 190 million were in Latin America and Caribbean (*New Internationalist*, May 2003);and (7) concern over deforestation and its ecological, social, and economic repercussions should call for increased efforts in conservation and improved forest management.

There should be a balance between national economic needs and increase in forest harvesting activities. Adequate forest management in the region could be achieved through clear land tenure, adequate information on proper management techniques, and sufficient institutional capacity and technical knowledge. There should be the development of non-wood products and adequate forest services, especially in adequately monitoring forest management plans and programs to foster the involvement of rural communities (the forest dependents, local communities, and indigenous groups) in forestry activities which should be encouraged, such as in Bolivia, Costa Rica, Chile, Honduras, and Peru (http://www.fao.org/docrep/W4343E. (For Latin America and the Caribbean, FAO Yearbook of Forest Products). Adequate and improved environmental and ecological education will reduce an individual's ecological impact since it has been recognized that the environmental crisis is less an environmental and technical problem than it is a behavioral and social one (Wackernagel and Rees, 1996).

CONCLUSIONS

This is an explanatory study using comparative model and descriptive statistics such as matrices, totals, ratios, and percentages, as well as regression analysis. The results and findings show that a country's total footprint, ecological balance, and carbon dioxide emissions (waste) are a function of its human development (welfare) position and affluence level in the world economy. Given the national biological capacity, it goes that the higher the total ecological footprint, thus, the lower the ecological balance. Therefore, as human development goes up, then environmental performance goes down.

Data analyses in the present study show that the higher the national populations, urban populations, and GDP per capita (affluence), the higher the *imbalaced* consumption, total footprint, carbon dioxide emissions, and the lower the ecological balance. The research findings also indicate that lack of improvements in human development index (human welfare) and lack of or declining affluence result in a lowering of a nation's ecological footprint position. This is not to say that improved human development (welfare) or affluence is not desirable but should be encouraged and pursued simultaneously with adequate public policy and sound practices for environmental sustainability. It is also found out that lower levels of literacy and urbanization correspond to lower levels of income, which allow for smaller material consumption. The higher domestic income inequality or intra-inequality in some Latin America and Caribbean countries has a negative or declining effect on per capita consumption levels that yield both low total and per capita ecological footprints.

The above analyses support the theorization and assertion that the global environmental stress is caused primarily by the increase in resource consumption (demand) and population size and growth. It is suggested also that sustainability will depend on such measures as greater emphasis on equity in international relationships, significant adjustment to prevailing terms of trade, increasing regional self-reliance, adequate population management initiatives, and policies to stimulate a massive increase in the material and energy efficiency of economic activities.

APPENDIX A: VARIABLE DEFINITIONS

Human Development Index (HDI): The HDI is a summary measure of human development (human welfare). It measures the average achievements of a country in three basic dimensions of human development (Global Footprint Network: *Africa's Ecological Footprint-2006 Factbook*, p. 89): a long and healthy life, as measured by life expectancy at birth; knowledge, as measured by the adult literacy rate (with two-thirds weight) and the combined primary, secondary and tertiary gross enrolment ratio (with one-third weight); and a decent standard of living, as measured by GDP per capita (PPP US$). Purchasing power parity (PPP) is a rate of exchange that accounts for price difference across countries, allowing international comparisons of real output and incomes. At the PPP US$ (as used in this study), PPP US$1 has the same purchasing power in the domestic economy as $1 has in the United States of America.

The high human development countries are countries with HDI of 0.800 and above, found mainly in North America, Western Europe and Australia, with Gross National Product (GNP) of $9,386 or more. The medium development countries are countries with HDI of 0.500–0.799, found in Eastern Europe, South-East Asia, Latin America and Caribbean, and North Africa, with per capita GNP of $766 and 9,385 (World Bank, 2005, p. 255). The low human development countries are countries with HDI below 0.500, found mainly in Sub-Saharan Africa except the countries of South Africa, Gabon, and Ghana, which are included in the medium human development, with per capita GNP of $765 or less (World Bank, 2005, p. 255). Data for this variable in this study is taken from UNDP's *Human Development Report 2004* and *Human Development Report 2005*.

Gross Domestic Product (GDP) per capita (PPP US$): GDP is converted to US dollars using the average official exchange rate reported by the International Monetary Fund (IMF). GDP alone does not capture the international relational characteristics as does the human development hierarchy of the world economy, which accounts for a country's relative socio-economic power and global dependence position in the modern world system. It is suggested elsewhere that GDP per capita is an inadequate measure of world-system position but a more appropriate indicator of domestic affluence or internal economic development (Burns, Kentor, and Jorgenson, 2003; Jorgenson, 2003; Dietz and Rosa, 1994). The Gross Domestic Product (GDP) per capita data for this study is taken from *Human Development Report 2007/2008*, table 1, and pp. 229–232. See also *World Development Report 2005*.

Domestic Income Inequality (Gini Index): The Gini index measures domestic income inequality of different countries, which had remained stable over a time with its impacts on other variables in the study (Bergesen and Bata, 2002; Jorgenson, 2003). Gini index measures the extent to which the distribution of income (or consumption) among individuals or households within a country deviates from a perfectly equal distribution (*Human Development Report, 2004*, p. 271)). It measures inequality over the entire distribution of income or consumption. A value of zero (0) represents perfect equality, and a value of hundred (100) represents perfect inequality. Data for domestic income inequality measured by Gini index are taken from World Bank, *World Development Report* (2005), table 2, pp. 258–259 and United Nations Development Report, *Human Development Report* 2007/2008, 15, pp. 281–284.

Urbanization Level (Urban Population as percent of Total Population): Cities are not self-contained entities and their concentration of intense economic processes and high levels of consumption both increase and stimulate their demands on resources. The cities have limited intrinsic biocapacity which undoubtedly must rely upon large hinterlands. Land consumed by urban regions is typically at least an order of magnitude greater than that contained within the usual political boundaries or the associated built-up area (The International Society for Ecological Economics and Island Press, 1994; Rees, 1992). The data are taken from *Human Development Report 2007/2008*, table 5, and pp. 243–246.

Literacy Rate: This variable refers to the percent of a nation's population over the age of fifteen (15) that can read and write in any language of their choice. Literate population generally has low domestic inequality and tends to consume more resources than their illiterate counterpart due to their higher income and urban living. High literate groups concentrate in urban areas where they consume more than their fair shares of biospheric resources. Higher levels of literacy correspond with higher incomes, which allow for greater consumption (Jorgenson, 2003). This is because literate populations are subject to increased consumerist ideologies and contextual images of good life through advertising (Princen et al, 2002), what Leslie Sklair (2001) and Jennifer Clapp (2002) labeled "cultural ideology of consumerism/consumption." The data for literacy rate is taken from the *Human Development Report 2007/2008*, table 1, pp. 229–232.

Population and Population Density: Apart from consumption, many have attributed to population as driving most of the sustainability problems (Palmer, 1998; Pimentel, 1996). Likewise, Dietz et al (2007) concluded that population size and affluence are the primary drivers of environmental impacts. Population growth and increases in consumption in many parts of the world have increased humanity's ecological burden on the planet. York et al (2003) indicated that population and affluence account for 95 percent of the variance in total footprints of countries. Others also see the ensuing human impact or footprint as a product of population, affluence (consumption), and technology (i.e. I = PAT (Ehrlich and Holdren, 1971; Holdren and Ehrlich, 1974; Hardin, 1991). Population as a variable in this study is taken from, *World Development Report 2005*, table 1, and pp. 256–259.

APPENDIX B: REGRESSION ANALYSIS TERMS

Correlation or co-relation: refers to the departure of two variables from independence or they are non-independent or redundant (Antonia D'Onofrio, 2001/2002; Richard Lowry, 1999–2008).

Collinearity: Refers to the presence of exact linear relationships within a set of variables, typically a set of explanatory (predictor) variables used in a regression-type model. It means that within the set of variables, some of the variables are (nearly) totally predicted by the other variables [(Rolf Sundberg, 2002). *Encyclopedia of Environmetrics*, edited by Abdel H. El-Shaarawi and Walter W. Piegorsch (Chichester: John Wiley & Sons, Ltd), Volume 1, pp. 365–366].

Partial Correlation Coefficients (r): When large, it means that there is no mediating variable (a third variable) between two correlated variables (Antonia D'Onofrio, 2001/2002).

Pearson's Correlation Coefficient (r): This is a measure of the strength of the association between two variables. It indicates the strength and direction of a linear relationship between two random variables. Value ranges from - to +1; –1.0 to –0.7 Strong negative association; –0.7 to –0.3 Weak negative association; –03 to +0.3 Little or no association; +0.3 to +0.7 Weak positive association; +0.7 to 1.0 Strong positive correlation (Brian Luke, "Pearson's Correlation Coefficient," Learning *From The Web.net*. Accessed Online on 5/30/2008).

Multiple "R": Indicates size of the correlation between the observed outcome variable and the predicted outcome variable (based on the regression equation).

"R^2" or Coefficient of Determination: Indicates the amount of variation (%) in the dependent scores attributable to all independent variables combined, and

ranges from 0 to 100 percent. It is a measure of model performance, summarizing how well the estimated Y values match the observed Y values.

"Adjusted R^2": The best estimate of R^2 for the population from which the sample was drawn. The Adjusted R-Squared is always a bit lower than the Multiple R-Squared value because it reflects model complexity (the number of variables) as it relates to the data.

R^2 and the *Adjusted R^2* are both statistics derived from the regression equation to quantify model performance (Scott and Pratt, 2009. *ArcUser*).

Standard Error of Estimate: Indicates the average of the observed scores around the predicted regression line.

Residuals: These are the unexplained portion of the dependent variable, represented in the regression equation as the random error term (E). The magnitude of the residuals from a regression equation is one measure of model fit. Large residuals indicate poor model fit. Residual = Observed – Predicted.

ANOVA: Decomposes the total sum of squares into regression (= explained) SS and residual (= unexplained) SS.

F-test in ANOVA represents the relative magnitude of explained to unexplained variation. If F-test is highly significant ($p = .000$), we reject the null-hypothesis that none of the independent variables predicts the effect (scores) in the population.

The "constant" represents the intercept in the equation and the coefficient in the column labeled by the independent variables.

REFERENCES

Aka, Ebenezer (2006). "Gender Equity and Sustainable Socio-Economic Growth and Development," *The International Journal of Environmental, Cultural, Economic & Social Sustainability*, Volume 1, Number 5, pp. 53–71.

Bergesen, Albert and Michelle Bata (2002). "Global and National Inequality: Are They Connected?" *Journal of World-System Research*, 8: 130–44.

Bergesen and Bartley (2000). "World-System and Ecosystem," in *A World-Systems Reader: New Perspectives on Gender, Urbanism, Culture, Indigenous Peoples, and Ecology*, edited by Thomas Hall (Lanham, MD: Rowman and Littlefield), pp. 307–22.

Bornschier, Volker and Christopher Chase-Dunn (1985). *Transnational Corporations and Underdevelopment* (New York: Praeger).

Brown, S., Jayant, S., Cannell, M., and Kauppi, P. (1996). "Mitigation of Carbon Emissions to the Atmosphere by Forest Management," *Commonwealth Forestry Review*, 75, 79–91.

Centro Internacional de Agricultura Tropical CIAT), United Nation Environment Program (UNEP), Center for International Earth Science Information Network (CIESIN), Columbia University, and the World Bank (2005). Latin America and Caribbean Population Database, Version 3. http://www.na.unep.net/datasets/datalist.php3.

City Mayors (2006). *The World's Largest Cities and Urban Areas*. http://www.citymayors.com/statistics/urban_2006.

Clapp, Jennifer (2002). "The Distancing of Waste: Over-consumption in a Global Economy," in *Confronting Consumption*, edited by T. Princen, M. Maniates, and K. Conca (Cambridge, MA: MIT Press) pp. 155–76.

Daly, H.E. (1986). *Beyond Growth: The Economics of Sustainable Development* (Boston, Massachusetts, USA: Beacon Press).

Dietz, Thomas and Eugene Rosa (1994). "Rethinking the Environmental Impacts of Population, Affluence, and Technology," *Human Ecology Review*, 1: 277–200.

Dietz, Thomas, Eugene Rosa, Yoor (2007). "Driving the Human Ecological Footprint." *Frontiers in Ecology and the Environment* (February).

D'Onofrio, Antonia (2001/2002). "Partial Correlation." Ed 710 Educational Statistics, Spring 2003. Accessed Online on 6/3/2008 at http://www2.widener.edu/.

Economic Commission for Latin America and Caribbean (CEPAC in Spanish) 2009–2010). *Economic Survey of Latin America and the Caribbean 2009–2010.*

Ehrlich and Holdren (1971). "Impacts of Population Growth," *Science*, 171, 1212–7.

Ferreira Francisco H and David de Ferranti et al (2004). "Inequality in Latin America: Breaking the History?" The World Bank. Washington, DC, USA.

Frank, Andre Gunder. See him for an in-depth discussion of Dependency Theory and Underdevelopment in Latin America.Some of his works include: *The Development of Underdevelopment*, 1966, MRP; *Capitalism and Underdevelopment in Latin America*, 1967; *Latin America: Underdevelopment or Revolution*, 1972; and *Theoretical Introduction to Five Thousand Years of World System History*, 1990, Review.

Girardet, Herbert (1996). "Giant Footprints," Accessed Online on 12/14/07, at http:// www.gdrc .org/uem/footprints/girardet.html.

Global Footprint Network: Africa's Ecological Footprint—2006 Fact book. Global Footprint Network, 1050 Warfield Avenue, Oakland, CA 94610, USA. http://www.footprintnetwork.org/Africa.

Global Footprint Network (2008). *The Ecological Footprint Atlas 2008*, October 28.

Goodwin, Paul B. (2007). Global Studies. *Latin America, Twelfth Edition* (Dubuque, Iowa: McGraw-Hill Contemporary Learning Series).

Hardin, G. (1991). "Paramount Positions in Ecological Economics," in Constanza, R. (editor), *Ecological Economics: The Science and Management of Sustainability* (New York: Columbia University Press).

Holdren and Ehrlich (1974). "Human Population and The Global Environment," *American Scientist*, 62: 282–92.

IPCC (2001). Intergovernmental Panel on Climatic Change.

Jorgenson, Andrew K. (2003). "Consumption and Environmental Degradation: A Cross-National Analysis of Ecological Footprint," *Social Problems*, Volume 50, No. 3, pp. 374–394.

Kentor, Jeffrey (2001). "The Long Term Effects of Globalization on Income Inequality, Population Growth, and Economic Development," *Social Problems*, 48: 435–55.

Kick, Edward L. (2000). "World-System Position, National Political Characteristics and Economic Development," Journal of Political and Military Sociology, (Summer), http://www.findarticles .com/p/articles/mi-qa3719.

Lowry, Richard (1999–2008). "Subchapter 3a. Partial Correlation." Accessed Online on 6/3/08 at http://faculty.vassar.edu/lowry/cha3a.html.

Luke, Brian T. "Pearson's Correlation Coefficient." Learning From The Web.net. Accessed Online on 5/30/2008.

New Internationalist (2003). Latin America and the Caribbean the Facts (May).

Palmer, A.R. (1998). "Evaluating Ecological Footprints," *Electronic Green Journal*. Special Issue 9 (December).

Pimentel, David (1996). "Impact of Population Growth on Food Supplies and Environment." American Association for the Advancement of Science (AAAS) (February, 9). See also GIGA DEATH, Accessed Online on 12/18/07, at http://dieoff.org/page13htm.

Princen, Thomas (2002). "Consumption and Its Externalities: Where Economy Meets Ecology," in *Confronting Consumption*, edited by T. Princen, M. Maniates, and K. Conca (Cambridge, MA: MIT Press) pp. 23–42.

Redefining Progress (2005). Footprints of Nations. 1904 Franklin Street, Oakland, California, 94612. See http://www.RedefiningProgress.org.

Rees, William E. (1992). "Ecological Footprint and Appropriated Carrying Capacity: What Urban Economics Leaves Out," Environment and Urbanization, Vol. 4, No. 2, (October). Accessed Online on 12/14/07, at http://eau.sagepub.com.

Sachs, Jeffrey D. (2007). "Can Extreme Poverty Be Eliminated"? *Developing World, 07/08* (Dubuque, IA: McGraw Hill0 pp. 10–14.

Scott, Lauren and Monica Pratt (2009). "An Introduction To Using Regression Analysis With Spatial Data." *ArcUser. The Magazine for ESRI Software Users* (Spring), pp. 40–43. Lauren Scott and Monica Pratt are ESRI Geo-processing Spatial Statistics Product Engineer and ArcUser Editor respectively.

Sklair, Leslie (2001). *The Transnational Capitalist Class* (Oxford, UK: Blackwell Press).

Solow, R. M. (1974). "The Economics of Resources or the Resources of Economics," *American Economics Review*, Vol. 64, pp. 1–14.

Sundberg, Rolf (2002). *Encyclopedia of Environmetrics*, edited by Abdel H. El-Shaarawi and Walter W. Piegorsch (Chichester: John Wiley & Sons, Ltd), Volume 1, pp. 365–366.

Suplee, D. (1998). "Unlocking the Climate Puzzle." *National Geographic*, 193 (5), 38–70.The international Society for Ecological Economics and Island Press (1994). "Investing in Natural Capital: The Ecological Approach to Sustainability." Accessed Online on 12/18/07, at http://www.dieoff.org/page13.htm.

United Nations (2009). *World Population Prospects: The 2008 Revision. The Highlights*. Population Division of the Department of Economic and Social Affairs of the United Nations Secretariat. United Nations, New York. See table 1.1.

United Nations Development Program (2004). *Human Development Report 2004 (HDR). Cultural Liberty in Today's Diverse World* (New York, N.Y: UNDP).

United Nations Development Program (2005). *Human Development Report (HDR) 2005. International Cooperation at a Crossroads: Aid, Trade and Security in an Unequal World* (New York, N.Y: UNDP).

United Nations Development Program (2007/2008). *Human Development Report 2007/2008. Fighting Climate Change: Human Solidarity in a Divided World.*

United Nations Development Program (2010). *Human Development Report 2010. The Real Wealth of Nations: Pathway to Human Development,* p. 187.

United Nations Educational, Scientific and Cultural Organization (UNESCO) (2006–2008). *International Hydrological Program (IHP) in Latin America and the Caribbean Report.*

United Nations, *World Population Prospects: The 2004 Revision*, March 2005.

Vitousek, P., P. Ehrlich, A. Ehrlich and P. Matson (1986). "Human Appropriation of the Products of Photosynthesis," *Bioscience*, Vol. 36, pp. 368–374.

Wackernagel, M. (1991). "Using 'Appropriated Carrying Capacity' as an Indicator: Measuring the Sustainability of a Community." Report for the UBC Task Force on Healthy and Sustainable Communities. UBC School of Community and Regional Planning, Vancouver, Canada.

Wackernagel and Rees (1996). *Our Ecological Footprint* (New Society Publisher).

World Bank (2005). *World Development Report 2005. A Better Investment Climate for Everyone* (New York, N.Y: A Co-publication of the World Bank and Oxford University Press).

World Bank (2010). *World Development Report 2010. Development and Climate Change*. Selected World Development Indicators, table 1, p. 379.

World Resources Institute (WRI) (1992). *World Resources, 1992–1993* (New York: Oxford University Press).

York, Richard, Eugene A. Rosa, and Thomas Dietz (2003). "Footprints on the Earth: The Environmental Consequences of Modernity." *American Sociological Review*, 68: 279–300.

Table 7.1. Regional Population, Gross Domestic Product (GDP), Total Footprint, Biological Capacity, and Ecological Balance by World Income Regions in 2005

World Income Table 1: Regions	*Regional Population (millions) (2005)*	*GDP PPP US$ Billions (2005)**	*Total Footprint Million gha (2005)*	*Biological Capacity Global gha (2005)*	*Ecological Balance Million gha (2005)*
Low Income	2370.6	5879.1	2377.2	2089.7	–287.5
% World	36.6	9.7	13.6	15.6	7.0
Middle Income	3097.9	22586.3	6787.0	6684.8	–102.2
% World	47.8	37.3	38.9	50.0	2.5
High Income	971.8	32680.7	6196.0	3561.5	–2634.5
% World	15.0	53.9	35.5	26.7	64.5
World Total	6475.6	60597.3	17443.6	13361.0	–4082.7

*Human Development Report (HDR) 2007/2008. *Fighting Climate Change: Human Solidarity in a Divided World*, table 14, p. 280.

Note: Percentages may not add up to 100 percent due to (i) rounding up; (ii) some countries that were not up to 1 million people in 2005 were not included; and (iii) some countries that had been at war by 2005 were not included.

Source: Global Footprint Network (2008). *The Ecological Footprint Atlas 2008*, Appendix F, table 2, p. 46.

Table 7.2. Total Population, Ecological Footprint, Biological Capacity, and Ecological Balance by World Regions in 2005

Region	*Total Population (million) 2005*	*Total Ecological Footprint (million gha) 2005*	*Total Biological Capacity (million gha) 2005*	*Total Ecological Balance (million gha) 2005*
Latin America and Caribbean	**553.0**	**1350.8**	**2655.7**	**1304.9**
% World	**8.5**	**7.7**	**19.9**	
Africa	902.0	1237.5	1627.1	389.56
% World	13.9	7.1	12.2	
Middle East and Central Asia	365.7	846.8	466.9	–379.9
% World	5.6	4.9	3.5	
Asia-Pacific	3562.1	5758.6	2923.3	–2835.3
% World	55.0	33.0	21.9	
North America	330.5	3037.8	2143.3	–894.5
% World	5.1	17.4	16.0	
Europe EU	487.3	2291.8	1128.2	–1163.6
% World	7.5	13.1	8.4	
Europe Non-EU	239.6	842.4	1391.6	549.2
% World	3.7	4.8	10.4	
World	**6475.6**	**17443.6**	**13361.0**	**–4082.7**

EU: European Union

Percentages may not add up to 100 percent due to (i) rounding up; (ii) some countries that were not up to 1 million people in 2005 were not included ; and (iii) some countries that had been at war by 2005 were not included.

Source: Global Footprint Network (2008). *The Ecological Footprint Atlas 2008*, Appendix F, table 2, pp. 46–50.

Table 7.3. Per Capita Ecological Footprint, Biological Capacity, and Ecological Balance by World Regions in 2005

Region	*Ecological Footprint (gha per person) 2005*	*Biological Capacity (gha per person) 2005*	*Ecological Balance (gha per person) 2005*
Latin America and Caribbean	**2.44**	**4.80**	**2.36**
Africa	1.37	1.80	0.43
Middle East and Central Asia	2.32	1.28	–1.04
Asia-Pacific	1.62	0.82	–0.80
North America	9.19	6.49	–2.71
Europe EU	4.69	2.32	–2.38
Europe Non-EU	3.52	5.81	2.29
World	**2.69**	**2.06**	**–0.63**

EU: European Union

Percentages may not add up to 100 percent due to: (i) rounding up; (ii) some countries not up to 1 million people in 2005 were not included; and (iii) some countries that had been at war by 2005 were not included.

Source: Global Footprint Network (2008). *The Ecological Footprint Atlas 2008*. Appendix F, table 1.

Table 7.4. Human Development and Affluence on Biological Capacity, Ecological Footprint and Balance in Latin America and Caribbean Countries

Countries	*Human Development Index (HDI) Value 2005*[1]	*GDP Per Capita PPP US($) 2005*[2]	*Carbon Dioxide Emissions Mil Tons 2004*[3]	*Total Footprint Million gha 2005*[4]	*Bio-Capacity Million gha 2005*[4]	*Ecological Balance Million gha 2005*[4]
Argentina	0.869	14280	141.7	95.2	315.1	220.0
Bolivia	0.695	2819	7.0	19.5	144.2	124.8
Brazil	0.800	8402	331.6	439.2	1353.8	914.6
Chile	0.867	12027	62.4	48.9	67.4	18.5
Colombia	0.791	7304	53.6	81.6	173.0	96.3
Costa Rica	0.846	10180	6.4	9.8	8.0	–1.9
Cuba	0.838	6000	25.8	19.9	11.8	–8.0
Dominican Republic	0.779	8217	19.6	13.2	7.1	–6.1
Ecuador	0.772	4341	29.3	29.1	28.3	–0.8
El Salvador	0.735	5255	6.2	11.1	5.0	–6.2
Guatemala	0.689	4568	12.2	19.0	16.2	–2.8
Haiti	0.529	1663	1.8	4.6	2.3	–2.3
Honduras	0.700	3430	7.6	12.8	13.5	0.68
Jamaica	0.736	4291	10.6	2.9	1.7	–1.20
Mexico	0.829	10751	437.8	361.9	178.4	–183.5
Nicaragua	0.710	3674	4.0	11.0	18.0	6.8
Panama	0.812	7605	5.7	10.3	11.3	1.0
Paraguay	0.755	4642	4.2	19.8	59.8	40.0
Peru	0.773	6039	31.5	43.8	112.5	68.6
Trinidad and Tobago	0.814	14603	32.5	2.8	2.7	–0.1
Uruguay	0.852	9962	5.5	19.0	36.4	17.4
Venezuela	0.792	6632	172.5	75.2	84.4	9.2
Latin America and Caribbean	**0.803**	**8417**	**1422.6**	**1350.7**	**2665.7**	**1304.9**

1. Human Development Report 2007/2008. *Fighting Climate Change: Human Solidarity in a Divided World.* Table 1: Human Development Index, pp. 229–232.
2. Human Development Report 2007/2008. *Fighting Climate Change: Human Solidarity in a Divided World.* Table 1: Human Development Index, pp. 229–232.
3. Human Development Report 2007/2008. *Fighting Climate Change: Human Solidarity in a Divided World.* Table 24: Carbon Dioxide Emissions and Stocks, pp. 310–313.
4. Global Footprint Network (2008). *The Ecological Footprint Atlas 2008, October 28*, tables 2 and 4, pp. 46–50 and pp. 55–58, respectively.

Table 7.5. Population, Ecostructural Factors, and Total Footprint Per Capita in Latin America and Caribbean Countries

Countries	*Population (Millions) 2005*[1]	*Population Density People per km^2 2003*[2]	*Urban Population (% of Total) 2005*[3]	*Adult Literacy Rates (%) by 2005*[4]	*Domestic Inequality (Gini Index) by 2005*[5]	*Total Footprint gha Per Capita 2005*[6]
Argentina	38.7	14	90.1	97.2	51.3	2.5
Bolivia	9.2	8	64.2	86.7	60.1	2.1
Brazil	186.4	21	84.2	88.6	57.0	2.4
Chile	16.3	21	87.6	95.7	54.9	3.0
Colombia	45.6	43	72.7	92.8	58.6	1.8
Costa Rica	4.3	78	61.7	94.9	49.8	2.3
Cuba	11.3	99[a]	75.5	99.8	—	1.8
Dominica Republic	8.9	181	66.8	87.0	51.6	1.5
Ecuador	13.2	47	62.8	91.0	53.6	2.2
El Salvador	6.9	315	59.8	80.6	52.4	1.6
Guatemala	12.6	114	47.2	69.1	55.1	1.5
Haiti	8.5	306	38.8	54.8	59.2	0.5
Honduras	7.2	62	46.5	80.0	53.8	1.8
Jamaica	2.7	244	53.1	79.9	45.5	1.1
Mexico	107.0	54	76.0	91.6	46.1	3.4
Nicaragua	5.5	45	59.0	76.7	43.1	2.1
Panama	3.2	40	70.8	91.9	56.1	3.2
Paraguay	6.16	14	58.5	93.5	58.4	3.2
Peru	28.0	21	72.6	87.9	52.0	1.6
Trinidad and Tobago	1.3	214[a]	72.0[a]	98.4	38.9	2.1
Uruguay	3.3	19	92.0	96.8	44.9	5.5
Venezuela	26.8	29	93.4	93.0	48.2	2.8
Latin America and Caribbean	**553.2**	**27**	**77.3**	**90.3**	—	**2.44**

a. Godwin, Paul B. (2007). *Global Studies. Latin America. Twelfth Edition* (Dubuque, Iowa: McGraw-Hill/ Contemporary Learning Series).

1. Human Development Report 2007/2008. *Fighting Climate Change: Human Solidarity in a Divided World*. Table 5: Demographic Trends, pp. 243–246. See also Global Footprint Network (2008). *The Ecological Footprint Atlas 2008, October 28*, pp. 48–49.
2. World Development Report 2005. *A Better Investment Climate for Everyone*. The World Bank. Table 1: Key Indicators of Development, pp. 256–259.
3. Human Development Report 2007/2008. *Fighting Climate Change: Human Solidarity in a Divided World*. Table 5: Demographic Trends, pp. 243–246. Note: Because data are based on national definitions of what constitutes a city or metropolitan area, cross country comparisons should be made with caution.
4. Human Development Report 2007/2008. *Fighting Climate Change: Human Solidarity in a Divided World*. Table 1: Human Development Index, pp. 229–232.
5. Human Development Report 2007/2008. *Fighting Climate Change: Human Solidarity in a Divided World*. Table 15: Inequality in Income or Expenditure, pp. 281–284. Note: A value of zero (0) represents absolute equality, and a value of hundred (100) absolute inequality.
6. Global Footprint Network (2008). *The Ecological Footprint Atlas 2008, October 28*, table 3, pp. 51–54.

Table 7.6. Descriptive Statistics

	N	*Minimum*	*Maximum*	*Mean*	*Std. Deviation*
Human Development Index	22	.529	.869	.77195	.077682
GDP Per Capita	22	1663	14603	7122.05	3607.976
Carbon Dioxide Emissions	22	1.8	437.8	64.068	113.8982
Total Footprint	22	2.8	439.2	61.391	113.4311
Bio-Capacity	22	1.7	1353.8	120.495	286.7224
Ecological Balance	22	−183.5	914.6	59.317	203.9197
Valid N (listwise)	22				

Table 7.7. Correlations

	Human Development Index	*GDP Per Capita*	*Carbon Dioxide Emissions*	*Total Footprint*	*Bio-Capacity*	*Ecological Balance*
Human Development Index Pearson Correlation	1	.795**	.319	.257	.168	.094
Sig. (2-tailed)		.000	.147	.249	.454	.676
N	22	22	22	22	22	22
GDP Per Capita Pearson Correlation	795**	1	.377	.282	.182	.099
Sig. (2-tailed)	.000		.084	.203	.417	.661
N	22	22	22	22	22	22
Carbon Dioxide Emissions Pearson Correlation	.319	.377	1	.942**	.637**	.372
Sig. (2-tailed)	.147	.084		.000	.001	.089
N	22	22	22	22	22	22
Total Footprint Pearson Correlation	.257	.282	.942**	1	.823**	.601**
Sig. (2-tailed)	.249	.203	.000		.000	.003
N	22	22	22	22	22	22
Bio-Capacity Pearson Correlation	.168	.182	.637**	.823**	1	.949**
Sig. (2-tailed)	.454	.417	.001	.000		.000
N	22	22	22	22	22	22
Ecological Balance Pearson Correlation	.094	.099	.372	.601**	.949**	1
Sig. (2-tailed)	.676	.661	.089	.003	.000	
N	22	22	22	22	22	22

**Correlation is significant at the 0.01 level (2-tailed).

Table 7.8. Variables Entered/Removed[a]

Model	*Variables Entered*	*Variables Removed*	*Method*
1	Carbon Dioxide Emissions		Stepwise (Criteria: Probability-of-F-to-remove <= .050, Probability-of-F-to-remove >= .100).
2	Bio-Capacity		Stepwise (Criteria: Probability-of-F-to-remove <= .050, Probability-of-F-to-remove >= .100).
3	Ecological Balance		Stepwise (Criteria: Probability-of-F-to-remove <= .050, Probability-of-F-to-remove >= .100).
4		Carbon Dioxide Emissions	Stepwise (Criteria: Probability-of-F-to-remove <= .050, Probability-of-F-to-remove >= .100).

a. Dependent Variable: Total Footprint

Table 7.9. Model Summary

Model	*R*	*R Square*	*Adjusted R Square*	*Std. Error of the Estimate*
1	.942[a]	.887	.881	39.1513
2	.985[b]	.970	.967	20.5788
3	1.000[c]	1.000	1.000	1.1169
4	1.000[d]	1.000	1.000	1.1016

a. Predictors: (Constant), Carbon Dioxide Emissions
b. Predictors: (Constant), Carbon Dioxide Emissions, Bio-Capacity
c. Predictors: (Constant), Carbon Dioxide Emissions, Bio-Capacity, Ecological Balance
d. Predictors: (Constant), Bio-Capacity, Ecological Balance

Table 7.10. ANOVA[e]

Model		*Sum of Squares*	*df*	*Mean Square*	*F*	*Sig.*
1	Regression	239542.387	1	239542.387	156.275	.000[a]
	Residual	30656.531	20	1532.827		
	Total	270198.918	21			
2	Regression	262152.692	2	131076.346	309.518	.000[b]
	Residual	8046.226	19	423.486		
	Total	270198.918	21			
3	Regression	270176.463	3	90058.821	72190.777	.000[c]
	Residual	22.455	18	1.248		
	Total	270198.918	21			
4	Regression	270175.861	2	135087.931	111319.674	.000[d]
	Residual	23.057	19	1.214		
	Total	270198.918	21			

a. Predictors: (Constant), Carbon Dioxide Emissions
b. Predictors: (Constant), Carbon Dioxide Emissions, Bio-Capacity
c. Predictors: (Constant), Carbon Dioxide Emissions, Bio-Capacity, Ecological Balance
d. Predictors: (Constant), Bio-Capacity, Ecological Balance
e. Dependent Variable: Total Footprint

Table 7.11. Coefficients[a]

Model	Unstandardized Coefficients		Standardized Coefficients		
	B	*Std. Error*	*Beta*	*t*	*Sig.*
1 (Constant)	1.314	9.632		.136	.893
Carbon Dioxide Emissions	.938	.075	.942	12.501	.000
2 (Constant)	–1.325	5.075		–.261	.797
Carbon Dioxide Emissions	.700	.051	.703	13.682	.000
Bio-Capacity	.148	.020	.375	7.307	.000
3 (Constant)	.225	.276		.815	.426
Carbon Dioxide Emissions	–.006	.009	–.006	–.694	.496
Bio-Capacity	1.007	.011	2.546	93.564	.000
Ecological Balance	–1.008	.013	–1.813	–80.199	.000
4 (Constant)	.193	.268		.718	.481
Bio-Capacity	1.000	.003	2.528	377.208	.000
Ecological Balance	–1.000	.004	–1.798	–268.217	.000

a. Dependent Variable: Total Footprint

Table 7.12. Descriptive Statistics

	N	*Minimum*	*Maximum*	*Mean*	*Std. Deviation*
Population	22	1.3	186.4	25.139	42.8444
Population Density	22	8	315	90.41	96.962
Urban Population	22	38.8	93.4	68.423	15.1661
Adult Literacy Rates	22	54.8	99.8	87.632	10.6981
Gini Index	21	38.9	60.1	51.933	5.7701
Total Footprint Per Capita	22	.5	5.5	2.273	1.0157
Valid N (listwise)	21				

Table 7.13. Correlations

	Population	*Population Density*	*Urban Population*	*Adult Literacy Rates*	*Gini Index*	*Total Footprint Per Capita*
Population Pearson Correlation	1	–.278	.367	.113	.139	.166
Sig. (2-tailed)		.211	.093	.616	.549	.606
N	22	22	22	22	21	22
Population Density Pearson Correlation	–.278	1	–.535*	–.540**	–.161	–.589
Sig. (2-tailed)	.211		.010	.009	.486	.004
N	22	22	22	22	21	22
Urban Population Pearson Correlation	.367	–.535*	1	.766**	–.239	.648**
Sig. (2-tailed)	.093	.010		.000	.296	.001
N	22	22	22	22	21	22
Adult Literacy Rates Pearson Correlation	.113	–.540**	.766**	1	–.262	.599**
Sig. (2-tailed)	.616	.009	.000		.251	.003
N	22	22	22	22	21	22
Gini Index Pearson Correlation	.139	–.161	–.239	–.262	1	–.238
Sig. (2-tailed)	.549	.486	.296	.251		.298
N	21	21	21	21	21	21
Total Footprint Per Capita Pearson Correlation	.116	–.589**	.648**	.599**	–.238	1
Sig. (2-tailed)	.606	.004	.001	.003	.298	
N	22	22	22	22	21	22

*Correlation is significant at the 0.05 level (2-tailed).
**Correlation is significant at the 0.01 level (2-tailed).

Table 7.14. Variables Entered/Removed[a]

Model	*Variables Entered*	*Variables Removed*	*Method*
1	Urban Population	.	Stepwise (Criteria: Probability-of-F-to-enter <= .050, Probability-of-F-to-remove >= .100).

a. Dependent Variable: Total Footprint Per Capita

Table 7.15. Model Summary

Model	*R*	*R Square*	*Adjusted R Square*	*Std. Error of the Estimate*
1	.667[a]	.444	.415	.7917

a. Predictors: (Constant), Urban Population

Table 7.16. ANOVA[b]

Model	*Sum of Squares*	*df*	*Mean Square*	*F*	*Sig.*
1 Regression	9.521	1	9.521	15.190	.001[a]
Residual	11.909	19	.627		
Total	21.430	20			

a. Predictors: (Constant), Urban Population
b. Dependent Variable: Total Footprint Per Capita

Table 7.17. Coefficients[a]

	Unstandardized Coefficients		*Standardized Coefficients*		
Model	*B*	*Std. Error*	*Beta*	*t*	*Sig.*
1 (Constant)	–.744	.799		–.932	.363
Urban Population	.045	.011	.667	3.897	.001

a. Dependent Variable: Total Footprint Per Capita

Figure 7.1. Map of Latin America and the Caribbean

Chapter Eight

Socio-Economic Determinants of National Ecological Footprints in Middle East and Central Asia Countries

A Multiple Regression Analysis of the Factorial Impacts

INTRODUCTION

In this study, Middle East and Central Asia or "Greater Middle East," is the crescent (concave) of territory stretching from Turkey in the northwest through the crescent trough of Saudi Arabia, Yemen, and Oman, traversing the newly independent states of Central Asia, and with a spur at the steppes of Kazakhstan in the northeast (see Map of the Study Area at the end of this chapter). The relative location of the area in the middle of Eastern Hemisphere makes it a frontier between the civilizations of Europe, eastern Asia, and Africa. Thus, the region is viewed as a "Land Island" (Spencer 2009, p. 5). Middle East and Central Asia cities were designed around access to water, God (or gods), and trade (Brunn, Hays-Mitchell, and Zeigler, 2012, p. 282). The region is the birthplace of the three monotheistic faiths of Judaism, Christianity, and Islam, with enduring stamp on the region, but widely associated with Islam. Most of the Arab countries are still connected to their former colonies. In British and U.S diplomatic parlance or phraseology, this area was formally divided into "Near," "Middle," and "Far" East.

The traditional role of this area, particularly the Middle East section, has been the exploitation of mineral resources vital to the global economy (oil); and also characterized by the rivalries of nations that regard the area as strategically important for their national interests (Spencer, 2009, p. 4). Middle East and Central Asia countries are one of the mix income economy regions of the world: Low Income Countries (LIC = $765 or less GNI per capita), Lower Middle Income Countries (LMC = $766–$3,035), Upper Middle Income Countries (UMC =

$3,036–$9,385), and High Income Countries (HIC = $9,386 or more) (World Development Report 2005, p. 255). Apart from Afghanistan, Kyrgyzstan, Tajikistan, Uzbekistan, and Yemen that were low-income countries in 2005; and Israel, Kuwait, and United Arab Emirates that were high-income countries, other countries of the region were either low-middle or upper-middle income (table 8.1). The human development index (HDI) value of the region in 2005 was 0.699, more than those of Yemen, Tajikistan, and Kyrgyzstan, reflecting their poor levels of development (table 8.1).

The gender-related development index (GDI) in the region follows the same national development hierarchy of the region. For example, among world countries, Yemen (0.472), Tajikistan (0.669), and Kyrgyzstan (0.692) ranked among the lowest (table 8.1). Gender-related development (GDI) is one of the indicators of human development developed by the United Nations, which addresses gender gaps in life expectancy, education, and income. It highlights inequalities in the areas of long and healthy life, knowledge, and decent standard of living between men and women (Human development Reports). Likewise, gender inequality index (GII) values of the region follows the same pattern of development, as Yemen (0.853), Afghanistan (0.797), Saudi Arabia (0.760), and Iraq (0.751) ranked highest among world countries. The GII shows the loss in human development due to inequality between female and male achievements in reproductive health, empowerment, and labor market. The index is designed to measure those aspects of human development that are eroded by gender inequality (see table 8.1).

Government or Public Social Spending as a percent of gross domestic product (GDP) is generally low, except in high income country of Israel and medium income countries of Turkey and Jordan (see table 8.2). Public expenditure exerts an effect on economic growth rate through positive externality in the productivity of the capital stocks. Investment in education and health is also investment in people and in the future. The level of public spending in different countries of the region would likely have effects on their economic growths and consumptions of natural capital.

As mix-income economies, Middle East and Central Asia countries belong to: low-income group that had in 2005 36.6 percent of the world population, 9.7 percent of the gross domestic product (GDP), 13.6 percent of the world total footprint, 15.6 percent of the world biological capacity, and 7.0 percent of the world ecological overshoot; middle-income group with 47.8 percent of the world population, 37.3 percent of the GDP, 38.9 percent of the total footprint, 50.0 percent of the biological capacity, and 2.5 percent of overshoot; and high-income group with 15.0 percent of the world population, 53.9 percent of the GDP, 35.5 percent of total footprint, 26.7 percent of biological capacity, and 64.5 percent of the overshoot (see table 8.3).

National and regional ecological footprints reflect the consumption patterns of nations and regions respectively, as well as the degree of consumption of life-support services (natural capital). The ecological footprint analysis (EFA) addresses the issues of human biological metabolism and humanity's industrial metabolism (Rees, 2012). Since Middle East and Central Asia countries belong to mixed-income economies where human development is relatively low, medium, and high with mixed ecological capacities and overshoots (table 8.3), the question is, what factors are responsible for or are driving the total and per capita footprints in those countries and the region?

FOCUS OF THE STUDY

Countries in Middle East and Central Asia (see also the map at the end of this chapter)

Given below are the 21 countries of Middle East and Central Asia for the study based on the availability of data and adequate population size of one million people or more in 2005, as represented in Global Footprint Network. *The Ecological Footprint Atlas 2008*. Large population is synonymous with increased consumption of natural capital, which increases anthropogenic (humanity's) ecological burden on the planet. The countries are: Afghanistan, Armenia, Azerbaijan, Georgia, Iran, Iraq, Israel, Jordan, Kazakhstan, Kuwait, Kyrgyzstan, Lebanon, Oman, Saudi Arabia, Syria, Tajikistan, Turkey, Turkmenistan, United Arab Emirates, Uzbekistan, and Yemen.

PURPOSE OF STUDY

The study analyzes and explains the national and regional variations in total footprints and per capita footprints associated with human development indexes and ecostructural factors [socioeconomic processes within nations such as urbanization, literacy rate, gross domestic product per capita (affluence), and domestic inequality (Gini index)] in Middle East and Central Asia. Progress towards sustainable development can be assessed using human development index (HDI) as an indicator of well-being, and the ecological footprint as a measure of demand on the biosphere. Table 8.4 exposes the inequitable distribution of Middle East and Central Asia ecological footprint, biological capacity, and ecological balance as compared to those of other regions of the world. From the table, the region had only 5.6 percent of world population in 2005 but with 3.5 percent of world's total biological capacity. The region also has 4.9 percent of world's total ecological footprint, as well as the world's lowest total ecological overshoot (–379.9 million global hectares), when compared to those of Asia-Pacific (–2835.3), European Union (–1163.6) and North America (–894.5) regions. Table 8.5 also shows that Middle East and Central Asia had in 2005 per capita ecological footprint of 2.32 (less than the world's 2.69); biological capacity of 1.28 per capita (about half of the world's 2.06); and per capita ecological overshoot of –1.04 global hectares, almost doubling that of the world (–0.83).

Using the tools of comparative model and regression analysis, the paper seeks to establish the key factors, among others, that highly influence and drive the footprints in different Middle East and Central Asia countries. The question is what type of secure environments should meet the needs of both people and natural environment in Middle East and Central Asia region? This constitutes an enormous challenge to policy makers, planners, and development experts. Thus, the study proffers some policies, recommendations, and solutions on how to reduce the footprints of Middle East and Central Asia nations, which also will help protect the natural environment and promote a more equitable and sustainable society.

THEORETICAL FRAMEWORK FOR THE STUDY

Within the biosphere, everything, including humans is interconnected, and sustainable development requires a good knowledge of human ecology. The human life-support functions of the ecosphere are maintained by nature's biocapacity which runs the risks of being depleted, the prevailing technology notwithstanding. Regardless of the humanity's mastery over the natural environment, it still remains a creature of the ecosphere and always in a state of *obligate dependency* on numerous biological goods and services (Rees, 1992, p. 123). Despite the above thesis, the prevailing economic mythology assumes a world in which carrying capacity is indefinitely expandable (Daly, 1986; Solow, 1974). The human species has continued to deplete, draw-down, and confiscate nature's biocapacity with reckless abandon. York et al (2003), indicated that population and affluence account for 95 percent of the variance in total footprints of countries. Total human impact on the ecosphere is given as: Population × Per capita impact (Ehrlich and Holdren, 1971; Holdren and Ehrlich, 1974); Hardin (1991). In other words, population size and affluence are the primary drivers of environmental impacts (Dietz et al, 1994; and 2007). The footprints of nations provide compelling evidence of the impacts of consumption, thus the need for humans to change their lifestyles. According to Palmer (1998), there are in order of decreasing magnitude, three categories of consumption that contribute enormously to our ecological footprints: wood products (53 percent), food (45 percent), and degraded land (2 percent). Degraded land includes land taken out of ecological availability by buildings, roads, parking lots, etc. Palmer also indicated that about 10 percent or more of earth's forests and other ecological land should be preserved in more or less pristine condition to maintain a minimum base for global biodiversity.

The environmental impacts of urban areas should be considered, because a rapidly growing proportion of world's population lives in cities; and more than one million people are added to the world's cities each week and majority of them are in developing countries of Africa, Latin America, and Southeast Asia (Wackernagel and Rees, 1996). In fact, today, more people live in cities than ever before. The reality is that the population of all urban regions and many whole nations already exceeded their territorial carrying capacities and depend on trade for survival. Of course, such regions are running an unaccounted ecological deficit; their populations are appropriating carrying capacity from elsewhere (Pimentel, 1996; Wackernagel and Rees, 1996; Girardet, 1996, accessed Online on 12/14/07; Rees, 1992; The International Society for Ecological Economics and Island Press, 1994; Vitouset et al, 1986; Wackernagel, 1991; WRI, 1992, p. 374). Undoubtedly, the rapid urbanization and increasing ecological uncertainty have implications for world development and sustainability. Cities are densely populated areas that have high ecological footprints, which leads to the perception of these populations as "parasitic," since these communities have little intrinsic biocapacity; and instead must rely upon large hinterlands. Land consumed by urban regions is typically at least an order of magnitude greater than that contained within the usual political boundaries or the associated built-up areas (The International Society for Ecological Economics and Island Press, 1994; Rees, 1992; 2012).

According to Rees (1992), every city is an entropic black hole drawing on the concentrated material resources and low-entropy production of a vast and scattered hinterlands many times the size of the city itself. In the same vain, Vitouset et al (1986) expressed that high density settlements "appropriate" carrying capacity from all over the globe, as well as from the past and the future (see also Wackernagel, 1991). In modern cities, resources flow through the urban system without much concern either about their origin, or about the destination of their wastes, thus, inputs and outputs are considered to be unrelated. The cities' key activities such as transport, electricity supply, heating, manufacturing and the provision of services depend on a ready supply of fossil fuels, usually from far-flung hinterlands than within their usual political boundaries or their associated built-up areas. Cities are not self-contained entities, and their concentration of intense economic processes and high levels of consumption both increase and stimulate their demands on resources. Cities occupy only 2 percent of the world's land surface, but use some 75 percent of the world resources, and release a similar percentage of waste (Girardet, 1996).

Like urbanization, energy footprint, from energy use and carbon dioxide emissions, is not subject to area constraints. Energy footprint is the area of forest that would be needed to sequester the excess carbon (as carbon dioxide) that is being added to the atmosphere by the burning of fossil fuels to generate energy for travel, heating, lighting, manufacturing, etc. Actually, the demand for energy defines modern cities more than any other single factor. The natural global systems of forests and oceans for carbon sequestration are not handling the human contributions fast enough, thus the Kyoto Conference of early 1998. According to Suplee (1998), only half of the carbon we generate burning fossil fuels can be absorbed in the oceans and existing terrestrial sinks. The oceans absorb about 35 percent of the carbon in carbon dioxide (Suplee, 1998), equivalent of 1.8 giga tons of carbon every year (IPCC, 2001), while the global forests under optimum management of existing forests could absorb about 15 percent of the carbon in the CO_2 produced from the burning of fossil fuels world-wide (Brown et al, 1996). The energy footprint is caused by the unsequestrated 50 percent in the atmosphere with the potentially troubling ecological consequences, such as rapid global warming and other environmental stresses. Carbon dioxide in the atmosphere will continue to increase unless humanity finds alternative energy sources of sufficient magnitude. It is in the humanity's best interest to get off its petroleum addiction and develop sustainable consumption habits.

High literate groups concentrate in urban areas where they consume more than their fair shares of biospheric resources. Literate population generally has low domestic inequality and tends to consume more resources than their illiterate counterpart due to their higher income and urban living. Higher levels of literacy correspond with higher incomes, which allow for greater consumption (Jorgenson, 2003). This is because literate populations are subject to increased consumerist ideologies and contextual images of good life through advertising (Princen et al, 2002); what Leslie Sklair (2001) and Jennifer Clapp (2002) labeled "cultural ideology of consumerism/consumption."

METHODOLOGY FOR THE STUDY

Unit of Analysis

The unit of analysis is "country" (individual Middle East and Central Asia countries), which should be used to reach defensible conclusions about the region.

Sample

To test for the national variations in total footprints and per capita footprints associated with human development index, Comparative Model Analysis and Stepwise Regression Analysis were used as the tools of analysis. Using Comparative Model, a sample of 21 countries out of many countries of the region was analyzed in tables 6–7 (see Global Footprint Network. *The Ecological Footprint Atlas 2008*; Spencer, William (2009), *Global Studies. The Middle East. Twelfth*

Edition). Step-wise Regression Analysis was also conducted using the sample. Each table represents Middle East and Central Asia countries as also represented by Global Footprint Network, *The Ecological Footprint Atlas 2008*, tables 2,3, and 4. These tables do not include countries that were less than one million people in 2005. The larger population of at least one million is chosen for the study because; population is a variable which affects the consumption rates and levels, therefore the ecological footprint.

Mode of Analysis

This is an explanatory study using Step-wise Regression, as well as Comparative Model and Descriptive Statistic, (such as matrices, totals, averages, ratios and percentages). Regression Analysis was performed to flesh out factors that highly impacted national footprints, as well as to strengthen and buttress the Comparative Model Analysis results. Potential technical problems are diagnosed such as the missing data on domestic inequality or Gini index data for most of the Middle East countries that are sensitive to them, as well as most data in Iraq and Afghanistan that have been in wars since 2000 decade; which did not actually alter the substantive conclusions, especially regarding the ecological footprint accounts and environmental sustainability of different Middle East and Central Asia countries. The national ecological footprint and ecological balance, as dependent variables, are explained recursively by the country's human development hierarchical category, population size, population density, urbanization, GDP per capita, domestic inequality (Gini index), literacy rate, government social spending (% of GDP), gender inequality index (GII), and gender-related development index (GDI), as the independent variables. The independent variables helped in explaining the varying levels of consumption and footprint among different human development indexes of the capitalist world economy (Bergesen and Bartley, 2000; Bergesen et al, 2002). It is hypothesized that: Human Development positions and Ecostructural Factors are likely to be responsible for the variations in the Ecological Footprints and Balances. The carbon dioxide emission levels (*Human Development Report*, 2007/2008, table 24, pp. 310–313) are included to depict the consumption, and environmental degradation impacts of different Middle East and Central Asia countries, which reflect their national ecological footprint and ecological balance levels.

HYPOTHESIS FOR THE STUDY

Null Hypothesis (H_0)

None of the independent variables predicts the national footprint effects. The independent variables do not influence national footprints.

Alternative Hypothesis (H_1)

Some (if not all) of the independent variables predict the national footprint effects.

Variables in the Study

The Dependent and Independent variables are selected on the basis of the theoretical themes and underpinnings, which indicate that national ecological footprint as dependent variable is explained by the country's human development (HDI) hierarchical category, population size, population density, urbanization level, government social spending (% of GDP), GDP per capita, domestic inequality (Gini Index), gender inequality index (GII), gender-related development index (GDI), and literacy rate, as the independent variable. The measures of the dependent and independent variables used in the present study are as follow:

Dependent Variables

The ecological footprint (or consumption) and ecological balance are measured in global hectares (1 hectare = 2.47 acres). A global hectare is 1 hectare of biologically productive space with world average productivity (Global Footprint Network: *Africa's Ecological Footprint-2006 Factbook*, pp. 82–90).

Independent Variables

The ecostructural mediating factors are most of the independent variables in this study (apart from human development index) that include, GDP per capita, urbanization (urban population as a percent of total population), literacy rate (%), domestic inequality (Gini index), government social spending, gender inequality index (GII), and gender-related development index (GDI) (see their definitions below).

Human Development Index (HDI): The HDI is a summary measure of human development (human welfare). It measures the average achievements of a country in three basic dimensions of human development (Global Footprint Network: *Africa's Ecological Footprint-2006 Factbook*, p. 89): a long and healthy life, as measured by life expectancy at birth; knowledge, as measured by the adult literacy rate (with two-thirds weight) and the combined primary, secondary and tertiary gross enrolment ratio (with one-third weight); and a decent standard of living, as measured by GDP per capita (PPP US$). Purchasing power parity (PPP) is a rate of exchange that accounts for price difference across countries, allowing international comparisons of real output and incomes. At the PPP US$ (as used in this study), PPP US$1 has the same purchasing power in the domestic economy as $1 has in the United States of America.

The high human development countries are countries with HDI of 0.800 and above, found mainly in North America, Western Europe and Australia, with Gross National Product (GNP) of $9,386 or more in 2005. The medium development countries are countries with HDI of 0.500–0.799, found in Eastern Europe, South-East Asia, Latin America and Caribbean, and North Africa, with per capita GNP of $766 and 9,385. The low human development countries are countries with HDI below 0.500, found mainly in Sub-Saharan Africa except the countries of South Africa, Gabon, and Ghana, which are included in the medium human development, with per capita GNP of $765 or less (World Bank, 2005, p. 255). Data for this variable in this study is taken from UNDP's *Human Development Report 2004* and *Human Development Report 2005*.

Gross Domestic Product (GDP) per capita (PPP US$): GDP is converted to US dollars using the average official exchange rate reported by the International Monetary Fund (IMF). GDP alone does not capture the international relational characteristics as does the human development hierarchy of the world economy, which accounts for a country's relative socio-economic power and global dependence position in the modern world system. It is suggested elsewhere that GDP per capita is an inadequate measure of world-system position but a more appropriate indicator of domestic affluence or internal economic development (Burns, Kentor, and Jorgenson, 2003; Jorgenson, 2003; Dietz and Rosa, 1994). The Gross Domestic Product (GDP) per capita data for this study is taken from *Human Development Report 2007/2008*, table 1, and pp. 229–232. See also *World Development Report 2005*.

Domestic Income Inequality (Gini Index): The Gini index measures domestic income inequality of different countries, which had remained stable over a time with its impacts on other variables in the study (Bergesen and Bata, 2002; Jorgenson, 2003). Gini index measures the extent to which the distribution of income (or consumption) among individuals or households within a country deviates from a perfectly equal distribution (*Human Development Report, 2004*, p. 271)). It measures inequality over the entire distribution of income or consumption. A value of zero (0) represents perfect equality, and a value of hundred (100) represents perfect inequality. Data for domestic income inequality measured by Gini index are taken from World Bank, *World Development Report* (2005), table 2, pp. 258–259 and United Nations Development Report, *Human Development Report* 2007/2008, 15, pp. 281–284.

Gender Inequality Index (GII): The index shows the loss in human development due to inequality between female and male achievements in reproductive health, empowerment, and labor market. It reflects women's disadvantage in those three dimensions. The index ranges from zero (0), which indicates that women and men fare equally, to one (1), which indicates that women fare as poorly as possible in all measured dimensions. The health dimension is measured by two indicators: maternal mortality ratio and the adolescent fertility rate. The empowerment dimension is measured by two indicators: the share of parliamentary seats held by each sex and by secondary and higher education attainment levels. The labor market dimension is measured by women's participation in the work force.

Gender Inequality Index is designed to measure the extent to which national achievements in these aspects of human development are eroded by gender inequality; also to provide empirical foundations for policy analysis and advocacy efforts. It can be interpreted as a percentage loss to potential human development due to shortfalls in the dimensions included. Countries with unequal distribution of human development also experience high inequality between women and men, and countries with high gender inequality also experience unequal distribution of human development (UNDP *Human Development Reports*). The Gender Inequality Index data for this study is taken from UNDP's *Human Development Report 2010___20th Anniversary Edition. The Real Wealth of Nations: Pathways to Human Development*. Table 4: Gender Inequality Index, pp. 156–160.

Gender-Related Development Index (GDI): The index is one of the indicators of human development developed by the United Nations. The GDI is considered a gender-sensitive extension of the Human Development Index (HDI), which addresses gender-gaps in life expectancy, education, and income. It highlights inequalities in the areas of long and healthy life, knowledge, and decent standard of living between women and men. It measures achievement in the same basic capabilities as the HDI, but takes note of inequality in achievement between women and men (*Human Development Reports*). The methodology used imposes a penalty for inequality, such that the GDI falls when the achievement levels of both women and men in a country go down or when the disparity in basic capabilities, the lower a country's GDI compared with its HDI. The GDI is simply the HDI discounted, or adjusted downwards, for gender inequality. Thus, if GDI goes down, HDI goes down, while GII goes up. The methodology used to construct the GDI could be used to assess inequalities not only between men and women, but also between other groups such as rich and poor, young and old, etc. (*Human Development Reports*). The Gender-Related Development Index data for this study is taken from UNDP's *Human Development Report 2010___20th Anniversary Edition. The Real Wealth of Nations: Pathways to Human Development*. Table 4: Gender Inequality Index, pp. 156–160.

Government Social Spending Per Capita: Spending by a government (federal, state, and local), municipality, or local authority, which covers such things as spending on healthcare, education, pensions, defense, welfare, interest, and other social services, and is funded by tax revenue, seigniorage, or government borrowing. Public expenditure exerts an effect on economic growth rate through the positive externality in the productivity of the capital stock. Investment in education and health is also investment in people and in the future (See, Education: Crisis Reinforces Importance of a Good Education, says OECD. Accessed Online on 9/23/2011, at http://www.oecd.org/document/21/0,3746. Data for this study were taken from: Human Development Report 2007/2008. *Fighting Climate Change: Human Solidarity in a Divided World.* Table 28: Gender-related Development Index, pp. 326–329; and Human Development Report 2010___20th Anniversary Edition. *The Real Wealth of Nations: Pathways to Human Development.* Table 4: Gender Inequality Index, pp. 156–160.

Urbanization Level (Urban Population as percent of Total Population): Cities are not self-contained entities and their concentration of intense economic processes and high levels of consumption both increase and stimulate their demands on resources. The cities have limited intrinsic biocapacity which undoubtedly must rely upon large hinterlands. Land consumed by urban regions is typically at least an order of magnitude greater than that contained within the usual political boundaries or the associated built-up area (The International Society for Ecological Economics and Island Press, 1994; Rees, 1992). The data are taken from *Human Development Report 2007/2008*, table 5, and pp. 243–246.

Literacy Rate: This variable refers to the percent of a nation's population over the age of fifteen (15) that can read and write in any language of their choice. Literate population generally has low domestic inequality and tends to consume more resources than their illiterate counterpart due to their higher income and urban living. High literate groups concentrate in urban areas where they consume more than their fair shares of biospheric resources. Higher levels of literacy correspond with higher incomes, which allow for greater consumption (Jorgenson, 2003). This is because literate populations are subject to increased consumerist ideologies and contextual images of good life through advertising (Princen et al, 2002), what Leslie Sklair (2001) and Jennifer Clapp (2002) labeled "cultural ideology of consumerism/consumption." The data for literacy rate is taken from the *Human Development Report 2007/2008*, table 1, pp. 229–232.

Population and Population Density: Apart from consumption, many have attributed to population as driving most of the sustainability problems (Palmer, 1998; Pimentel, 1996). Likewise, Dietz et al (2007) concluded that population size and affluence are the primary drivers of environmental impacts. Population growth and increases in consumption in many parts of the world have increased humanity's ecological burden on the planet. York et al (2003) indicated that population and affluence account for 95 percent of the variance in total footprints of countries. Others also see the ensuing human impact or footprint as a product of population, affluence (consumption), and technology (i.e. I = PAT (Ehrlich and Holdren, 1971; Holdren and Ehrlich, 1974; Hardin, 1991). Population as a variable in this study is taken from, *World Development Report 2005*, table 1, and pp. 256–259.

COMPARATIVE MODEL ANALYSIS (RESULTS AND DISCUSSIONS)

The results and findings in this study are discussed under subtitles that include, "Per Capita Ecological Footprint, Biological Capacity, and Ecological Balance by World Regions," "Human Development and Affluence on Biological Capacity, Ecological Footprint and Balance in Middle East and Central Asia Countries," and "Population and Ecostructural Factors on Total Footprint Per Capita in Middle East and Central Asia Countries."

PER CAPITA ECOLOGICAL FOOTPRINT, BIOLOGICAL CAPACITY, AND ECOLOGICAL BALANCE BY WORLD REGIONS

Table 8.5 shows a comparison of per capita ecological footprint, biological capacity, and ecological balance by world regions. The world regions include, Middle East and Central Asia, Latin America and Caribbean, Africa, Asia-Pacific, North America, Europe EU, Europe Non-EU. Analysis of table 8.5 exposes the inequitable distribution of Middle East and Central Asia nations' ecological footprints as compared to those of other world regions. The table shows a comparison of each region's footprint with its bio-capacity depicting whether that region has an ecological reserve or is running a deficit (overshoot). From the table, North America with its considerable per capita biological capacity in 2005 (6.49 gha) had the largest per capita footprint (9.19 gha) and the largest per capita ecological deficit or overshoot (–2.71 gha); followed by Europe EU (Western Europe) with per capita footprint of 4.69 and per capita ecological deficit of –2.38; followed by Europe Non-EU (Central and Eastern Europe), 3.52 and ecological reserve of 2.29; Latin America and Caribbean, 2.44 and ecological balance or reserve of 2.36; and Middle East and Central Asia, 2.32 and –1.04 respectively. The lowest per capita ecological footprint of 1.37 was in Africa with per capita ecological reserve of 0.43; followed by Asia-Pacific, 1.62 and ecological deficit of –1.04. The highest per capita ecological reserve is found in Latin America and Caribbean (2.36), followed by Europe Non-EU (2.29), and Africa (0.43). Middle East and Central Asia per capita ecological footprint (2.32), biological capacity (1.28), and ecological balance (–1.04) were less than the world averages in 2005 of 2.69, 2.06, and –0.63 respectively. A regional footprint per capita that is less than the world average and per capita ecological balance more than the world average may denote sustainability at the global level (Living Planet Report, 2006); all were found in varying proportions in developing countries in 2005, especially in Africa, Middle East and Central Asia, and Latin America and Caribbean regions.

HUMAN DEVELOPMENT AND AFFLUENCE ON BIOLOGICAL CAPACITY, ECOLOGICAL FOOTPRINT AND BALANCE IN MIDDLE EAST AND CENTRAL ASIA COUNTRIES

Table 8.6 presents the national comparisons for both dependent (Ecological footprint and Ecological balance) and some of the independent (Human development or welfare, GDP per capita or affluence, and Carbon dioxide emissions) variables in Middle East and Central Asia countries. The table shows that a country's total footprint, ecological balance, and carbon dioxide emissions (waste) are a function of its human development position and affluence level in the world economy. Given the national biological capacity measured in global hectares, generally the higher the total ecological footprint, the lower the ecological balance and vice versa. Table 8.6 assesses the progress towards sustainability development in Middle East and Central Asia by using human development index (HDI) as an indicator of national well-being, and the ecological footprint and balance as measures of national demand on the biosphere and biological capacity respectively. The sustainability quotient is depicted by any country with low footprint and high human development levels. In fact, table 8.6 shows that, among Middle East and Central Asia countries, improvements in the human development indexes (human welfare) have resulted in the worsening of the countries' positions on their ecological footprints and balances. Table 8.6 shows that Middle East and Central Asia countries are among the high human development (HDI of 0.800 and above and per capita GNP of $9,386 or more) or medium human development (HDI of 0.500–0.799 and per capita GNP of between $766 and $9,385), even Yemen that is one of the poorest countries of the world had HDI of more than 0.500 (World Bank, 2005, p. 255).

From table 8.6 and based on the existing biological capacity, it becomes apparent that the higher the HDI (human welfare), GDP per capita (affluence) i.e. consumption; the higher the carbon dioxide emissions and ecological footprint. For example, Israel shows HDI of 0.932, GDP per capita ($25,864), carbon dioxide emission (71.2 million tons), total footprint (32.6 global hectares), biological capacity (2.7 global hectares), and ecological balance of –29.9 global hectares. Actually, Israel "appropriates" most (about 11 times) of its carrying capacity from "elsewhere." Kuwait shows HDI of 0.891, GDP per capita ($26,321), carbon dioxide emission (99.3), and total footprint (23.9). United Arab Emirates shows HDI of 0.868, GDP per capita ($25,514), carbon dioxide emission (149.1), and total footprint (42.5). Likewise, Saudi Arabia shows HDI of 0.812, GDP per capita ($15,711), carbon dioxide emission of (308.2), and total footprint of 64.5; Turkey shows HDI of 0.775, GDP per capita ($8407), carbon dioxide emission of 226.0, and total footprint of 198.6; and Iran shows HDI of 0.759, GDP per capita ($7968), carbon dioxide emission (433.3), and total footprint of 186.0. On the other hand, Yemen shows the lowest HDI of 0.508,

GDP per capita($930), carbon dioxide emission (21.1), total footprint (19.1), biological capacity (12.3), and ecological balance (–6.9). Likewise, Kyrgyzstan shows HDI of 0.696, GDP per capita ($1,927), carbon dioxide emission (5.7), total footprint (5.8), biological capacity (8.7), and ecological balance (3.0). The table shows that most of the oil exporting countries of Middle East have high carbon dioxide emissions, high footprints, and high negative ecological balances (overshoots), probably due to mining of petroleum and flaring of natural gas, as well as excessive environmental degradations. Analysis of table 8.6 has shown that, given a nation's biological capacity, its levels of human development (welfare), GDP per capita (affluence), and carbon dioxide emissions (waste) help to explain its total ecological footprint (consumption or demand) and ecological balance (reserve or deficit).

POPULATION, ECOSTRUCTURAL FACTORS, AND TOTAL FOOTPRINT PER CAPITA IN MIDDLE EAST AND CENTRAL ASIA COUNTRIES

Table 8.7 shows the effects of population and ecostructural factors (socioeconomic processes) on the total ecological footprints of nations. A nation's total population size and population density, as well as its urban population, literacy rate, and domestic inequality (Gini index), lead to its ecological footprint difference and variation vis-à-vis other nations. The above indicators constitute the structural causes of the ecological footprint variations among Middle East and Central Asia countries. Table 8.7 shows that a nation with large population, high rates of urbanization and literacy, and low domestic inequality, has high ecological footprint, and vice versa. Large population level, high population density, high level of urbanization, high level of literacy, and low domestic inequality promote domestic consumption of biospheric resources.

Cities are not self contained entities and have little intrinsic biological capacity; their concentration of intense economic processes and high levels of consumption both increase and stimulate their demands on bio-resources. Thus, cities are parasitic and in most cases must rely upon large hinterlands for their supplies which increase their footprint impacts. High density settlements appropriate carrying capacity from all over the globe, as well as from the past and the future (Vitousek et al, 1986; Wackernagel, 1991). High levels of urbanization correspond to higher levels of consumption, higher levels of energy use and carbon dioxide emissions (waste), and thus, higher energy footprints and ecological footprints. Biospheric resources are consumed at higher levels in urban areas that generally contain high literate groups with higher incomes, which allow for greater material consumption (Jorgenson, 2003; Princen et al, 2002; Sklair, 2001; Clapp, 2002).

According to table 8.7, Turkey (Istanbul), Saudi Arabia (Riyadh), and Iran (Tehran) are nations with large populations and large cities. They have high urban populations as percent of their total populations. From the table, high population density and high literacy rate depict high per capita footprints; while low domestic inequality does not necessarily result in high footprint per capita. Thus, the level of consumption, spurred by high population density, high level of urbanization, large urban population, and high level of adult literacy are the major driving force of environmental sustainability and ecological footprints.

Tables 8.6 and 8.7 depict mix-income human development countries with mix-levels of urbanization, population density, literacy, and domestic inequality in 2005. These correspond to mix-income economies, which allow for mix-levels of material consumption and footprint. The findings here parallel the theorization above, which emphasizes that the low consumption levels and the accompanying low ecological footprints and high ecological balances (reserves) and vice versa, are based on the nations' positions in the world capitalist economy and system (Bergesen and Bartley, 2000; Bergesen et al, 2002). Most of the Middle East and Central Asia countries were once colonized by the medieval sovereign nation states of England, France, Italy, and recently Soviet Union (Brunn, Hays-Mitchell, and Zeigler, 2012, pp. 284–285). The region is where the process of underdevelopment, economic stagnation, and dependent industrialization limited their populations to consuming a greater share of unprocessed bio-productive resources, rather than industrial processed finished products (see Frank, Gunder for more discussion on dependency theory; see also Brunn, et al, 2012, p. 284). The mix domestic income inequality or intra-inequality has mix effects on per capita consumption levels (Bornschier and Chase-Dunn, 1985; Kick, 2000; Kentor, 2001; and Jorgenson, 2003) that yield both mix total and mix per capita ecological footprints.

REGRESSION ANALYSIS (RESULTS AND DISCUSSIONS)

Regression analysis allows the modeling, examining, and exploring of relationships and can help explain the factors behind the observed relationships or patterns. It is also used for predictions. Regression equation, on the other hand, utilizes the Ordinary Least Square (OLS) technique, is the mathematical formula that is applied to the explanatory variables to best predict the dependent variable one is trying to model. Each independent variable or explanatory variable is associated with a regression coefficient describing the strength and the sign of that variable's relationship to the dependent variable.

Table 8.8 depicts the descriptive statistics of the data in tables 8.1 and 8.6. The measure of variability or dispersion, from table 8, such as standard deviation signifies how homogeneous (in terms of low levels) Middle East and Central

Asia countries are in terms of Human Development Index (HDI), Gender-related Development Index (GDI), and Gender Inequality Index (GII); and how heterogeneous they are (in terms of high levels) in terms of GDP Per Capita, Carbon dioxide Emissions, and to some extent Total Footprint. The same is also true in table 8.14, where low variability is associated with Public Expenditure on Health, Public Expenditure on Education, Gini Index, Total Footprint Per Capita, and Adult Literacy Rates; and high variability with Population, Urban Population and Population Density. Table 8.9 shows the Pearson Correlation (2-tailed) of the data from table 8.6. From table 8.9, there is a strong correlation or relationship (at the 0.01 level) between Human Development Index and GDP Per Capita; Gender-related Development Index and Gender Inequality Index; Bio-Capacity, Carbon Dioxide Emissions, Total Footprints, and Ecological Balance; Total Footprint and Ecological Balance; and Ecological Balance, Carbon dioxide Emissions and Bio-Capacity. Likewise, table 8.15 shows a strong correlation (at 0.05 levels, 2-tailed) between Population Density and Urban Population; Population and Domestic Inequality; Public Expenditure on Health, Urban Population, and Domestic Inequality; Domestic Inequality, Population, Urban Population, Public Expenditure on Health, and Total Footprint Per Capita; and at 0.01 level between Urban Population and Total Footprint Per Capita; and Adult Literacy and Public Expenditure on Education.

Table 8.11 is the Regression model summary, which shows the Multiple R and R-Squared or Coefficient of Determination very high and significant. For example, about 94 percent of the dependent variable (Total Footprint) is accounted for by the independent variable, Bio-Capacity; and 100 percent of Total Footprint is explained by Bio-Capacity and Ecological Balance. These are entered in Stepwise Regression Model in table 8.10. The Multiple R is very large (100 percent), which indicates size of the correlation between the observed outcome variables and the predicted outcome variables (based on the regression equation).

Tables 8.16 and 8.17 show the Multiple R and R-Squared between dependent variable (Total Footprint Per Capita) and the independent variables: Population Density and Population. Nevertheless, their relationships as well as predictability and accountability are equally as high and significant as found in table 8.11. From table 8.17, Model Summary, about 95 percent of the variations in Total Footprint Per Capita is explained or accounted for by Population Density and Population. Tables 8.12 and 8.18 depict the analysis of variance (ANOVA) and F-Tests. The F-tests show how strong or significant the selected independent variables in the regression models are in predicting the dependent variable. From table 8.12, the F values are high and the residual (observed – predicted) values are very small. Since the F-Tests are high and significant, we reject the Null Hypothesis that none of the independent variables (predictors) predict the National footprint and per capita footprint effects. It is worthy to note that the results in table 8.18 are equally as strong as those in table 8.12

and very significant. Using the B columns of the Unstandardized Coefficients in tables 8.13 and 8.19, the regression model equations are constructed from tables 8.1, 8.2, 8.6 and 8.7 (see below).

For tables 8.1 and 8.6: Human Development Index, GDP Per Capita, Gender-related Development Index, Gender Inequality Index, Carbon Dioxide Emissions, Bio-Capacity, Ecological Balance Factors on Total Footprints, the significant explanatory (predictor) variables are: Bio-Capacity and Ecological Balance (see table 8.13). The Regression Equation is: Total Footprints = –630–5.219Human Development Index+3.167 GDP Per Capita+5.571Gender-Related Development Index+.721Gender Inequality Index+Carbon Dioxide Emissions Bio-Capacity-Ecological Balance.

For tables 8.2 and 8.7: Population, Population Density, Urban Population, Adult Literacy, Public Expenditure on Health, Public Expenditure on Education, Domestic Inequality on Total Footprints Per Capita (see table 8.19). The Regression Equation is = 0.312 + 0.013Population Density + 0.021Population.

From tables 8.13 and 8.19, Total National Footprints in Middle East and Central Asia is predicted or explained by Bio-Capacity, Ecological Balance, Population Density, Population, Human Development Index, GDP Per Capita, Gender-Related Development Index, Gender Inequality Index, and Carbon Dioxide Emissions.

POLICY RECOMMENDATIONS

Middle East and Central Asia region is classified by the United Nations Development Reports as a mix income economy, therefore policy recommendations will address and reflect environmental issues in high, medium, and low income countries. The way additional population or material growth can be sustained, without ravaging biodiversity and ultimately destroying the ecological basis of human life is through *cutbacks in resource or natural capital consumption* by the wealthy nations of Israel, Kuwait, and United Arab Emirates; for equity considerations. The above action is necessary to vacate the ecological space necessary to justify growth in low income countries of Afghanistan, Kyrgyzstan, Tajikistan, Uzbekistan, and Yemen. *Population management* is very essential and necessary since population and population density are major factors of national footprints in this region. This could be achieved through the education and empowerment of women, who are major players in reproduction, economy, environmental sustainability, and biodiversity conservation (Aka, 2006). Population should also be dispersed to small and intermediate-sized cities rather than concentrating in major cities, such as Istanbul, Turkey; Riyadh, Saudi Arabia, Tehran, Iran, and others, that are largely responsible for environmental unsustainability in the region. Apart from Israel, Gender

Inequality is high and public expenditures are low in health and education. Investment in education and health in low income countries is also investment in people and in the future, with their positive externalities in the productivity of the capital stock. Adequate environmental and ecological education, adequate technologies, and the willingness to behave appropriately (conserve resources), will undoubtedly reduce individual's ecological impacts, thus the national footprints in Middle East and Central Asia region.

CONCLUSIONS

This is an explanatory study using comparative model analysis, regression analysis, and descriptive statistical analysis. The results and the findings show that a country's total footprint, ecological balance, and carbon dioxide emissions (waste) are a function of its human development (welfare) position and affluence in the world economy. As human development and income go up, the environmental performance goes down. Middle East and Central Asia region is a mix income economy region with mixed national footprint and environmental problems, recommendations, and solutions. The present study shows that, the higher the national populations, population densities, and GDP per capita (affluence), the higher the *imbalanced* consumption of resources (bio-capacity or natural capital), as well as the total footprints.

The above analysis supports the theorization and assertion that the global environmental stress is caused primarily by the increase in resource consumption (demand) and population size and growth. The regional ecological sustainability calls for: (a) gender empowerment; (b) population management, especially redistribution to smaller centers; and (c) the already rich countries to reduce their consumptive appetite and eco-footprints to create the ecological space needed for a justifiable growth in the impoverished countries.

REFERENCES

Aka, Ebenezer (2006). "Gender Equity and Sustainable Socio-Economic Growth and Development," *The International Journal of Environmental, Cultural, Economic & Social Sustainability*, Volume 1, Number 5, pp. 53–71.

Bergesen, Albert and Michelle Bata (2002). "Global and National Inequality: Are They Connected?" *Journal of World-System Research*, 8: 130–44.

Bergesen and Bartley (2000). "World-System and Ecosystem," in *A World-Systems Reader: New Perspectives on Gender, Urbanism, Culture, Indigenous Peoples, and Ecology*, edited by Thomas Hall (Lanham, MD: Rowman and Littlefield), pp. 307–22.

Bornschier, Volker and Christopher Chase-Dunn (1985). *Transnational Corporations and Underdevelopment* (New York: Praeger).

Brown, S., Jayant, S., Cannell, M., and Kauppi, P. (1996). "Mitigation of Carbon Emissions to the Atmosphere by Forest Management," *Commonwealth Forestry Review*, 75, 79–91.

Brunn, Stanley, Maureen Hays-Mitchell, and Donald Zeigler (eds.) (2012). *Cities of the World: World Regional Urban Development. Fifth Edition* (Lanham, Boulder, New York, Toronto, Plymouth, UK: Rowman & Littlefield Publishers, Inc.).

Clapp, Jennifer (2002). "The Distancing of Waste: Over-consumption in a Global Economy," in *Confronting Consumption*, edited by T. Princen, M. Maniates, and K. Conca (Cambridge, MA: MIT Press) pp. 155–76.

Daly, H.E. (1986). *Beyond Growth: The Economics of Sustainable Development* (Boston, Massachusetts, USA: Beacon Press).

Dietz, Thomas and Eugene Rosa (1994). "Rethinking the Environmental Impacts of Population, Affluence, and Technology," *Human Ecology Review*, 1: 277–200.

Dietz, Thomas, Eugene Rosa, Yoor (2007). "Driving the Human Ecological Footprint." *Frontiers in Ecology and the Environment* (February).

Ehrlich and Holdren (1971). "Impacts of Population Growth," *Science*, 171, 1212–7.

Frank, Andre Gunder. See him for an in-depth discussion of Dependency Theory and Underdevelopment. Some of his works include: *The Development of Underdevelopment*, 1966, MRP; *Capitalism and Underdevelopment in Latin America*, 1967; *Latin America: Underdevelopment or Revolution*, 1972; and *Theoretical Introduction to Five Thousand Years of World System History,* 1990, Review.

Girardet, Herbert (1996). "Giant Footprints," Accessed Online on 12/14/07, at http:// www.gdrc .org/uem/footprints/girardet.html.

Global Footprint Network: Africa's Ecological Footprint—2006 Fact book. Global Footprint Network, 1050 Warfield Avenue, Oakland, CA 94610, USA. http://www.footprintnetwork.org/ Africa.

Global Footprint Network (2008). *The Ecological Footprint Atlas 2008*, October 28.

Hardin, G. (1991). "Paramount Positions in Ecological Economics," in Constanza, R. (editor), *Ecological Economics: The Science and Management of Sustainability* (New York: Columbia University Press).

Holdren and Ehrlich (1974). "Human Population and The Global Environment," *American Scientist*, 62: 282–92.

IPCC (2001). Intergovernmental Panel on Climatic Change.

Jorgenson, Andrew K. (2003). "Consumption and Environmental Degradation: A Cross-National Analysis of Ecological Footprint," *Social Problems*, Volume 50, No. 3, pp. 374–394.

Kentor, Jeffrey (2001). "The Long Term Effects of Globalization on Income Inequality, Population Growth, and Economic Development," *Social Problems*, 48: 435–55.

Kick, Edward L. (2000). "World-System Position, National Political Characteristics and Economic Development," Journal of Political and Military Sociology, (Summer), http://www.findarticles .com/p/articles/mi-qa3719.

Iiving Planet Report, 2006.

Palmer, A.R. (1998). "Evaluating Ecological Footprints," *Electronic Green Journal*. Special Issue 9 (December).

Pimentel, David (1996). "Impact of Population Growth on Food Supplies and Environment." American Association for the Advancement of Science (AAAS) (February, 9). See also GIGA DEATH, Accessed Online on 12/18/07, at http://dieoff.org/page13htm.

Princen, Thomas (2002). "Consumption and Its Externalities: Where Economy Meets Ecology," in *Confronting Consumption*, edited by T. Princen, M. Maniates, and K. Conca (Cambridge, MA: MIT Press) pp. 23–42.

Redefining Progress (2005). Footprints of Nations. 1904 Franklin Street, Oakland, California, 94612. See, http://www.RedefiningProgress.org.

Rees, William E. (1992). "Ecological Footprint and Appropriated Carrying Capacity: What Urban Economics Leaves Out," Environment and Urbanization, Vol. 4, No. 2, (October). Accessed Online on 12/14/07, at http://eau.sagepub.com.

Rees, William E. (2012). "Ecological Footprint, Concept of," in Simon Levin (ed.). *Encyclopedia of Biodiversity (2nd Edition)*.

Sklair, Leslie (2001). *The Transnational Capitalist Class* (Oxford, UK: Blackwell Press).

Solow, R. M. (1974). "The Economics of Resources or the Resources of Economics," *American Economics Review*, Vol. 64, pp. 1–14.

Spencer, William (ed.) (2009). *Global Studies: The Middle East* (Boston: McGraw Hill, Higher Education).

Suplee, D. (1998). "Unlocking the Climate Puzzle." *National Geographic*, 193 (5), 38–70.

The international Society for Ecological Economics and Island Press (1994). "Investing in Natural Capital: The Ecological Approach to Sustainability." Accessed Online on 12/18/07, at http://www.dieoff.org/page13.htm.

United Nations Development Program (2004). *Human Development Report 2004 (HDR). Cultural Liberty in Today's Diverse World* (New York, N.Y: UNDP).

United Nations Development Program (2005). *Human Development Report (HDR) 2005. International Cooperation at a Crossroads: Aid, Trade and Security in an Unequal World* (New York, N.Y: UNDP).

United Nations Development Program (2007/2008). *Human Development Report 2007/2008. Fighting Climate Change: Human Solidarity in a Divided World* (New York, NY: UNDP).

United Nations Development Program (2010). *Human Development Report 2010. The Real Wealth of Nations: Pathway to Human Development* (New York, NY: UNDP), p. 187.

Vitousek, P., P. Ehrlich, A. Ehrlich and P. Matson (1986). "Human Appropriation of the Products of Photosynthesis," *Bioscience*, Vol. 36, pp. 368–374.

Wackernagel, M. (1991). "Using 'Appropriated Carrying Capacity' as an Indicator: Measuring the Sustainability of a Community." Report for the UBC Task Force on Healthy and Sustainable Communities. UBC School of Community and Regional Planning, Vancouver, Canada.

Wackernagel and Rees (1996). *Our Ecological Footprint* (New Society Publisher).

World Bank (2005). *World Development Report 2005. A Better Investment Climate for Everyone* (New York, N.Y: A Co-publication of the World Bank and Oxford University Press).

World Bank (2010). *World Development Report 2010. Development and Climate Change*. Selected World Development Indicators, table 1, p. 379.

World Resources Institute (WRI) (1992). *World Resources, 1992–1993* (New York: Oxford University Press).

York, Richard, Eugene A. Rosa, and Thomas Dietz (2003). "Footprints on the Earth: The Environmental Consequences of Modernity." *American Sociological Review*, 68: 279–300.

Table 8.1. Gender Related Development and Gender Inequality Indexes in Middle East and Central Asia Countries

Countries	*Economy Class (2005)*[1]	*HDI Rank*	*Human Development Index (HDI) Value (2005)*[2]	*GDI Rank*	*Gender Related Development Index (GDI) Value (2005)*[3]	*HDI Rank minus GDI Rank*	*Rank*	*Gender Inequality Index (GII) Value (2008)*[4]
Afghanistan	LIC	—	—	—	—	—	134	0.797
Armenia	LMC	83	0.775	75	0.772	8	66	0.570
Azerbaijan	LMC	93	0.746	67	0.743	26	62	0.553
Georgia	LMC	96	0.754	—	—	—	71	0.597
Iran	LMC	94	0.759	84	0.750	10	98	0.674
Iraq	LMC	—	—	—	—	—	123	0.751
Israel	HIC	23	0.932	21	0.927	2	28	0.332
Jordan	LMC	83	0.773	80	0.760	3	76	0.616
Kazakhstan	LMC	73	0.794	65	0.792	8	67	0.575
Kuwait	HIC	33	0.891	32	0.884	1	43	0.451
Kyrgyzstan	LIC	116	0.696	102	0.692	14	63	0.560
Lebanon	UMC	88	0.772	81	0.759	7	—	—
Oman	UMC	58	0.814	67	0.788	–9	—	—
Saudi Arabia	UMC	61	0.812	70	0.783	–9	128	0.760
Syria	LMC	108	0.724	96	0.710	12	103	0.687
Tajikistan	LIC	122	0.673	106	0.669	16	65	0.568
Turkey	LMC	84	0.775	79	0.763	5	77	0.621
Turkmenistan	LMC	109	0.713	—	—	—	—	—
United Arab Emirates	HIC	39	0.868	43	0.855	–4	45	0.464
Uzbekistan	LIC	113	0.702	98	0.699	15	—	—
Yemen	LIC	153	0.508	136	0.472	17	138	0.853
Middle East and Central Asia	**Mix Income Economy**		**0.699**					**0.699**[5]

1. World Development Report 2005. *A Better Investment for Everyone*. Page 225. Classification of Economies by Region and Income, FY 2005. World Bank Data. LIC = Low Income, $765 or less GNI per capita in 2005; LMC = Lower Middle Income, $766–3,035; UMC = Upper Middle Income, $3,036–9,385; and HIC = High Income, $9,386 or more. \
2. Human Development Report 2007/2008. *Fighting Climate Change: Human Solidarity in a Divided World*. Table 1: Human Development Index, pp. 229–232.
3. Human Development Report 2007/2008. *Fighting Climate Change: Human Solidarity in a Divided World*. Table 28: Gender-related Development Index, pp. 326–329.
4. Human Development Report 2010__20th Anniversary. *The Real Wealth of Nations: Pathways to Human Development*. Table 4: Gender Inequality Index, pp. 156–160.
5. See also, *The Real Wealth of Nations: Pathways to Human Development*, P. 160. The figure is for Arab States only, without Central Asia countries.

Table 8.2. Public Spending on Health and Education in Middle East and Central Asia Countries

Countries	*Public Expenditure on Health (% of GDP) 2004*	*Public Expenditure on Education (% of GDP) 2002–2005*
Afghanistan	—	—
Armenia	1.4	3.2
Azerbaijan	0.9	2.5
Georgia	1.5	2.9
Iran	3.2	4.7
Iraq	—	—
Israel	6.1	6.9
Jordan	4.7	4.9
Kazakhstan	2.3	2.3
Kuwait	2.2	5.1
Kyrgyzstan	2.3	4.4
Lebanon	3.2	2.6
Oman	2.4	3.6
Saudi Arabia	2.5	6.8
Syria	2.2	—
Tajikistan	1.0	3.5
Turkey	5.7	3.7
Turkmenistan	3.3	—
United Arab Emirates	2.0	1.3
Uzbekistan	2.4	—
Yemen	1.9	9.6

Source: Human Development Report 2007/2008. *Fighting Climate Change: Human Solidarity in a Divided World.* Table 19: Priorities in Spending, pp. 294–297.

Table 8.3. Regional Population, Gross Domestic Product (GDP), Total Footprint, Biological Capacity, and Ecological Balance by World Income Regions in 2005

World Income Regions	*Regional Population (millions) (2005)*	*GDP PPP US$ Billions (2005)**	*Total Footprint Million gha (2005)*	*Biological Capacity Global gha (2005)*	*Ecological Balance Million gha (2005)*
Low Income	2370.6	5879.1	2377.2	2089.7	–287.5
% World	36.6	9.7	13.6	15.6	7.0
Middle Income	3097.9	22586.3	6787.0	6684.8	–102.2
% World	47.8	37.3	38.9	50.0	2.5
High Income	971.8	32680.7	6196.0	3561.5	–2634.5
% World	15.0	53.9	35.5	26.7	64.5
World Total	6475.6	60597.3	17443.6	13361.0	–4082.7

*Human Development Report (HDR) 2007/2008. *Fighting Climate Change: Human Solidarity in a Divided World,* table 14, p. 280.

Note: Percentages may not add up to 100 percent due to (i) rounding up; (ii) some countries that were not up to 1 million people in 2005 were not included; and (iii) some countries that had been at war by 2005 were not included.

Source: Global Footprint Network (2008). *The Ecological Footprint Atlas 2008,* Appendix F, table 2, p. 46.

Table 8.4. Total Population, Ecological Footprint, Biological Capacity, and Ecological Balance by World Regions in 2005

Region	*Total Population (million) 2005*	*Total Ecological Footprint (million gha) 2005*	*Total Biological Capacity (million gha) 2005*	*Total Ecological Balance (million gha) 2005*
Middle East and Central Asia	365.7	846.6	466.9	–379.9
% World	5.6	4.9	3.5	
Africa	902.0	1237.5	1627.1	389.56
% World	13.9	7.1	12.2	
Latin America and Caribbean	553.0	1350.8	2655.7	1304.9
% World	8.5	7.7	19.9	
Asia-Pacific	3562.1	5758.6	2923.3	–2835.3
% World	55.0	33.0	21.9	
North America	330.5	3037.8	2143.3	–894.5
% World	5.1	17.4	16.0	
Europe EU	487.3	2291.8	1128.2	–1163.6
% World	7.5	13.1	8.4	
Europe Non-EU	239.6	842.4	1391.6	549.2
% World	3.7	4.8	10.4	
World	**6475.6**	**17443.6**	**13361.0**	**–4082.7**

Note: EU: European Union. Percentages may not add up to 100 percent due to (i) rounding up; (ii) some countries that were not up to 1 million people in 2005 were not included; and (iii) some countries that had been at war by 2005 were not included.

Source: Global Footprint Network (2008). *The Ecological Footprint Atlas 2008*, Appendix F, table 2, pp. 46–50.

Table 8.5. Per Capita Ecological Footprint, Biological Capacity, and Ecological Balance by World Regions in 2005

Region	*Ecological Footprint (gha per person) 2005*	*Biological Capacity (gha per person) 2005*	*Ecological Balance (gha per person) 2005*
Middle East and Central Asia	**2.32**	**1.28**	**–1.04**
Africa	1.37	1.80	0.43
Latin America and Caribbean	2.44	4.80	2.36
Asia-Pacific	1.62	0.82	–0.80
North America	9.19	6.49	–2.71
Europe EU	4.69	2.32	–2.38
Europe Non-EU	3.52	5.81	2.29
World	**2.69**	**2.06**	**–0.63**

EU: European Union

Percentages may not add up to 100 percent due to: (i) rounding up; (ii) some countries not up to 1 million people in 2005 were not included; and (iii) some countries that had been at war by 2005 were not included.

Source: Global Footprint Network (2008). *The Ecological Footprint Atlas 2008*. Appendix F, table 1.

Table 8.6. Human Development and Affluence on Biological Capacity, Ecological Footprint and Balance in Middle East and Central Asia Countries

Countries	*Human Development Index (HDI) Value 2005*[1]	*GDP Per Capita PPP US ($) 2005*[2]	*Carbon Dioxide Emissions Mil. Tons 2004*[3]	*Total Footprint Million gha 2005*[4]	*Bio-Capacity Million gha 2005*[4]	*Ecological Balance Million gha 2005*[4]
Afghanistan	—	—	—	14.3	21.8	7.5
Armenia	0.775	4945	3.6	4.3	2.5	–1.9
Azerbaijan	0.746	5016	31.3	18.2	8.6	–9.6
Georgia	0.754	3365	3.9	4.8	7.9	3.1
Iran	0.759	7968	433.3	186.0	98.5	–87.6
Iraq	—	2900[b]	—	38.4	8.0	–30.5
Israel	0.932	25864	71.2	32.6	2.7	–29.9
Jordan	0.773	5530	16.5	9.7	1.6	–8.2
Kazakhstan	0.794	7857	200.2	50.0	63.5	13.5
Kuwait	0.891	26321	99.3	23.9	1.4	–22.5
Kyrgyzstan	0.696	1927	5.7	5.8	8.7	3.0
Lebanon	0.772	5584	16.3	11.0	1.5	–9.5
Oman	0.814	15602	30.9	12.0	6.6	–5.5
Saudi Arabia	0.812	15711	308.2	64.5	31.3	–33.2
Syria	0.724	3808	68.4	39.6	16.1	–23.2
Tajikistan	0.673	1356	5.7	4.6	3.6	–1.0
Turkey	0.775	8407	226.0	198.6	120.9	–77.7
Turkmenistan	0.713	3838	41.7	18.7	17.8	–0.9
United Arab Emirates	0.868	25514	149.1	42.5	4.8	–37.7
Uzbekistan	0.702	2063	137.8	48.2	27.2	–21.1
Yemen	0.508	930	21.1	19.1	12.3	–6.9
Middle East and Central Asia	**0.699**[a]	**—**	**1348.6**[a]	**846.8**	**466.9**	**–379.9**

1. Human Development Report 2007/2008. *Fighting Climate Change: Human Solidarity in a Divided World.* Table 1: Human Development Index, pp. 229–232.
2. Human Development Report 2007/2008. *Fighting Climate Change: Human Solidarity in a Divided World.* Table 1: Human Development Index, pp. 229–232.
3. Human Development Report 2007/2008. *Fighting Climate Change: Human Solidarity in a Divided World.* Table 24: Carbon Dioxide Emissions and Stocks, pp. 310–313.
4. Global Footprint Network (2008). *The Ecological Footprint Atlas 2008, October 28*, tables 2 and 4, pp. 46–50 and pp. 55–58, respectively.

a. For Arab States only.
b. Spencer, William (2009). *Global Studies: The Middle East, Twelfth Edition* (New York: McGraw Hill), pp. 72–84.

Table 8.7. Population, Ecostructural Factors, and Total Footprint Per Capita in Middle East and Central Asia Countries

Countries	*Population (Millions) 2005*[1]	*Population Density People Per Km^2 2003*[2]	*Urban Population (% of Total) 2005*[3]	*Adult Literacy Rates (%) by 2005*[4]	*Domestic Inequality (Gini Index) by 2005*[5]	*Total Footprint gha Per Capita 2005*[6]
Afghanistan	29.9[a]	—	—	28.0[d]	—	0.5
Armenia	3.0	108	64.1	99.4	33.8	1.4
Azerbaijan	8.4	95	51.5	98.8	36.5	2.2
Georgia	4.5	74	52.2	100.0	40.4	1.1
Iran	69.4	41	66.9	82.4	43.0	2.7
Iraq	28.8[a]	—	75[c]	74.1[d]	—	1.3
Israel	6.7	324	91.6	97.1	39.2	4.9
Jordan	5.5	60	82.3	91.1	38.8	1.7
Kazakhstan	15.2	6	57.3	99.5	33.9	3.4
Kuwait	2.7	134	98.3	93.3	—	8.9
Kyrgyzstan	5.2	26	35.8	98.7	30.3	1.1
Lebanon	4.0	440	86.6	88.3	—	3.1
Oman	2.5	—	71.5	81.4	—	4.7
Saudi Arabia	23.6	10	81.0	82.9	—	2.6
Syria	18.9	95	50.6	80.8	—	2.1

(continued)

Table 8.7. (*continued*)

Countries	*Population (Millions) 2005*[1]	*Population Density People Per Km*^{2} *2003*[2]	*Urban Population (% of Total) 2005*[3]	*Adult Literacy Rates (%) by 2005*[4]	*Domestic Inequality (Gini Index) by 2005*[5]	*Total Footprint gha Per Capita 2005*[6]
Tajikistan	6.6	45	24.7	99.5	32.6	0.7
Turkey	73.0	92	67.3	87.4	43.6	2.7
Turkmenistan	4.8	10	46.2	98.8	40.8	3.9
United Arab Emirates	4.1	—	76.7	88.7	—	9.5
Uzbekistan	26.6	62	36.7	99.4	36.8	1.8
Yemen	21.1	36	27.3	54.1	33.4	0.9
Middle East and Central Asia	**364.5**	**28**[b]	**55.1**[e]	**70.3**[e]	**—**	**2.3**

1. Human Development Report 2007/2008. *Fighting Climate Change: Human Solidarity in a Divided World.* Table 5: Demographic Trends, pp. 243–246. See also Global Footprint Network (2008). *The Ecological Footprint Atlas 2008, October 28*, pp. 48–49.
2. World Development Report 2005. *A Better Investment Climate for Everyone*. The World Bank. Table 1: Key Indicators of Development, pp. 256–259.
3. Human Development Report 2007/2008. *Fighting Climate Change: Human Solidarity in a Divided World.* Table 5: Demographic Trends, pp. 243–246. Note: Note: Because data are based on national definitions of what constitutes a city or metropolitan area, cross country comparisons should be made with caution.
4. Human Development Report 2007/2008. *Fighting Climate Change: Human Solidarity in a Divided World.* Table 1: Human Development Index, pp. 229–232.
5. Human Development Report 2007/2008. *Fighting Climate Change: Human Solidarity in a Divided World.* Table 15: Inequality in Income or Expenditure, pp. 281–284. Note: A value of zero (0) represents absolute equality, and a value of hundred (100) absolute inequality.
6. Global Footprint Network (2008). *The Ecological Footprint Atlas 2008, October 28*, table 3, pp. 51–54.

a. Global Footprint Network (2008). *The Ecological Footprint Atlas 2008, October 28*, table 2, p. 47.
b. Middle East and North Africa.
c. Spencer, William (2009). *Global Studies: The Middle East, Twelfth Edition* (New York: McGraw Hill), pp. 72–84.
d. Human Development Report 2007/2008. *Fighting Climate Change: Human Solidarity in a Divided World.* Table 1a: Basic Indicators for Other UN Member States, 233.
e. Arab States only and Central Asia not included.

Table 8.8. Descriptive Statistics

	N	*Minimum*	*Maximum*	*Mean*	*Std. Deviation*
Human Development Index	19	.508	.932	.76217	.090720
GDP Per Capita	20	930	26321	8725.30	8420.708
Gender-Related Development Index	17	.472	.927	.75400	.098941
Gender Inequality Index	17	.332	.853	.61324	.132503
Carbon Dioxide Emissions	19	3.6	433.3	98.437	119.1094
Total Footprint	21	4.3	198.6	40.324	53.4153
Bio-Capacity	21	1.4	120.9	22.252	32.5966
Ecological Balance	21	−87.6	13.5	−18.086	25.7581
Valid N (listwise)	14				

Table 8.9. Correlations

	Human Development Index	*GDP Per Capita*	*Gender-Related Development Index*	*Gender Inequality Index*	*Carbon Dioxide Emissions*	*Total Footprint*	*Bio-Capacity*
Human Development Index							
Pearson Correlation	1	.828**	.995**	–.758**	.238	.104	–.017
Sig. (2-tailed)		.000	.000	.001	.327	.673	.944
N	19	19	17	15	19	19	19
GDP Per Capita							
Pearson Correlation	.828**	1	.796**	–.633**	.258	.086	–.079
Sig. (2-tailed)	.000		.000	.009	.285	.718	.741
N	19	20	17	16	19	20	20
Gender-Related Development Index	.995**	.796**	1	–.793**	.209	.084	–.021
Sig. (2-tailed)	.000	.000		.001	.420	.749	.937
N	17	17	17	14	17	17	17
Gender Inequality Index							
Pearson Correlation	–.758**	–.633**	–.793**	1	.218	.118	.201
Sig. (2-tailed)	.001	.009	.001		.435	.653	.440
N	15	16	14	17	15	17	17

Carbon Dioxide Emissions							
Pearson Correlation	.238	.258	.209	.218	1	.835**	.775**
Sig. (2-tailed)	.327	.285	.420	.435		.000	.000
N	19	19	17	15	19	19	19
Total Footprint							
Pearson Correlation	.104	.086	.084	.118	.835**	1	.934**
Sig. (2-tailed)	.673	.718	.749	.653	.000		.000
N	19	20	17	17	19	21	21
Bio-Capacity							
Pearson Correlation	–.017	–.079	–.021	.201	.775**	.934**	1
Sig. (2-tailed)	.944	.741	.937	.440	.000	.000	
N	19	20	17	17	19	21	21
Ecological Balance							
Pearson Correlation	–.243	–.285	–.209	.011	–.768**	–.892**	–.672**
Sig. (2-tailed)	.317	.224	.421	.967	.000	.000	.001
N	19	20	17	17	19	21	21

**Correlation is significant at the 0.01 level (2-tailed).

Table 8.10. Variables Entered/Removed[b]

Model	*Variables Entered*[a]	*Variables Removed*	*Method*
1	Ecological Balance, Gender Inequality Index, GDP Per Capita, Bio-Capacity, Carbon Dioxide Emissions, Human Development Index, Gender-Related Development Index	.	Enter

a. All requested variables entered.
b. Dependent Variable: Total Footprint

Table 8.11. Model Summary[c]

				Change Statistics				
Model	*R*	*R Square*	*Adjusted R Square*	*R Square Change*	*F Change*	*df1*	*df2*	*Sig. F Change*
1	.943[a]	.889	.880	.889	96.276	1	12	.000
2	1.000[b]	1.000	1.000	.111	455424.895	1	11	.000

a. Predictors: (Constant), Ecological balance, Gender Inequality Index, GDP Per Capita, Bio-Capacity, Carbon Dioxide Emissions, Human Development Index, Gender Related Index.
b. Dependent Variable: Total Footprint

Table 8.12. ANOVA[b]

Model		*Sum of Squares*	*df*	*Mean Square*	*F*	*Sig.*
1	Regression	51629.865	7	7375.695	342668.541	.000[a]
	Residual	.129	6	.022		
	Total	51629.994	13			

a. Predictors: (Constant), Ecological Balance, Gender Inequality Index, GDP Per Capita, Bio-Capacity, Carbon Dioxide Emissions, Human Development Index, Gender-Related Development Index
b. Dependent Variable: Total Footprint

Table 8.13. Coefficients[a]

Model	*Unstandardized Coefficients*		*Standardized Coefficients*		
	B	*Std. Error*	*Beta*	*t*	*Sig.*
1 (Constant)	–.630	1.220		–.516	.624
Human Development Index	–5.219	17.394	–.009	–.300	.774
GDP Per Capita	3.167E-6	.000	.000	.149	.886
Gender-Related Development Index	5.571	16.918	.010	.329	.753
Gender Inequality Index	.721	1.791	.001	.403	.701
Carbon Dioxide Emissions	.000	.001	–.001	–.446	.671
Bio-Capacity	1.000	.002	.620	439.371	.000
Ecological Balance	–1.001	.003	–.465	–398.205	.000

a. Dependent Variable: Total Footprint

Table 8.14. Descriptive Statistics

	N	*Minimum*	*Maximum*	*Mean*	*Std. Deviation*
Population	21	2.5	73.0	17.357	20.1929
Population Density	17	6	440	97.53	115.065
Urban Population	20	24.7	98.3	62.180	21.3162
Adult Literacy	21	28.0	100.0	86.843	17.6135
Pub Exp Health	19	.9	6.1	2.695	1.4308
Pub Exp Education	16	1.3	9.6	4.250	2.0979
Domestic Inequality (Gini)	12	30.3	43.6	37.433	4.2547
Total Footprint Per Capita	20	.5	9.5	2.890	2.4887
Valid N (listwise)	10				

Table 8.15 Correlations

	Population	*Population Density*	*Urban Population*	*Adult Literacy*	*Pub Exp. Health*	*Pub Exp. Education*	*Domestic Inequality (Gini)*	*Total Footprint Per Capita*
Population								
Pearson Correlation	1	–.214	–.024	–.317	.384	.155	.590*	–.223
Sig. (2-tailed)		.409	.921	.161	.105	.567	.043	.344
N	21	17	20	21	19	16	12	20
Population Density								
Pearson Correlation	–.214	1	.548*	.053	.362	–.099	.165	.340
Sig. (2-tailed)	.409		.023	.840	.153	.736	.607	.197
N	17	17	17	17	17	14	12	16
Urban Population								
Pearson Correlation	–.024	.548*	1	–.001	.487*	–.062	.588*	.595**
Sig. (2-tailed)	.921	.023		.996	.034	.820	.044	
N	20	17	20	20	19	16	12	
Adult Literacy								
Pearson Correlation	–.317	.053	–.001	1	–.036	–.652**	.006	.224
Sig. (2-tailed)	.161	.840	.996		.884	.006	.986	.343
N	21	17	20	21	19	16	12	20

Pub Exp. Health								
Pearson Correlation	.384	.362	.487*	−.036	1	.271	.592*	.133
Sig. (2-tailed)	.105	.153	.034	.884		.310	.043	.599
N	19	17	19	19	19	16	12	18
Pub Exp. Education								
Pearson Correlation	.155	−.099	−.062	−.652**	.271	1	−.134	−.208
Sig. (2-tailed)	.567	.736	.820	.006	.310		.711	.457
N	16	14	16	16	16	16	10	15
Domestic Inequality								
Pearson Correlation	.590*	.165	.588*	.006	.592*	−.134	1	.601*
Sig. (2-tailed)	.043	.607	.044	.986	.043	.711		.039
N	12	12	12	12	12	10	12	12
Total Footprint								
Pearson Correlation	−.223	.340	.595**	.224	.133	−.208	.601*	1
Sig. (2-tailed)	.344	.197	.007	.343	.599	.457	.039	
N	20	16	19	20	18	15	12	20

*Correlation is significant at the 0.05 level (2-tailed).
**Correlation is significant at the 0.01 level (2-tailed).

Table 8.16. Variables Entered/Removed[a]

Model	*Variables Entered*	*Variables Removed*	*Method*
1	Population Density	.	Stepwise (Criteria: Probability-of-F-to-enter <= .050, Probability-of-F-to-remove >= .100).
2	Population	.	Stepwise (Criteria: Probability-of-F-to-enter <= .050, Probability-of-F-to-remove >= .100).

a. Dependent Variable: Total Footprint Per Capita

Table 8.17. Model Summary

					Change Statistics				
Model	*R*	*R Square*	*Adjusted R Square*	*Std. Error of the Estimate*	*R Square Change*	*F Change*	*df1*	*df2*	*Sig. F Change*
1	.847[a]	.717	.681	.7132	.717	20.240	1	8	.002
2	.953[b]	.908	.881	.4350	.191	14.509	1	7	.007

a. Predictors: (Constant), Population Density
b. Predictors: (Constant), Population Density, Population

Table 8.18. ANOVA[c]

Model		Sum of Squares	df	Mean Square	F	Sig.
1	Regression	10.295	1	10.295	20.240	.002[a]
	Residual	4.069	8	.509		
	Total	14.364	9			
2	Regression	13.040	2	6.520	34.462	.000[b]
	Residual	1.324	7	.189		
	Total	14.364	9			

a. Predictors: (Constant), Population Density
b. Predictors: (Constant), Population Density, Population
c. Dependent Variable: Total Footprint Per Capita

Table 8.19. Coefficients[a]

Model		Unstandardized Coefficients		Standardized Coefficients		
		B	Std. Error	Beta	t	Sig.
1	(Constant)	.829	.334		2.479	.038
	Population Density	.012	.003	.847	4.499	.002
2	(Constant)	.312	.245		1.275	.243
	Population Density	.013	.002	.922	7.917	.000
	Population	.021	.005	.444	3.809	.007

a. Dependent Variable: Total Footprint Per Capita

Figure 8.1. Map of Middle East and Central Asia

Chapter Nine

Socio-Economic Determinants of National Ecological Footprints in Asia-Pacific Countries

A Multiple Regression Analysis of the Factorial Impacts

INTRODUCTION

Asia and the Pacific Countries include sub-regions: North-East Asia, South-East Asia, South-West Asia, and Central Asia (Global Footprint Network (2009). In this study, Asia-Pacific Countries do not include Central Asia Region Countries such as: Armenia, Azerbaijan, Georgia, Kyrgyzstan, Russian Federation, Tajikistan, Turkey, Turkmenistan, and Uzbekistan. The Asia-Pacific Study Area does not also include countries that are less than one million people during the study period, such as: Bhutan, Maldives, American Samoa, Fiji, Kiribati, Marshal Islands, Micronesia Fed. Sts., Northern Mariana Islands, Palau, Samoa, Solomon Islands, Timor-Leste, Tonga, and Vanuatu.

FOCUS OF STUDY

The study area is included in table 9.1 at the end of this chapter. See also the area map at the end of this chapter.

Currently more than 4.2 billion people live in Asia-Pacific region, consisting sixty-one percent (61%) of the world's population. The population of Asia and the Pacific region increased from 3.3 billion in 1990, to 3.8 billion in 2000, to 4 billion in 2005, and 4.2 billion in 2010 (See United Nations *Statistical Yearbook for Asia and the Pacific*, 2011). The region includes the only two countries in the world, China and India, which have populations exceeding 1billion. The three most populous countries in the region are China (1.3 billion), India (1.1billion), and Indonesia (226 million) (United Nations Development Program. *Human Development Report 2007/2008*, table 5, pp. 243–246). The population growth rate in the Asia-Pacific region has been steadily declining over the last two decades. The region's population growth rates have declined 1.5 percent in the early 1990s to 1.0 percent in 2000 due to declining birth rates across the region and stabilization in death rates over the last two decades (*Statistical Yearbook for Asia and the Pacific*, 2011). The most rapid decline of up to 40 percent was recorded in Cambodia, the Laos People's Republic, and Singapore. Nonetheless, the world population growth rate has been also declining but at a slower rate, averaged slightly at 1.2 percent over a similar period. In China and India that have populations exceeding 1 billion, the growth rates fell to 0.5 percent and 1.2 percent, respectively in 2010.

As expected, fertility rate decline in the region follows the birth rate decline, and currently, the region-wide fertility rate is equal to the "replacement rate" at 2.1 percent, with considerable transformations in the population structure (*Statistical Yearbook for Asia and the Pacific*, 2011). For example, according to the Yearbook, the proportion of the elderly (aged 65 years and above) in Asia-Pacific region increased from 5.3 percent in 1990 to 7.0 percent in 2010, representing a 34 percent increase in their share of the total population, and numbered 294 million. The elderly population is projected to reach 1.3 billion by 2050, constituting almost 25 percent of the total regional population (United Nations, 2009). According to the Statistical Yearbook (2011), in Japan, China, and some other countries in the region, one third (1/3) of the population is expected to be over the age of 60 years by 2050; and should constitute significant challenges to governments and other stakeholders in the region. Population aging will undoubtedly have significant and pervasive social, economic, and political implications.

Thus, planning for the future should give priority considerations to the aging societies in the region. Other remarkable demographic observations in the region include the sex ratio structure and the nature of life expectancy that is highly varied among countries and between genders. Asia-Pacific region has the highest ratio of boys to girls in the world (that hovers just below 100), with population sex ratio of 104 (men per 100 women) in 2010. Life expectancy in the region is highly varied with ranges from 48 years in Afghanistan to 86 years in Japan for women; and for men, 47 years and 79 years respectively in Afghanistan and Japan in 2005–2010 (see *Statistical Yearbook for Asia and the Pacific*, 2011).

Compared to other regions of the world in which urbanized populations overtook their rural counterparts during the second half of 2000 decade, the Asia-Pacific urbanized population is still less than 50 percent. According to Yearbook statistics, during the period the urbanized world population rose from 49 percent in 2005 to 51 percent in 2010, with annual growth rate of 1.9 percent. Relatively, Asia-Pacific region is second to Africa in the least percent of population living in urban areas; as well as having the fastest annual growth rate. For example, 43 percent of Asia and the Pacific population live in urban areas with average annual growth rate of 2.0 percent. The Pacific region of Asia-Pacific is the most urbanized with 71 percent of the population living in cities and towns, although less than 25 percent in Federated States of Micronesia, Papua New Guinea, Samoa, Solomon Island, and Tonga (*Statistical Yearbook for Asia and the Pacific*, 2011, Demographic Trend, p. 1). Asia-Pacific high income countries have an average urbanized population of 75 percent, while the less developed countries (LDCs) of the region have an average of 27 percent. Conversely, countries with the fastest urban population growth rates are also those with the fastest levels of urbanization.

In this region rapid economic growth is closely linked with urbanization that had encouraged rural to urban migration. Thus, Rural-to-Urban migration is caused by economic development in the urban areas, as well as some "push" factors the rural areas such as the inability of households to sustain livelihoods in rural areas, conflicts, natural disasters, and environmental changes such as desertification and saltwater intrusion. Other factors in urban growth also include the regional natural increase of population with growth rate of 2 percent and the reclassification of rural areas as urban. Indeed, Asia-Pacific region contains a majority of world's 21 mega-cities (population exceeding 10 million) in 2010. For example, 12 mega cities are in Asia, including 7 of the largest 10 cities (United Nations Department of Economic and Social Affairs, Population Division, 2010) (see table 9.2).

Asia-Pacific region economy has been growing since the last three decades. In the region, the Pacific Island developing economies are currently recording the highest average annual growth rates of over 2 percent. In fact, growth rates of over 2 percent have been recorded in China, Papua New Guinea, Singapore, Solomon Islands, Timor-Leste, and Vanuatu. More so, poverty and overall inequality have been declining since 1990 decade. For example, the numbers of the poverty-stricken (living on less than PPP $1.25 per day) declined from about 1.6 billion in 1990 to 0.9 billion in 2008. Since the 1990s, inequality has increased in some countries such as: Bangladesh, Cambodia, Nepal, and Sri Lanka. Countries with Gini Index above 40 in 2005 include: Cambodia, China, Nepal, Papua New Guinea, Philippines, Singapore, Sri Lanka, and Thailand. However, inequality also decreased in the 1990s in some countries such as: Indonesia, Malaysia, and Thailand. An important feature in the region is that poverty in urban areas has declined faster than the rural poverty (World Bank, *Povcal Net*. Available here at: http;//go.worldbank.org/WE8P118250). For example, poverty has been reduced faster in urban areas of the most populous countries of the region, namely, China, India, and Indonesia; and in fact, urban poverty has been virtually eliminated in China (see table 9.3). Greater poverty in rural areas is due to the fact that: greater proportion of population in the region lives in rural areas; incidence of poverty tends to be higher in rural than urban areas; and the effects of climate-change related impacts such as, floods, droughts, desertification, and soil erosion, which trigger rural-to-urban migration and increase food insecurity in cities.

PURPOSE OF STUDY

The study analyzes and explains the national and regional variations in total footprints and per capita footprints associated with human development indexes and ecostructural factors in Asia-Pacific region. Ecostructural factors are socioeconomic processes within nations e.g., urbanization, literacy rate, gross domestic product per capita (affluence), and domestic inequality (Gini Index). National Ecological Footprints reflect the consumption patterns of nations, as well as the consumption of life-support services (natural capital). Ecological Footprint Analysis addresses the issue of human biological metabolism and humanity's industrial metabolism (Rees, 2012). Progress towards sustainable development can be assessed using human development index (HDI) as an indicator of well-being, and ecological footprint as a measure of demand on the biosphere. Table 9.4 exposes the inequitable distribution of Asia-Pacific nations' ecological footprints as compared to those of other regions of the world. From this table, Asia-Pacific region had 55 percent of world population in 2005 and with about 22 percent of world's total biological capacity. The region also has 33 percent of world's total ecological footprint, as well as the world's highest total ecological offshoots (–2835.3 million global hectares). Table 9.5 also shows that Asia-Pacific had in 2005 per capita ecological footprint of 1.62 (less than the world's 2.69); biological capacity of 0.82 per capita (2/5 the world's 2.06); and per capita negative ecological balance of –0.80 global hectares, one of the lowest in the world. Using the tools of comparative model and regression analysis, the paper seeks to establish the key factors, among others, that highly influence and drive the ecological footprints in different Asia-Pacific countries and region. The question is what factors are responsible for or are driving the total and per capita footprints in these Asia-Pacific countries. What type of secure environments should meet the needs of both people and natural environments in these mix income consumption countries? The study also proffers some policies, recommendations, and solutions on how to reduce the footprints of Asia-Pacific region and nations, which also will help protect the national and regional environments, as well as promote a more equitable and sustainable society.

SOME SOCIO-ECONOMIC PROCESSES IN ASIA-PACIFIC REGION AND COUNTRIES: THE ECONOMY

Although the economy of Asia-Pacific region and countries has been doing well in recent decades, nevertheless, the financial crisis of second half of the 2000 decade resulted in negative GDP growth rates in the countries that depend on exports. Countries in Asia and the Pacific are vulnerable to both domestic and external shocks. Such domestic shock elements include, changing needs of aging populations and evolving requirement for public services, social welfare policies, and policies of infrastructure investment. Low (LIC) and lower middle income (LMC) countries where the economies are led by domestic demand buoyed the region and kept the regional average growth positive at 0.5 percent (*Statistical Yearbook for Asia and the Pacific*, 2011). Majority of the countries in the region are of low incomes and belong to developing countries of the world with GDP per capita of below $7,567 PPP US$ (United Nations Development Program. *Human Development Report*, 2010, table 16, p. 210). As a matter of fact, the average GDP per capita for the region is about $6604 PPP US$; and only a few countries are of upper middle income (Thailand) and high income (Malaysia, Republic of Korea, New Zealand, Singapore, Japan, and Australia). For different economy classes and their respective countries in Asia-Pacific region that include, Low Income (LIC), Lower Middle Income (LMC), Upper Middle Income (UMC), and High Income (HIC), see table 9.6; and for the classification of world economies by region and income, see World Development Report, 2005: *A Better Investment Climate for Everyone*, p. 255. Large trade surpluses, which accumulate foreign exchange, played an important role in supporting economic growth in East and South-East Asia but not in South Asia. Since 2005, South East Asia achieved fiscal surplus through trade, such as in Viet Nam, Malaysia, Philippines, and Thailand. The ASEAN Free Trade Area (AFTA) is a unifying factor in the region that has united Indonesia, Malaysia, Philippines, Singapore, and Thailand in a commitment to abolish or reduce tariffs on all goods except some agricultural products (*Statistical Yearbook of Asia and the Pacific*, 2011, p. 4). In the Pacific, according to the Yearbook, fiscal surpluses are from tourism earnings and remittances. South and South-West Asia countries recorded fiscal deficit stemming from increases in public expenditure and the underlying structural factor of relatively low revenues of less than 15 percent of GDP, particularly in India and Sri Lanka (p. 4).

Table 9.7 shows that Low Income Region has about 37 percent of the world population. It has also about 10 percent of the world GDP, about 14 percent of total footprints, about 16 percent of world biological capacity, and 7 percent of world ecological balance overshoots. Likewise, Middle Income Region has about 48 percent of the world population. It has also 37 percent of the world GDP, about 39 percent of the total footprint, 50 percent of world biological capacity, and about 2.5 percent of world ecological balance overshoots. The High Income Region has only 15 percent of the world population, with the highest GDP (53.9 percent), and highest ecological balance overshoot (64.5 percent). Since Asia-Pacific countries largely belong to low income region where human development is relatively low, though with the second lowest ecological balance overshoot when compared to Middle Income Region and High Income Region (See table 9.7), the question is: what factors are responsible for or are driving the total and per capita footprints.

WOMEN ISSUES AND DEVELOPMENT

A great deal of gender inequality exists in Asia-Pacific region in terms of gender related development (see table 9.6). According to Asia-Pacific Development indexes in 2011, female participation in the Asia and Pacific labor force remained at 65 employed women per 100 men employed men from 1991 to 2009. More so, women's access to land and property is still "very limited" in some Asia and Pacific countries, such as Bangladesh, India, Sri Lanka, Papua New Guinea, and Mongolia (United Nations, 2011. Asia-Pacific Development—Did You Know?). Women are underrepresented in national and local politics in almost all Asia and Pacific countries, although in Nepal and New Zealand women occupied 30 percent or more seats in their national parliaments during the second half of last decade. In Asia-Pacific region, women composed only 18 percent of the regional research and development (R & D) work force at the middle of 2000 decade, lower than in Africa (33 percent), Latin America and the Caribbean (45 percent), and Europe (34 percent) (see United Nations, 2011. Asia-Pacific Development—Did You Know? p.3). From table 9.6, about half of the Asia-Pacific countries are within the LIC bracket; has low gender related development index value (GDI), ranked low, which is less than the regional average (0.719); and are within medium and high human development index value (HDI). Thus, Asia-Pacific region belongs to mix income economy, according to United Nations *World Development Reports* and *Human Development Reports*, over the years. Likewise, about half of the countries have high gender inequality index value, which is higher than the regional average (0.546).

In Asia-Pacific region, just like in other societies and cultures, poverty impacts women and men unequally, differently, and in disfavor of women due to a number of factors. Such factors include, among others, biased macroeconomic and institutional structures, discriminatory laws and customs, and societal attitudes make it more likely that women will fall into and remain in poverty. Women are particularly vulnerable to exploitation, discrimination and violence, thereby exacerbating their experiences of hardship in many different areas of their lives and presenting them with multiple obstacles to escaping poverty (*United Nations Statistical Yearbook for Asia and the Pacific*, 2011). According to the *Statistical Yearbook* (p. 2), poverty denies women opportunities and the ability to live healthy, long, and productive lives; to participate in decision making; to enjoy basic rights and freedoms, such as access to clean drinking water and sanitation; or even to receive adequate respect and dignity in societies, giving their usually lower status than men.

EDUCATION

During the last decade most Asian and Pacific countries made substantial progress in education by bringing children into school. Between 2000 and

2005, the Asia-Pacific experienced an increase in net enrolment rate (NER) from 86 percent to 89 percent, but since 2005 the rate of increase has slowed down (*EFA Global Monitoring Report, 2011*, p. 8). The Asia-Pacific NER is slightly lower than that of Latin America and the Caribbean at 94 percent, but significantly higher than that of Africa at 77 percent. Public-Sector spending for education in Asia and the Pacific shows mixed trends, with some countries increasing spending while others are cutting spending (see table 8). Public expenditure on education is one indication of the political priority of education in national policy. Public expenditure as a percentage of GDP of most countries in the region remains below the recommended 6 percent threshold (UNESCO and CONFINTEA VI, 2011). In most Asia and Pacific countries, public spending in education in 2005 ranged from about 1 percent in Myanmar (1.3%) to more than 6 percent of GDP in Malaysia (6.2%) and New Zealand (6.5%) (table 9.8). According to the UNESCO document (p. 3), Asia- Pacific region includes the largest number of illiterate adults of any region in the world, as the region by recent data was home to 518 million of the 793 million illiterate adults worldwide, with 416 million of them in South and South-West Asia alone. In fact, female illiterate adults continue to outnumber males and composed 65 percent of the region-wide total, similar to the proportion twenty years ago (64 percent).

HEALTH

Healthcare financing is the key in sustaining and developing healthcare systems that aim at improving human health, because a healthy nation is a wealthy nation. Public resources for healthcare programs in the Asia-Pacific region fall far short of the level needed for ensuring equitable access to essential services. Total expenditure on health as a percentage of GDP in the region was also mixed as in education e.g. in 2004, as small as 0.3 percent in Myanmar and 0.4 percent in Pakistan, to as large as 6.5 percent in Australia and New Zealand, and also 6.3 percent in Japan (see table 9.8). Total expenditure on health includes governmental spending from tax-funded health budgets and social health insurance funds, private health insurance, out-of-pocket spending on personal healthcare (including medicines) and external financing by international partners. Countries in the region exhibit large differences in the mobilization of healthcare financial resources, which is worst in the poor developing countries of the region. Government commitment to health in terms of total health expenditure as a percentage of gross domestic product (GDP) varies in the region. Even in some countries health was de-prioritized in the allocation of public resources, especially from 2000 to 2005, e.g. in Bangladesh, China, the Lao People's Democratic Republic, Myanmar, Mongolia, Papua New Guinea, and Philippines (UNESCO and CONFINTEA VI, p. 2).

ENVIRONMENT

Cities contribute to climate change and are also affected by climate change. For example, about 54 percent of the Asian and Pacific urban population lives in low-lying coastal zones or deltas that are highly vulnerable to sea-level rises, storm-water surges, and flooding, even deforestation; and such cities include: Dhaka, Bangkok, Ho Chi Minh City, Jakarta, Kolkata, Shanghai, and Manila, among others (UN-HABITAT, 2008). A majority of the cities are also indirectly affected by climate-change-related impacts in rural areas such as floods, draughts, desertification, and soil erosion, which increase food insecurity in cities and provide another "push" factor for rural-to-urban migration. These cities (Kolkata, Shanghai, Dhaka, Manila, and Jakarta), according to international ranking, are among the largest 30 urban agglomerations in 2010 (United Nations Department of Economics and Social Affairs, 2010) (see also, table 9.2). Moreover, the poor that contribute least to climate change in most world regions and cultures tend to suffer most from the negative impacts, whether they live in urban or rural areas, and Asia-Pacific region is not an exception. According to World Bank *Povcal Net*, most of these cities are located in countries where the rural proportion of total population is high, for example, India (74.5 percent), (China, 72.6 percent), and Indonesia (69.4 percent); and also where rural population below poverty line is high such as in China (74.1 percent), Indonesia (54.1 percent), and India (52.5 percent) (see also, table 9.a)

In 2006, the world's cities generated an estimated 67 percent of primary energy demand and 71 percent of energy-related global greenhouse gas emissions (International Energy Agency, 2008). In fact, China's largest 35 cities contributed 40 percent of its energy-related carbon dioxide emissions (Dhakal, 2010). Based on United Nations report, Asia and Pacific untreated wastes contribute as much as 75 billion tons of carbon dioxide to the atmosphere every year (United Nations ESCAP et al, 2007). In the second half of 2000 decade, Asia and the Pacific was by far the major energy producer among the world's regions accounting for 46 percent of global production, but as a consumer it ranked as the second most frugal after Africa (40 percent of world average) at 74 percent of the world average (*Statistical Yearbook for Asia and the Pacific*, 2011, p. 1). Per capita footprints in Asia and Pacific developing countries remain relatively low, but growing rapidly, compared with those of developed countries. Carbon dioxide emissions in the region have been increasing, between 2000 and 2008, almost twice as fast as the global average (5.4 percent change per annum as compared to 2.8 percent); and China was the single largest emitter of greenhouse gases worldwide, emitting 6.5 billion tons of CO_2 (0.4 billion tons more than all of North America (*Statistical Yearbook for Asia and the Pacific*, 2011, pp. 1–8). According to the Yearbook (p.1), within the region, Brunei Darussalam is the highest emitter of greenhouse gases at 20 tons per capita of CO_2, followed closely by Australia at 19 tons, while China emits 4.9 tons per capita. Nonetheless, it has been noted that, since the 1990s due to some policies and reformed instituted in the region by the large carbon-emitting economies (UN-ESCAP, 2010). For example, such countries that have recently instigated policies and reforms to reduce their CO_2 intensity by improving energy efficiency in various sectors and increasing the use of renewable energy include China, India, Indonesia, Marshall Islands, Maldives, Mongolia, Papua New Guinea, The Republic of Korea, and Singapore). They have also introduced voluntary targets to reduce CO_2 emissions or reduce the consumption of fossil fuels.

In many regions and cultures of the world, forests impact many aspects of economic and social development; and forests and the people depending on them are under increasing pressures because of land-use changes due to agriculture (e.g. for bio-fuel production), human settlements, unsustainable logging, and inefficient soil management. As noted elsewhere, economic activities that are related to forests influence the life of 1.6 billion people globally, and forests play a major role in the mitigation and attenuation of the effects of climate change (ASEAN Biodiversity Outlook, 2010). During the past two decades, in the case of biodiversity, protected areas, and forests, both primary forest and total forest cover expanded in Asia and Pacific region. In the same period, two-thirds (2/3) of countries in the region experienced an increase in the number of threatened species, and South-East lost nearly one seventh (1/7) of its forest cover (*Statistical Yearbook for Asia and the Pacific*, 2011). According to the Yearbook (p.1), deforestation is highest in low-income countries; and deforestation and forest degradation account for up to 20 percent of global greenhouse gas emissions that contribute to global warming. In the Asia-Pacific region, the South-East is being severely deforested because the growing population depends heavily on timber for livelihood, wood for fuel, and new land to convert into agricultural and industrial estates (*ASEAN Biodiversity Outlook, 2010*).

THEORETICAL FRAMEWORK FOR THE STUDY

Within the biosphere, everything is interconnected, including humans, thus sustainable development requires a good knowledge of human ecology. The human life-support functions of the ecosphere are maintained by nature's biocapacity which runs the risks of being depleted, the prevailing technology notwithstanding. Regardless of the humanity's mastery over the natural environment, it still remains a creature of the ecosphere and always in a state of *obligate dependency* on numerous biological goods and services (Rees, 1992, p. 123). Despite the above *dependency* thesis, the prevailing economic mythology assumes a world in which carrying capacity is indefinitely expandable (Daly, 1986; Solow, 1974). The human species has continued to deplete, draw-down, and confiscate nature's biocapacity with reckless abandon. York et al (2003), indicated that population and affluence account for 95 percent of the variance in total footprints of countries. Thus, large human population all over the world and their excessive consumption of the scarce natural resources are responsible for the national footprint of nations. Thus, total human impact on the ecosphere is given as: population × per capita impact (Ehrlich and Holdren, 1971; Holdren and Ehrlich, 1974); Hardin (1991). In other words, population size and affluence are the primary drivers of environmental impacts (Dietz et al, 2007). The footprints of nations provide compelling evidence of the impacts of consumption, thus the need for humans to change their lifestyles and conserve scarce natural capital. According to Palmer (1998), there are in order of decreasing magnitude, three categories of consumption that contribute enormously to our ecological footprints: wood products (53 percent), food (45 percent), and degraded land (2 percent). Degraded land includes land taken out of ecological availability by buildings, roads, parking lots, recreation, businesses, and industries. Palmer also indicated that about 10 percent or more of earth's forests and other ecological land should be preserved in more or less pristine condition to maintain a minimum base.

The environmental impacts of urban areas should be considered because a rapidly growing proportion of world's population lives in cities, and more than one million people are added to the world's cities each week and majority of them are in developing countries of Africa, Latin America, and Southeast Asia (Wackernagel and Rees, 1996). The reality is that the populations of urban regions of many nations had already exceeded their territorial carrying capacities and depend on trade for survival. Of course, such regions are running an unaccounted ecological deficit; their populations are appropriating and meeting their carrying capacity from elsewhere (Pimentel, 1996; Wackernagel and Rees, 1996; Girardet, 1996, accessed Online on 12/14/07; Rees, 1992; The International Society for Ecological Economics and Island Press, 1994; Vitouset et al, 1986; Wackernagel, 1991; WRI, 1992, p. 374). Undoubtedly, the rapid urbanization occurring in many regions and the increasing ecological uncertainty have implications for world development and sustainability. Cities are densely populated areas that have high ecological footprints which leads to the perception of these populations as "parasitic," since these communities have little intrinsic biocapacity, and instead, must rely upon large hinterlands. Land consumed by urban regions is typically at least an order of magnitude greater than that contained within the usual political boundaries or the associated built-up areas (The International Society for Ecological Economics and Island Press, 1994; Rees, 1992).

According to Rees (1992), every city is an entropic black hole drawing on the concentrated material resources and low-entropy production of a vast and scattered hinterlands many times the size of the city itself. In the same vein, Vitouset et al (1986) asserted that high density settlements "appropriate" or augment their carrying capacity from all over the globe, as well as from the past and the future (see also Wackernagel, 1991). In modern industrial cities, resources flow through the urban system without much concern either about their origins, or about the destination of their wastes, thus, inputs and outputs are considered to be unrelated. The cities' key activities such as transportation, provision of electricity supply, heating, manufacturing and the provision of socio-economic services depend on a ready supply of fossil fuels, usually from far-flung hinterlands than within their usual political boundaries or their associated built-up areas. Cities are not self-contained entities, and their concentration of intense economic processes and high levels of consumption both increase and stimulate their demands on resources. Cities occupy only 2 percent of the world's land surface, but use some 75 percent of the world resources, and release a similar percentage of waste (Girardet, 1996).

Like urbanization, energy footprint, created from energy use and carbon dioxide emissions, is not subject to area constraints. Energy footprint is the area of forest that would be needed to sequester the excess carbon (as carbon dioxide) that is being added to the atmosphere by the burning of fossil fuels to generate energy for travel, heating, lighting, manufacturing, recreation, among other uses. Actually, the demand for energy defines modern cities more than any other single factor. Cities contain enormous concentration of economic activities that consume enormous quantities of energy. The natural global systems of forests and oceans for carbon sequestration are not handling the human carbon contributions fast enough, thus the Kyoto Conference of early 1998 (see UNDP, Human Development Report, 2007/2008, pp. 314–317, for the Status of Major International Environmental Treaties). According to Suplee (1998), only half of the carbon humans generate burning fossil fuels can be absorbed in the oceans and existing terrestrial sinks. The oceans absorb about 35 percent of the carbon in carbon dioxide (Suplee, 1998), equivalent of 1.8 giga tons of carbon every year (IPCC, 2001), while the global forests under optimum management of existing forests could absorb about 15 percent of the carbon in the CO_2 produced from the burning of fossil fuels world-wide (Brown et al, 1996). The energy footprint is caused by the un-sequestrated 50 percent in the atmosphere with the potentially troubling ecological consequences, such as rapid global warming and other environmental stresses, including climate change. Carbon dioxide in the atmosphere will continue to increase unless humanity finds alternative energy sources of sufficient magnitude. It is in the humanity's best interest to get off its petroleum addiction (control and minimize fossil fuel use) and develop sustainable consumption habits.

Literacy affects the consumption of natural capital resources. Highly literate groups concentrate in urban areas where they consume more than their fair shares of biospheric resources. Literate populations generally have lower rates of domestic inequality and tend to consume more resources than their illiterate counterparts due to their higher incomes and higher standards of urban living. Furthermore, higher levels of literacy correspond with higher incomes, which allow for greater consumption (Jorgenson, 2003). This is because literate populations are subject to increased consumerist ideologies and contextual images of good life through advertising (Princen et al, 2002); what Sklair (2001) and Clapp (2002) labeled "cultural ideology of consumerism/consumption."

METHODOLOGY FOR THE STUDY

Unit of Analysis

The unit of analysis is "country" (individual Asia-Pacific countries).

Sample

To test for the national variations in total footprints and per capita footprints associated with human development index, Comparative Model Analysis and Stepwise Regression Analysis were used as the tools of analysis. Using comparative model, a sample of 21 countries out of many countries and subregions was analyzed in tables 9.9–9.10 (see Global Footprint Network. *The Ecological Footprint Atlas 2008*; http://www.unescap.org/stat/data/syb2011). In this study, Asia-Pacific Countries do not include Central Asia Region Countries such as: Armenia, Azerbaijan, Georgia, Kyrgyzstan, Russian Federation, Tajikistan, Turkey, Turkmenistan, and Uzbekistan; and does not also include countries that are less than one million people during the study period, such as: Bhutan, Maldives, American Samoa, Fiji, Kiribati, Marshal Islands, Micronesia Fed. Sts., Northern Mariana Islands, Palau, Samoa, Solomon Islands, Timor-Leste, Tonga, and Vanuatu. Step-wise Regression Analysis was also conducted using the sample. Each table represents Asia-Pacific countries as also represented by Global Footprint Network, *The Ecological Footprint Atlas 2008*, tables 2,3, and 4. The larger population of at least one million is chosen for the study because; population is a variable which affects the consumption rates and levels, therefore the ecological footprint.

Mode of Analysis

This is an explanatory study using Step-wise Regression, as well as Comparative Model, and Descriptive Statistics such as matrices, totals, averages, ratios and percentages. Regression Analysis was performed to flesh out factors that highly impacted national footprints, as well as to strengthen the comparative model analysis results. Potential technical problems are diagnosed that might affect the validity of the result of the analysis such as the missing data (e.g., domestic inequality or Gini index data for Myanmar) or the excluded countries mentioned above, which did not alter the substantive conclusions, especially regarding the ecological footprint accounts and environmental sustainability of different Asia-Pacific countries. The national ecological footprint and ecological balance, as dependent variables, are explained recursively by the country's human development hierarchical category, population size, population density, urbanization, GDP per capita, domestic inequality (Gini index), and literacy rate, as the independent variables. The independent variables mediated in explaining the varying levels of consumption among different human development indexes of the capitalist world economy. It is hypothesized that Human Development positions and Ecostructural Factors are likely to be responsible for the variations in the Ecological Footprints and Balances. The carbon dioxide emission levels (UNDP, *Human Development Report*, 2007/2008, table 24, pp. 310–313) are included to depict the consumption and environmental degradation impacts of different Asia-Pacific countries that reflect their national ecological footprint and ecological balance levels.

HYPOTHESIS FOR THE STUDY

Null Hypothesis (H_0)

None of the independent variables predicts the national footprint effects. The independent variables do not influence national footprints.

Alternative Hypothesis (H_1)

Some (if not all) of the independent variables predict the national footprint effects.

Variables

Dependent and independent variables are selected on the basis of the theoretical themes and underpinnings described above. The measures of the dependent and independent variables used in the present study are as follow:

Dependent Variables

The ecological footprint or consumption and ecological balance are measured in global hectares (1 hectare = 2.47 acres). A global hectare is 1 hectare of biologically productive space with world average productivity (Global Footprint Network: *Africa's Ecological Footprint-2006 Factbook*, pp. 82–90).

Independent Variables

The ecostructural mediating factors are most of the independent variables in this study that include, population, GDP per capita, urbanization (urban population as a percent of total population), literacy rate (%), and domestic inequality (Gini index) (See Appendix A for Variable definitions).

COMPARATIVE MODEL ANALYSIS (RESULTS AND DISCUSSIONS)

The results and findings in this study are discussed under subtitles that include, "Per Capita Ecological Footprint, Biological Capacity, and Ecological Balance by World Regions," "Human Development and Affluence on Biological Capacity, Ecological Footprint and Balance in Asia-Pacific Countries," and "Population and Ecostructural Factors on Total Footprint Per Capita in Asia-Pacific Countries."

PER CAPITA ECOLOGICAL FOOTPRINT, BIOLOGICAL CAPACITY, AND ECOLOGICAL BALANCE BY WORLD REGIONS IN 2005

Different regions of the world have different per capita ecological footprints, biological capacities, and ecological balances. Table 9.5 shows a comparison of per capita ecological footprint, biological capacity, and ecological balance by world regions. The world regions include Asia-Pacific, Africa, Latin America and Caribbean, Middle East and Central Asia, North America, Europe EU, Europe Non-EU. Analysis of table 9.5 exposes the inequitable distribution of Asia-Pacific nations' ecological footprints as compared to those of other world regions. Furthermore, this table shows a comparison of each region's footprint with its biocapacity depicting whether that region has an ecological reserve or is running a deficit. Also from the table, one may find that North America, with its considerable per capita biological capacity in 2005 (6.49 gha) had the largest per capita footprint (9.19 gha) and the largest per capita ecological deficit or overshoot (–2.71 gha); followed by Europe EU (Western Europe) with per capita footprint of 4.69 and per capita ecological deficit of –2.38; followed by Europe Non-EU (Central and Eastern Europe), 3.52 and ecological reserve of 2.29; Latin America and Caribbean, 2.44 and ecological balance or reserve of 2.36; and Middle East and Central Asia, 2.32 and –1.04 respectively. The lowest per capita ecological footprint of 1.37 was in Africa with per capita ecological reserve of 0.43; followed by Asia-Pacific, 1.62 and ecological deficit of –0.80.

The highest per capita ecological reserve is found in LAC (2.36), followed by Europe Non-EU (2.29), and Africa (0.43). Asia-Pacific per capita ecological footprint (1.62), biological capacity (0.82) were less than the world average of 2.69 and 2.06 respectively in 2005, and the ecological balance (–0.80) was more than the world average (–0.63) in the same period. A regional footprint per capita that is less than the world average and per capita ecological balance more than the world average may denote sustainability at the global level (*Living Planet Report, 2006*). The less than world's average in footprint per capita and more than world's average in per capita ecological balance were found in developing countries in 2005, especially in Africa, Middle East and Central Asia, and Latin America and Caribbean regions, but not in Asia-Pacific that was less than in both footprint per capita and ecological balance.; thus not ecologically sustainable.

HUMAN DEVELOPMENT AND AFFLUENCE ON BIOLOGICAL CAPACITY, ECOLOGICAL FOOTPRINT AND BALANCE IN ASIA-PACIFIC COUNTRIES

Table 9.9 presents the national comparisons for both dependent (Ecological footprint and Ecological balance) and some of the independent (Human development or welfare, GDP per capita or affluence, and Carbon dioxide emissions) variables in Asia-Pacific. The table shows that a country's total footprint, ecological balance, and carbon dioxide emissions (waste) are a function of its human development position and affluence level in the world economy. Given the national biological capacity measured in global hectares, the higher the total ecological footprint; the lower the ecological balance and vice versa. Table 9.9 assesses the progress towards sustainability development in Asia-Pacific by using human development index (HDI) and GDP per capita as indicators of national well-being, and the ecological footprint and balance as measures of national demand on the biosphere and biological capacity respectively. The sustainability quotient is depicted by any country with low footprint and high human development levels. In fact, table 9.9 shows that, among Asia-Pacific countries, improvements in the human development indexes (human welfare) have resulted in the worsening of the countries' positions on their ecological footprints and balances, such as in Japan, China, India, Korea Republic, and Indonesia. Table 9.9 shows that Asia-Pacific countries are among the low income countries of less than per capita GNP of $766), although a few countries are of medium human development (HDI of 0.500–0.799) and medium per capita GNP (between $766 and $9,385), as well as high human development (0.800 and above) and high per capita GNP ($9,386 or more), e.g. Australia, Japan, Korea Republic, Malaysia, New Zealand, and Singapore (See table 6; World Bank, 2005, p. 255).

From table 9.9 and based on the existing biological capacity, it becomes apparent that the higher the HDI (human welfare), GDP per capita (affluence) i.e. consumption, the higher the carbon dioxide emissions, ecological footprint, and ecological deficit or offshoot. For example, Japan shows HDI of 0.953, GDP per capita ($31,267), carbon dioxide emission (1257.2 million tons), total footprint (626.6 global hectares), biological capacity (77.2 global hectares), and ecological balance of –549.4 global hectares. Actually, Japan "appropriates" more than 600 percent of its carrying capacity from "elsewhere." However, China and India have low HDI and low GDP per capita (as developing countries), as well as high carbon dioxide emissions and large ecological deficits or overshoots. Their high carbon dioxide emissions (wastes), high total footprints, and large ecological deficits are likely due to having large number of polluting industries, many large and mega cities, and large national populations. Nonetheless, New Zealand and Singapore, although with low national populations, have high HDI and high GDP per capita, but with low carbon dioxide emissions and with low ecological deficits or positive ecological balance. Likewise, Australia has high HDI, high GDP per capita, but moderately high carbon dioxide emissions and positively ecological balance. New Zealand, Australia, and Singapore, although highly industrialized, but might have instituted adequate policies and programs dealing with carbon dioxide emissions. On one hand, and as expected, Cambodia, Lao PDR, Nepal, Papua New Guinea, and Sri Lanka show the lowest HDI, GDP per capita, carbon dioxide emission, total footprint, biological capacity, and ecological balance. In fact, Asia-Pacific depicts of countries with "multiple colors" of HDI, GDP per capita, carbon dioxide emissions, total footprints, bio-capacities, and ecological balances. Subsequently, analysis of table 9 shows that, given a nation's biological capacity, its levels of human development (welfare), GDP per capita (affluence), and carbon dioxide emissions (waste), they help to explain, most of the time but not always, its total ecological footprint (consumption or demand) and ecological balance (reserve or deficit).

POPULATION, ECOSTRUCTURAL FACTORS, AND TOTAL FOOTPRINT PER CAPITA IN ASIA-PACIFIC COUNTRIES

Table 9.10 shows the effects of population and ecostructural factors (socioeconomic processes) on the total ecological footprints of nations. A nation's total population size and population density, as well as its urban population, literacy rate, and domestic inequality (Gini index), lead to its ecological footprint difference and variation vis-à-vis other nations. The above indicators constitute the structural causes of the ecological footprint variations among Asia-Pacific countries. Table 10 shows that a nation with large population, high rates of urbanization and literacy, and low domestic inequality, has high ecological footprint, and vice versa. Large population level, high population density, high level of urbanization, high level of literacy, and low domestic inequality promote domestic consumption of biospheric resources.

Cities are not self contained entities and have little intrinsic biological capacity; therefore their concentration of intense economic processes and high levels of consumption both increase and stimulate their demands on bioresources. Thus, cities are parasitic and in most cases must rely upon large hinterlands for their supplies which increase their footprint impacts. Correspondingly, high density settlements appropriate carrying capacity from all over the globe, as well as from the past and the future (Vitousek et al, 1986; Wackernagel, 1991). According to international ranking of cities, Asia and the Pacific countries are the locations of the 30 largest urban agglomerations in the world (United Nations Department of Economic and Social Affairs, Population Division, 2010) (see also table 9.2). High levels of urbanization correspond to higher levels of consumption, higher levels of energy use and carbon dioxide emissions (waste), and thus, higher energy footprints and ecological footprints. As mentioned earlier, biospheric resources are consumed at higher levels in urban areas that generally contain high literate groups with higher incomes, which allow for greater material consumption (Jorgenson, 2003; Princen et al, 2002; Sklair, 2001; Clapp, 2002).

According to table 9.10, Singapore (100), Australia (88.2), New Zealand (86.2), Republic of Korea (80.8), Malaysia (67.3), and Japan (65.8) have high urban population as percent of their total populations. China (Shanghai, Beijing, Chongqing, Shenzhen, and Guangzhou, Guangdong), India (Delhi, Mumbai, and Kolkata), Japan (Tokyo, Osaka-Kobe), Bangladesh (Dhaka), Pakistan (Karachi), Philippines (Manila), Russian Federation (Moscow), Turkey (Istanbul), Republic of Korea (Seoul), and Indonesia (Jakarta) are nations with large population and large cities (see table 9.2). Table 9.10 also shows that Australia (99), Japan (99), Republic of Korea (99), New Zealand (99), Mongolia (97.8), Philippines (92.6), Thailand (92.6), Singapore (92.5), China (90.9), Sri Lanka (90.7), Indonesia (90.4), Viet Nam (90.3), Mongolia (89.9), Myanmar (89.9), and Malaysia (88.7) have high adult literacy rates. The table also shows that Australia (35.2), Japan (24.9), Republic of Korea (3.7), New Zealand (36.4), and Mongolia (32.8) with low domestic inequality, have high per capita footprints. What may be deduced from table 9.10 is that, high population density does not depict high footprint per capita but the opposite. For example, Bangladesh (1061), Singapore (6967), India (358), Philippines (273), Sri Lanka (297), and Viet Nam (250), except Republic of Korea (485), all showed low footprints per capita that are less than 1.6, the region's average. Thus, the level of consumption, and not population density, spurred by large urban population, high levels of adult literacy, and low levels of inequality are the major driving forces of environmental sustainability and ecological footprints.

Tables 9.9 and 9.10 depict medium-high human development countries or higher level developing countries with medium-high levels of urbanization, illiteracy, and domestic inequality in 2005. These correspond to low and medium-high levels of income, which allow for low and medium levels of material consumption and footprints. The findings here parallel the theorization above, which emphasizes that the low consumption levels and the accompanying low ecological footprints and high ecological balances (reserves) and vice versa, are based on the nations' positions in the world capitalist economy and system. Most of the Asia-Pacific countries were once colonized by the medieval sovereign nation states of Britain, Portugal, Dutch, Holland, and France, thus the spoken languages in the region (Brunn et al, 2012) . The region is where the process of underdevelopment, economic stagnation, and dependent industrialization limited their populations to consuming a greater share of unprocessed bio-productive resources, rather than industrial processed finished products (see Frank, Andre Gunder (1966, 1967, 1972, 1990) for more discussions on dependency theory in Latin America). The high domestic income inequality or intra-inequality has a negative or declining effect on per capita consumption levels (Bornschier and Chase-Dunn, 1985; Kick, 2000; Kentor, 2001; and Jorgenson, 2003) that yield both low total and per capita ecological footprints.

REGRESSION ANALYSIS AND REGRESSION EQUATION

On one hand regression analysis allows the modeling, examining, and exploring of relationships and can help explain the factors behind the observed relationships or patterns. It is also used for predictions. Regression equation, on the other hand, utilizes the Ordinary Least Square (OLS) technique. This mathematical formula when applied to the explanatory variables is best used to predict the dependent variable that one is attempting to model. Each independent variable or explanatory variable is associated with a regression coefficient describing the strength and the sign of that variable's relationship to the dependent variable. A regression equation might look like the one given below where Y is the dependent variable, the Xs are the explanatory variables, and the Bs are regression coefficients:

$$Y = B_0 + B_1 X_1 + B_2 X_2 + \text{-----} B_n X_n + E \text{ (Random Error Term/Residuals)}$$

See Appendix B for Regression Analysis Terms.

The row of "unstandardized coefficients" or "Bs" gives us the necessary coefficient values for the multiple regression models or equations.

From table 9.9, *Human Development and Affluence on Biological Capacity, Ecological Footprint and Balance in Latin America and Caribbean Countries*, the dependent variable is Total Footprint, and the independent variables are Human Development Index (HDI), GDP Per Capita, Carbon Dioxide Emissions, Bio-Capacity, and Ecological Balance. Also, from table 9.10, *Population, Ecostructural Factors, and Total Footprint Per Capita in Latin America and Caribbean Countries*, the dependent variable is Total Footprint, and independent variables are Population, Population Density, Urban Population, Literacy Rates, and Gini Index.

REGRESSION ANALYSIS (RESULTS AND DISCUSSIONS)

Table 9.11 depicts the descriptive statistics of the data in table 9.9. The measures of variability or dispersion, from table 9.11, expressed using standard deviation and variance, signify how homogeneous (in terms of low levels) Asia-Pacific countries are in terms of Human Development Index (HDI) and how heterogeneous they are (in terms of high levels) in terms of GDP Per Capita, Carbon dioxide Emissions, Total Footprint, Biocapacity, and Ecological Balance. The same is also true in table 9.17 where descriptive statistics of the data in tables 9.8 and 9.10 show low levels of variability associated with Public Expenditure on Health, Total Footprint Per Capita, Gini Index, and Public Expenditure on Education; and high variability with Population and Population Density. Table 9.12 shows the Pearson Correlation (1-tailed test) of the data from table 9.9. From table 9.12, there is a strong correlation or relationship (at the 0.01 level) between Human Development Index and GDP Per Capita; Total Footprint and Carbon Dioxide Emissions, Bio-Capacity, Ecological Balance; Bio-Capacity and Carbon Dioxide Emissions; Carbon Dioxide Emissions and Ecological Balance; and Ecological Balance and Bio-Capacity. Likewise, table 9.18 shows the Pearson Correlation a strong correlation (at 0.05 levels, 1-tailed) Total Footprint and Urban Population, Adult Literacy, Public Expenditure on Health; Urban Population and Adult Literacy Rates, Public Expenditure on Health; and Adult Literacy and Public.

Table 9.14 is the Regression model summary, which shows the Multiple R and R-Squared or Coefficient of Determination are very high and significant. For example, about 99 percent of the dependent variable (Total Footprint) is accounted for by the independent variable, Carbon Dioxide Emissions; about 100 percent of Total Footprint is explained by Carbon Dioxide Emissions and Human Development Index. These are entered in Stepwise Regression Model in table 9.13. The Multiple R is very large (100 percent), which indicates size of the correlation between the observed outcome variables and the predicted outcome variables (based on the regression equation).

Tables 9.19 and 9.20 show the Multiple R and R-Squared between dependent variable (Total Footprint Per Capita) and the independent variables (Public Expenditure on Health, and Urban Population). Nevertheless, their relationships as well as predictability and accountability are also as high and significant as found in table 9.14. From table 9.20, the Model Summary, about 90 percent of the variations in Total Footprint Per Capita is explained or accounted for by Public Expenditure on Health; and about 96 percent of Total Footprint Per Capita is explained by Public Expenditure on Health and Urban Population. Tables 9.15 and 9.21 depict the analysis of variance (ANOVA) and F-Tests. The F-tests show how strong or significant the selected independent variables in the regression models are in predicting the dependent variable. From table 9.15, the F values are high and the residual (observed – predicted) values are small. Since the F-Tests are high and significant, the Null Hypothesis may be rejected by saying that none of the independent variables (predictors) predict the National footprint and per capita footprint effects. It is worthy to note that the results in table 9.21 are as strong as those in table 9.15 and also significant.

The "*Coefficients*[a]" table of the step-wise regression process is used to construct the regression model equation. Using the B columns of the Unstandardized Coefficients in tables 9.16 and 9.22, the regression model equations are constructed from tables 9.9 and 9.10 (including table 9.8).

a. For table 9.9: Human Development, GDP Per Capita, Carbon Dioxide Emissions, Bio-Capacity Factors on Total Footprint, the significant explanatory (predictor) variables are: Carbon Dioxide Emissions and Human Development Index. *The Regression Equation is: Total Footprint = 185.341+0.564 Carbon Dioxide Emissions –237.034 Human Development Index + E (residuals).*
b. Also for table 9.10: Population, Population Density, Urban Population, Literacy Rates, Domestic Inequality (Gini Index) on Total Footprint Per Capita, including Public Expenditure on Health and Education (table 9.8), the significant explanatory (predictor) variables are Public Expenditure on Health and Urban Population. *The Regression Equation is: Total Footprint Per Capita = –0.964 + 0.756 Public Expenditure on Health + 0.034 Urban Population + E (residuals).*

Therefore, given the factors A and B above, *Total National Footprint in Asia-Pacific is predicted or explained by Carbon Dioxide Emissions, Human Development Index, Public Expenditure on Health and Urban Population.*

POLICY RECOMMENDATIONS

Asia and the Pacific region is classified by the United Nations Development Reports as a mix income economy, therefore policy recommendations will address and reflect issues in high, medium, and low income countries.

The above analyses support the theorization and assertion that the global environmental stress is caused primarily by increase in resource consumption (demand) and human population size and growth. The analyses indicate that large human population all over the world and their excessive consumption of the scarce natural resources are responsible for the national footprint of nations. Absolute population growth in Asia-Pacific countries and their rapid growing world largest cities are exerting tremendous pressures on natural resources in the region. This region in 2005 had total bio-capacity, which was approximately half the size of its total footprint and per capita footprint (see tables 9.4 and 9.5). Its net demand on the planet is more than, if not double, it's available capacity. However, as a mix income economy (per capita income ranging from $765 to $9,385 and above in 2005) and relatively medium-high human development region of the world capitalist economy; resource consumption is relatively modest and increasing, as well as taking its toll on bio-capacity's forest resources. Wealth (growth) can be good for the environment, only if public policy and technologies encourage sound practices and the necessary investments are made in environmental sustainability (Sachs, 2007). More affluent countries among them (e.g. Australia, Japan, Republic of Korea, New Zealand, Singapore, and Malaysia) can reduce consumption and still improve their quality of life, if adequate resource management policies are in place. Large nations with rapid economic growth, such as China and India should also embark on population and pollution management. Therefore, long-term investment will be required in many areas, including education, technology, conservation, urban and family planning, and forest management systems. Innovative approaches to meeting human needs without environmental destruction are called for as well as a shift in belief that greater well-being necessarily entails more consumption. The way additional population or material growth can be sustained, without ravaging biodiversity and ultimately destroying the ecological basis of human life is through cutbacks in resource consumption by the wealthy Australia, Japan, New Zealand and Republic of Korea, for equity considerations. The above action is necessary to vacate the ecological space necessary to justify growth in low income countries of Cambodia, Lao PDR, Mongolia, and Papua New Guinea.

Population management in populous countries such as China and India is very essential and necessary since excessive consumption of natural capital is a major factor of national footprints. This could be achieved through the education and empowerment of women, who are major players in reproduction, economy, environmental sustainability, and biodiversity conservation (Aka, 2006). Population should also be dispersed to small and intermediate-sized cities rather than concentrating in major mega-cities such as Tokyo, Delhi; Mumbai, Shanghai, Kolkata, Dhaka, Karachi, Beijing, Manila, and others, that are largely responsible for carbon dioxide emissions, environmental un-sustainability, and climate change. Apart from Australia, Japan, Korea Republic, New Zealand and

Singapore, Gender Inequality is high and public expenditures are low in health and education in low income countries. Investment in education and health is also investment in people and in the future, with their positive externalities in the productivity of the capital stock. Adequate environmental and ecological education, especially in coastal and delta regions, and the willingness to behave, will reduce individual's ecological impacts. Appropriate policies and programs are required to address the socio-economic adjustments and needs of increasing ageing and older persons in Asia-Pacific countries in both rural and urban areas.

CONCLUSIONS

This is an explanatory study using comparative model analysis, regression analysis, and descriptive statistical analysis. The results and the findings show that a country's total footprint, ecological balance, and carbon dioxide emissions (waste) are a function of its human development (welfare) position and affluence in the world economy. As human development and income go up, the environmental performance goes down. Asia-Pacific region is a mix income economy region with mixed national footprint problems, recommendations, and solutions.

The present study shows that, the higher the national population, urban population, public expenditure on health, and human development index (development), the higher the *imbalanced* consumption of resources, as well as the total footprints. The above analysis supports the theorization and assertion that the global environmental stress is caused primarily by the increase in resource consumption (demand) and population size and growth. The regional ecological sustainability calls for the already rich countries to reduce their consumptive appetite and eco-footprints to create the ecological space needed for a justifiable growth in the impoverished countries (Rees, 2012).

APPENDIX A: VARIABLE DEFINITIONS

Human Development Index (HDI): The HDI is a summary measure of human development (human welfare). It measures the average achievements of a country in three basic dimensions of human development (Global Footprint Network: *Africa's Ecological Footprint-2006 Factbook*, p. 89): a long and healthy life, as measured by life expectancy at birth; knowledge, as measured by the adult literacy rate (with two-thirds weight) and the combined primary, secondary and tertiary gross enrolment ratio (with one-third weight); and a decent standard of living, as measured by GDP per capita (PPP US$). Purchasing power parity (PPP) is a rate of exchange that accounts for price difference across countries, allowing international comparisons of real output and incomes. At the PPP US$

(as used in this study), PPP US$1 has the same purchasing power in the domestic economy as $1 has in the United States of America.

Gross Domestic Product (GDP) per capita (PPP US$): GDP is converted to US dollars using the average official exchange rate reported by the International Monetary Fund (IMF). GDP alone does not capture the international relational characteristics as does the human development hierarchy of the world economy, which accounts for a country's relative socio-economic power and global dependence position in the modern world system. It is suggested elsewhere that GDP per capita is an inadequate measure of world-system position but a more appropriate indicator of domestic affluence or internal economic development (Burns, Kentor, and Jorgenson, 2003; Jorgenson, 2003; Dietz and Rosa, 1994). The Gross Domestic Product (GDP) per capita data for this study is taken from UNDP, *Human Development Report 2007/2008*, table 1, and pp. 229–232. See also World Bank (2005), *World Development Report 2005*.

Domestic Income Inequality (Gini Index): The Gini index measures domestic income inequality of different countries, which had remained stable over a time with its impacts on other variables in the study (Bergesen and Bata, 2002; Jorgenson, 2003). Gini index measures the extent to which the distribution of income (or consumption) among individuals or households within a country deviates from a perfectly equal distribution (UNDP, *Human Development Report, 2004*, p. 271)). It measures inequality over the entire distribution of income or consumption. A value of zero (0) represents perfect equality, and a value of hundred (100) represents perfect inequality. Data for domestic income inequality measured by Gini index are taken from World Bank, *World Development Report* (2005), table 2, pp. 258–259 and United Nations Development Program, *Human Development Report* 2007/2008, 15, pp. 281–284.

Gender Inequality Index (GII): The index shows the loss in human development due to inequality between female and male achievements in reproductive health, empowerment, and labor market. It reflects women's disadvantage in those three dimensions. The index ranges from zero (0), which indicates that women and men fare equally, to one (1), which indicates that women fare as poorly as possible in all measured dimensions. The health dimension is measured by two indicators: maternal mortality ratio and the adolescent fertility rate. The empowerment dimension is measured by two indicators: the share of parliamentary seats held by each sex and by secondary and higher education attainment levels.

The labor market dimension is measured is measured by women's participation in the work force. Gender Inequality Index is designed to measure the extent to which national achievements in these aspects of human development are eroded by gender inequality; also to provide empirical foundations for policy analysis and advocacy efforts. It can be interpreted as a percentage loss to potential human development due to shortfalls in the dimensions included.

Countries with unequal distribution of human development also experience high inequality between women and men, and countries with high gender inequality also experience unequal distribution of human development (UNDP *Human Development Reports*). The Gender Inequality Index data for this study is taken from UNDP's *Human Development Report 2010___20th Anniversary Edition. The Real Wealth of Nations: Pathways to Human Development*. Table 4: Gender Inequality Index, pp. 156–160.

Gender-Related Development Index (GDI): The index is one of the indicators of human development developed by the United Nations. The GDI is considered a gender-sensitive extension of the Human Development Index (HDI), which addresses gender-gaps in life expectancy, education, and income. It highlights inequalities in the areas of long and healthy life, knowledge, and decent standard of living between women and men. It measures achievement in the same basic capabilities as the HDI, but takes note of inequality in achievement between women and men (UNDP, *Human Development Reports*). The methodology used imposes a penalty for inequality, such that the GDI falls when the achievement levels of both women and men in a country go down or when the disparity in basic capabilities, the lower a country's GDI compared with its HDI. The GDI is simply the HDI discounted, or adjusted downwards, for gender inequality. Thus, if GDI goes down goes down, HDI goes down, while GII goes up. The methodology used to construct the GDI could be used to assess inequalities not only between men and women, but also between other groups such as rich and poor, young and old, etc. (UNDP, *Human Development Reports*). The Gender-Related Development Index data for this study is taken from UNDP's *Human Development Report 2010___20th Anniversary Edition. The Real Wealth of Nations: Pathways to Human Development*. Table 4: Gender Inequality Index, pp. 156–160.

Urbanization Level (Urban Population as percent of Total Population): Cities are not self-contained entities and their concentration of intense economic processes and high levels of consumption both increase and stimulate their demands on resources. The cities have limited intrinsic biocapacity which undoubtedly must rely upon large hinterlands. Land consumed by urban regions is typically at least an order of magnitude greater than that contained within the usual political boundaries or the associated built-up area (The International Society for Ecological Economics and Island Press, 1994; Rees, 1992). The data are taken from UNDP, *Human Development Report 2007/2008*, table 5, and pp. 243–246.

Literacy Rate: This variable refers to the percent of a nation's population over the age of fifteen (15) that can read and write in any language of their choice. Literate population generally has low domestic inequality and tends to consume more resources than their illiterate counterpart due to their higher income and urban living. High literate groups concentrate in urban areas where they consume more than their fair shares of biospheric resources. Higher levels of literacy correspond with higher incomes, which allow for greater consumption (Jorgenson, 2003). This is because literate populations are subject to increased consumerist ideologies and contextual images of good life through advertising (Princen et al, 2002), what Leslie Sklair (2001) and Jennifer Clapp (2002) labeled "cultural ideology of consumerism/consumption." The data for literacy rate is taken from the UNDP, *Human Development Report 2007/2008*, table 1, pp. 229–232.

Population and Population Density: Apart from consumption, many have attributed to population as driving most of the sustainability problems (Palmer, 1998; Pimentel, 1996). Likewise, Dietz et al (2007) concluded that population size and affluence are the primary drivers of environmental impacts. Population growth and increases in consumption in many parts of the world have increased humanity's ecological burden on the planet. York et al (2003) indicated that population and affluence account for 95 percent of the variance in total footprints of countries. Others also see the ensuing human impact or footprint as a product of population, affluence (consumption), and technology (i.e. I = PAT (Ehrlich and Holdren, 1971; Holdren and Ehrlich, 1974; Hardin, 1991). Population as a variable in this study is taken from, World Bank (2005), *World Development Report 2005*, table 1, and pp. 256–259.

Government Social Spending Per Capita: Spending by a government (federal, state, and local), municipality, or local authority, which covers such things as spending on healthcare, education, pensions, defense, welfare, interest, and other social services, and is funded by tax revenue, seigniorage, or government borrowing. Public expenditure exerts an effect on economic growth rate through the positive externality in the productivity of the capital stock. Investment in education and health is also investment in people and in the future (See, Education: Crisis Reinforces Importance of a Good Education, says OECD. Accessed Online on 9/23/2011, at http://www.oecd.org/document/21/0,3746. Data for this study were taken from: UNDP, *Human Development Report, 2007/2008. Fighting Climate Change: Human Solidarity in a Divided World.* Table 28: Gender-related Development Index, pp. 326–329; and UNDP, *Human Development Report 2010___20th Anniversary Edition. The Real Wealth of Nations: Pathways to Human Development*. Table 4: Gender Inequality Index, pp. 156–160.

APPENDIX B: REGRESSION ANALYSIS TERMS

Correlation or co-relation: refers to the departure of two variables from independence or they are non-independent or redundant (Antonia D'Onofrio, 2001/2002; Richard Lowry, 1999–2008).

Collinearity: Refers to the presence of exact linear relationships within a set of variables, typically a set of explanatory (predictor) variables used in a regression-type model. It means that within the set of variables, some of the variables are (nearly) totally predicted by the other variables [(Rolf Sundberg, 2002). *Encyclopedia of Environmetrics*, edited by Abdel H. El-Shaarawi and Walter W. Piegorsch (Chichester: John Wiley & Sons, Ltd), Volume 1, pp. 365–366].

Partial Correlation Coefficients (r): When large, it means that there is no mediating variable (a third variable) between two correlated variables (Antonia D'Onofrio, 2001/2002).

Pearson's Correlation Coefficient (r): This is a measure of the strength of the association between two variables. It indicates the strength and direction of a linear relationship between two random variables. Value ranges from - to +1; –1.0 to –0.7 Strong negative association; –0.7 to –0.3 Weak negative association; –03 to +0.3 Little or no association; +0.3 to +0.7 Weak positive association; +0.7 to 1.0 Strong positive correlation (Brian Luke, "Pearson's Correlation Coefficient," Learning *From The Web.net*. Accessed Online on 5/30/2008).

Multiple "R": Indicates size of the correlation between the observed outcome variable and the predicted outcome variable (based on the regression equation).

"R^2" or Coefficient of Determination: Indicates the amount of variation (%) in the dependent scores attributable to all independent variables combined, and ranges from 0 to 100 percent. It is a measure of model performance, summarizing how well the estimated Y values match the observed Y values.

"Adjusted R^2": The best estimate of R^2 for the population from which the sample was drawn. The Adjusted R-Squared is always a bit lower than the Multiple R-Squared value because it reflects model complexity (the number of variables) as it relates to the data.

R^2 and the *Adjusted R^2* are both statistics derived from the regression equation to quantify model performance (Scott and Pratt, 2009. *ArcUser*).

Standard Error of Estimate: Indicates the average of the observed scores around the predicted regression line.

Residuals: These are the unexplained portion of the dependent variable, represented in the regression equation as the random error term (E). The magnitude of the residuals from a regression equation is one measure of model fit. Large residuals indicate poor model fit. Residual = Observed – Predicted.

ANOVA: Decomposes the total sum of squares into regression (= explained) SS and residual (= unexplained) SS.

F-test in ANOVA represents the relative magnitude of explained to unexplained variation. If F-test is highly significant (p = .000), we reject the null-hypothesis that none of the independent variables predicts the effect (scores) in the population.

The "constant" represents the intercept in the equation and the coefficient in the column labeled by the independent variables.

REFERENCES

Aka, Ebenezer (2006). "Gender Equity and Sustainable Socio-Economic Growth and Development," *The International Journal of Environmental, Cultural, Economic & Social Sustainability*, Volume 1, Number 5, pp. 53–71.

ASEAN Biodiversity Outlook (2010). ASEAN Cooperation on Environment. Association of South-East Asian Nations.

Bornschier, Volker and Christopher Chase-Dunn (1985). *Transnational Corporations and Underdevelopment* (New York: Praeger).

Brown, S., Jayant, S., Cannell, M., and Kauppi, P. (1996). "Mitigation of Carbon Emissions to the Atmosphere by Forest Management," *Commonwealth Forestry Review*, 75, 79–91.

Brunn, et al (2012). *Cities of the World. World Regional Urban Development, 5th Edition* (Lanham, Boulder, New York, Toronto, Plymouth/UK: Rowman & Littlefield Publishers, Inc.).

Burns, Thomas J., Jeffery Kentor, and Andrew K. Jorgenson (2003). "Trade Dependence, Pollution, and Infant Mortality in Less Developed Countries," pp. 14–28, in *Crises and Resistance in the 21st Century World-System*, edited by Wilma A. Dunaway (Westport, CT: Pager).

Clapp, Jennifer (2002). "The Distancing of Waste: Over-consumption in a Global Economy," in *Confronting Consumption*, edited by T. Princen, M. Maniates, and K. Conca (Cambridge, MA: MIT Press) pp. 155–76.

Daly, H.E. (1986). *Beyond Growth: The Economics of Sustainable Development* (Boston, Massachusetts, USA: Beacon Press).

Dhakal, Shobhakar (2010). "GHG Emissions from Urbanization and Opportunities for Urban Carbon Mitigation," *Current Opinion in Environmental Sustainability*, Vol. 2, No. 4 (October 2010), pp. 277–283. Accessible from: www.sciencedirect.com.

Dietz, Thomas and Eugene Rosa (1994). "Rethinking the Environmental Impacts of Population, Affluence, and Technology," *Human Ecology Review*, 1: 277–200.

Dietz, Thomas, Eugene Rosa, Yoor (2007). "Driving the Human Ecological Footprint." *Frontiers in Ecology and the Environment* (February).

D'Onofrio, Antonia (2001/2002). "Partial Correlation." Ed 710 Educational Statistics, Spring 2003. Accessed Online on 6/3/2008 at http://www2.widener.edu/.

Education for all (EFA) Global Monitoring Report (2011). Statistical table 5 (Central Asia, East Asia and the Pacific, and South and West Asia), p. 8.

Ehrlich and Holdren (1971). "Impacts of Population Growth," *Science*, 171, 1212–7.

Frank, Andre Gunder (1966, 1967, 1972, 1990). See him for an in-depth discussion of Dependency Theory and Underdevelopment in Latin America. Some of his works include: *The Development of Underdevelopment*, 1966, MRP; *Capitalism and Underdevelopment in Latin America*, 1967; *Latin America: Underdevelopment or Revolution*, 1972; and *Theoretical Introduction to Five Thousand Years of World System History,* 1990, Review.

Girardet, Herbert (1996). "Giant Footprints," Accessed Online on 12/14/07, at http:// www.gdrc .org/uem/footprints/girardet.html.

Global Footprint Network (2006): *Africa's Ecological Footprint—2006 Fact book*. Global Footprint Network, 1050 Warfield Avenue, Oakland, CA 94610, USA. http://www.footprintnetwork .org/Africa.

Global Footprint Network (2006). *Living Planet Report, 2006*.

Global Footprint Network (2008). *The Ecological Footprint Atlas 2008*, October 28.

Global Footprint Network (2009). *Ecological Footprint Atlas 2009*, November 24, 2009, pp. 48–53.

Hardin, G. (1991). "Paramount Positions in Ecological Economics," in Constanza, R. (editor), *Ecological Economics: The Science and Management of Sustainability* (New York: Columbia University Press).

Holdren and Ehrlich (1974). "Human Population and The Global Environment," *American Scientist*, 62: 282–92.

Human Development Report (2007/2008). *Fighting Climate Change: Human Solidarity in a Divided World*. Table 5: Demographic Trend, pp. 243–246. See also Global Footprint Network (2008). *The Ecological Footprint Atlas 2008*, October 28, p. 48.

Human Development Report (2010). 20th Anniversary Edition. *The Real Wealth of Nations: Pathways to Human Development.*

International Energy Agency (2008). World Energy Outlook (Paris, 2008). Available at: www.worldenergyoutlook.org/2008.asp.

IPCC (2001). Intergovernmental Panel on Climatic Change.

Jorgenson, Andrew K. (2003). "Consumption and Environmental Degradation: A Cross-National Analysis of Ecological Footprint," *Social Problems*, Volume 50, No. 3, pp. 374–394.

Kentor, Jeffrey (2001). "The Long Term Effects of Globalization on Income Inequality, Population Growth, and Economic Development," *Social Problems*, 48: 435–55.

Kick, Edward L. (2000). "World-System Position, National Political Characteristics and Economic Development," Journal of Political and Military Sociology, (Summer), http://www.findarticles.com/p/articles/mi-qa3719.

Living Planet Report (2006). Global Footprint Network.

Lowry, Richard (1999–2008). "Subchapter 3a. Partial Correlation." Accessed Online on 6/3/08 at http://faculty.vassar.edu/lowry/cha3a.html.

Luke, Brian T. "Pearson's Correlation Coefficient." Learning From The Web.net. Accessed Online on 5/30/2008.

Palmer, A.R. (1998). "Evaluating Ecological Footprints," *Electronic Green Journal*. Special Issue 9 (December).

Pimentel, David (1996). "Impact of Population Growth on Food Supplies and Environment." American Association for the Advancement of Science (AAAS) (February, 9). See also GIGA DEATH, Accessed Online on 12/18/07, at http://dieoff.org/page13htm.

Princen, Thomas (2002). "Consumption and Its Externalities: Where Economy Meets Ecology," in *Confronting Consumption*, edited by T. Princen, M. Maniates, and K. Conca (Cambridge, MA: MIT Press) pp. 23–42.

Redefining Progress (2005). Footprints of Nations. 1904 Franklin Street, Oakland, California, 94612. See http://www.RedefiningProgress.org.

Rees, William E. (1992). "Ecological Footprint and Appropriated Carrying Capacity: What Urban Economics Leaves Out," Environment and Urbanization, Vol. 4, No. 2, (October). Accessed Online on 12/14/07, at http://eau.sagepub.com.

Rees, William E. (2012). "Ecological Footprint, Concept of," in Simon Levin (ed.). *Encyclopedia of Biodiversity (2nd Edition)*.

Sachs, Jeffrey D. (2007). "Can Extreme Poverty Be Eliminated"? *Developing World, 07/08* (Dubuque, IA: McGraw Hill0 pp. 10–14.

Scott, Lauren and Monica Pratt (2009). "An Introduction To Using Regression Analysis With Spatial Data." *ArcUser. The Magazine for ESRI Software Users* (Spring), pp. 40–43. Lauren Scott and Monica Pratt are ESRI Geo-processing Spatial Statistics Product Engineer and ArcUser Editor respectively.

Sklair, Leslie (2001). *The Transnational Capitalist Class* (Oxford, UK: Blackwell Press).

Solow, R. M. (1974). "The Economics of Resources or the Resources of Economics," *American Economics Review*, Vol. 64, pp. 1–14.

Sundberg, Rolf (2002). *Encyclopedia of Environmetrics*, edited by Abdel H. El-Shaarawi and Walter W. Piegorsch (Chichester: John Wiley & Sons, Ltd), Volume 1, pp. 365–366.

Suplee, D. (1998). "Unlocking the Climate Puzzle." *National Geographic*, 193 (5), 38–70.

The international Society for Ecological Economics and Island Press (1994). "Investing in Natural Capital: The Ecological Approach to Sustainability." Accessed Online on 12/18/07, at http://www.dieoff.org/page13.htm.

United Nations (2009). *World Population Prospects: The 2008 Revision. The Highlights*. Population Division of the Department of Economic and Social Affairs of the United Nations Secretariat. United Nations, New York. See table 1.1.

United Nations (2009). *Urban Agglomerations 2009*. United Nations Publications, Sale No. E. 10. XIII.7. Available at: http://esa.un.org/undp/wup/Documents/WUP2009.

United Nations (2011). *Statistical Yearbook for Asia and the Pacific*.

United Nations (2011). *Statistical Yearbook for Asia and the Pacific*. "Economic Growth," p. 4. Available at: http://www.unescap.org/stat/data/syb2011.

United Nations (2011). *Statistical Yearbook for Asia and the Pacific*. "Energy Supply and Use." Available at: http://www.unescap.org/stat/data/syb2011/II-Environment/Energy-supply-and-use.asp.

United Nations (2011). *Statistical Yearbook for Asia and the Pacific*. 1.1. Population, p. 147.

United Nations (2011). *Statistical Yearbook for Asia and the Pacific*. "Did You Know?" p. 3.

United Nations Department of Economic and Social Affairs, Population Division (2010). *World Urbanization Prospects, the 2009 Revision*. ESA/P/WP/215, New York. Available at: http://www.nescap.org/stat/data/syb2011/I-People/Urbanization.asp.

United Nations Development Program (2004). *Human Development Report 2004 (HDR). Cultural Liberty in Today's Diverse World* (New York, N.Y: UNDP).

United Nations Development Program (2007/2008). *Human Development Report 2007/2008. Fighting Climate Change: Human Solidarity in a Divided World.*

United Nations Development Program (2010). *Human Development Report 2010. The Real Wealth of Nations: Pathway to Human Development*, p. 187.

UN-ESCAP (2010). "Preview Green Growth, Resources and Resilience, Environmental Sustainability in Asia and the Pacific." Available at: http://www.unescap.org/esd/environment/flagpubs/GGRAP/.

United Nations ESCAP, et al (2007). "Report: Regional Seminar & Study Visit on Community-based Solid Waste Management," Quy Nhon City, 15–16, December 2007. Accessible at: http://www.housing-the-urban-poor.Net/Doc/SWMreport.pdf.

UNESCO and CONFINTEA VI (2011). *Harnessing the Power and Potential of Adult Learning and Education for Viable Future: Belem Framework for Action*, Para.14a, p. 5.

UN-HABITAT (2008/2009). *State of the World's Cities 2008/2009: Harmonious Cities*, HS/1031/08E (Nairobi, 2008).

Vitousek, P., P. Ehrlich, A. Ehrlich and P. Matson (1986). "Human Appropriation of the Products of Photosynthesis," *Bioscience*, Vol. 36, pp. 368–374.

Wackernagel, M. (1991). "Using 'Appropriated Carrying Capacity' as an Indicator: Measuring the Sustainability of a Community." Report for the UBC Task Force on Healthy and Sustainable Communities. UBC School of Community and Regional Planning, Vancouver, Canada.

Wackernagel and Rees (1996). *Our Ecological Footprint* (New Society Publisher).

World Bank (2005). *World Development Report 2005. A Better Investment Climate for Everyone* (New York, N.Y: A Co-publication of the World Bank and Oxford University Press).

World Bank, *Povcal Net*. Available at: http://go.worldbank.org/WE8P118250.

World Resources Institute (WRI) (1992). *World Resources, 1992–1993* (New York: Oxford University Press).

York, Richard, Eugene A. Rosa, and Thomas Dietz (2003). "Footprints on the Earth: The Environmental Consequences of Modernity." *American Sociological Review*, 68: 279–300.

Table 9.1. The Asia-Pacific Study Area Countries and Sub-Regions

Country	*Asia-Pacific Sub-Region*
Australia	Pacific
Bangladesh	South and South-West Asia
Cambodia	South-East Asia
China	East and North-East Asia
India	South and South-West Asia
Indonesia	South-East Asia
Japan	East and North-East Asia
Korea, Republic of	East and North-East Asia
Lao PDR	South-East Asia
Malaysia	South-East Asia
Mongolia	East and North-East Asia
Myanmar	South-East Asia
Nepal	South and South-West Asia
New Zealand	Pacific
Pakistan	South and South-West Asia
Papua New Guinea	Pacific
Philippines	South-East Asia
Singapore	South-East Asia
Sri Lanka	South and South-West Asia
Thailand	South-East Asia
Viet Nam	South-East Asia

Source: http://www.unescap.org/stat/data/syb2011/II-Environment/Air-pollution-and-cllmate-change. See also, Brunn, et al (2012). *Cities of the World, 5th Edition.*

Table 9.2. The Largest 30 Urban Agglomerations in Asia and the Pacific Countries by International Ranking, 2010

World Rank Order	*Country*	*Urban Agglomeration*	*Population (Millions)*
1	Japan	Tokyo	36.67
2	India	Delhi	22.16
4	India	Mumbai (Bombay)	20.04
7	China	Shanghai	16.58
8	India	Kolkata (Calcutta)	15.55
9	Bangladesh	Dhaka	14.65
10	Pakistan	Karachi	13.12
13	China	Beijing	12.39
15	Philippines	Manila	11.63
16	Japan	Osaka-Kobe	11.34
19	Russian Federation	Moskva (Moscow)	10.55
20	Turkey	Istanbul	10.52
22	Republic of Korea	Seoul	9.77
23	China	Chongqing	9.40
24	Indonesia	Jakarta	9.21
26	China	Shenzhen	9.01
28	China	Guangzhou, Guangdong	8.88

Source: United Nations Department of Economic and Social Affairs, Population Division (2010). *World Urbanization Prospects, the 2009 Revision.* ESA/P/WP/215. New York. Available from: Urbanization—Statistical Yearbook for Asia and the Pacific 2011, pp. 2–3. Available Online at: http://www.unescap.org/stat/data/syb2011/I-People/Urbanization.asp

Table 9.3. Proportion of the Rural and Urban Population below Poverty Line of PPP$1.25 per Day* in Three Most Populous Countries of Asia-Pacific Region: China, India, and Indonesia

Country	*Year Data Collected*	*Rural Population Below Poverty Line*	*Urban Population Below Poverty Line*	*Rural Proportion of Total Population*
China	1990	74.1	23.4	72.6
	2005	26.1	1.7	59.6
India	1994	52.5	40.8	74.5
	2005	43.8	36.2	71.3
Indonesia	1990	57.1	47.8	69.4
	2005	24.0	18.7	51.9

*World Bank, *Povcal Net*. Available here at: http://go.worldbank.org/WE8P1I8250

Table 9.4. Total Population, Ecological Footprint, Biological Capacity, and Ecological Balance by World Regions in 2005

Region	*Total Population (Million) 2005*	*Total Ecological Footprint (million gha) 2005*	*Total Biological Capacity (million gha) 2005*	*Total Ecological Balance (million gha) 2005*
Asia-Pacific	3562.1	5758.6	2923.3	–2835.3
% World	55.0	33.0	21.9	
Africa	902.0	1237.5	1627.1	389.6
% World	13.9	7.1	12.2	
Latin America and Caribbean	553.0	1350.8	2655.7	1304.9
% World	8.5	7.7	19.9	
Middle East and Central Asia	365.7	846.6	466.9	–379.9
% World	5.6	4.9	3.5	
North America	330.5	3037.8	2143.3	–894.5
% World	5.1	17.4	16.0	
Europe EU	487.3	2291.8	1128.2	–1163.6
% World	7.5	13.1	8.4	
Europe Non-EU	239.6	842.4	1391.6	549.2
% World	3.7	4.8	10.4	
World	6475.6	17443.6	13361.0	–4082.7

EU: European Union

Percentages may not add up to 100 percent due to (i) rounding up; (ii) some countries that were not up to 1 million people in 2005 were not included; and (iii) some countries that had been at war by 2005 were not included.

Source: Global Footprint Network (2008). *The Ecological Footprint Atlas 2008*, Appendix F, table 2, pp. 46–50.

Table 9.5. Per Capita Ecological Footprint, Biological Capacity, and Ecological Balance by World Regions in 2005

Region	*Ecological Footprint (gha per person) 2005*	*Biological Capacity (gha per person) 2005*	*Ecological Balance (gha per person) 2005*
Asia-Pacific	**1.62**	**0.82**	**–0.80**
Africa	1.37	1.80	0.43
Latin America and Caribbean	2.44	4.80	2.36
Middle East and Central Asia	2.32	1.28	–1.04
North America	9.19	6.49	–2.71
Europe EU	4.69	2.32	–2.38
Europe Non-EU	3.52	5.81	2.29
World	**2.69**	**2.06**	**–0.63**

EU: European Union

Percentages may not add up to 100 percent due to: (i) rounding up; (ii) some countries not up to 1 million people in 2005 were not included; and (iii) some countries that had been at war by 2005 were not included.

Source: Global Footprint Network (2008). *The Ecological Footprint Atlas 2008*. Appendix F, table 1.

Table 9.6. Gender Related Development and Gender Inequality Indexes in Asia-Pacific Countries

Countries	*Economy Class (2005)*[1]	*Human Development Index (HDI) Value (2005)*[2]	*HDI Rank*	*Gender Related Development Index (GDI) Value (2005)*[3]	*GDI Rank*	*HDI Rank Minus GDI Rank*	*Gender Inequality Index (GII) Value (2008)*[4]	*GII Rank*
Australia	HIC	0.962	3	0.960	2	1	0.296	18
Bangladesh	LIC	0.547	140	0.539	121	19	0.734	116
Cambodia	LIC	0.598	131	0.594	114	17	0.672	95
China	LMC	0.777	81	0.7776	73	8	0.405	38
India	LIC	0.619	128	0.600	113	15	0.748	122
Indonesia	LMC	0.728	107	0.721	94	13	0.680	100
Japan	HIC	0.953	8	0.942	13	–5	0.273	12
Korea Rep.	HIC	0.921	26	0.910	26	0	0.310	20
Lao PDR	LIC	0.601	130	0.593	115	15	0.650	88
Malaysia	UMC	0.811	63	0.802	58	5	0.493	50
Mongolia	LIC	0.700	114	0.695	100	14	0.523	57
Myanmar	LIC	0.583	132	—	—	—	—	—
Nepal	LIC	0.534	142	0.520	128	14	0.716	110
New Zealand	HIC	0.943	19	0.935	18	1	0.320	25

(continued)

Table 9.6. *(continued)*

Countries	*Economy Class (2005)*[1]	*Human Development Index (HDI) Value (2005)*[2]	*HDI Rank*	*Gender Related Development Index (GDI) Value (2005)*[3]	*GDI Rank*	*HDI Rank Minus GDI Rank*	*Gender Inequality Index (GII Value (2008)*[4]	*GII Rank*
Pakistan	LIC	0.551	136	0.525	125	11	0.721	112
Papua New Guinea	LIC	0.530	145	0.529	124	21	0.784	133
Philippines	LMC	0.771	90	0.768	77	13	0.623	78
Singapore	HIC	0.922	25	—	—	—	0.255	10
Sri Lanka	LMC	0.743	99	0.735	89	10	0.599	72
Thailand	LMC	0.781	78	0.779	71	7	0.586	69
Viet Nam	LIC	0.733	105	0.732	91	14	0.530	58
Asia-Pacific	**Mix Income Economy**	**0.729**[a]		**0.719**[b]			**0.546**[c]	

a. Asia-Pacific minus Bhutan and Korea DPR countries.
b. Asia-Pacific countries minus Bhutan, Korea DPR, Myanmar, and Singapore.
c. Asia-Pacific countries minus Bhutan, Korea DPR, and Myanmar.

1. World Development Report 2005. *A Better Investment for Everyone*. Page 225. Classification of Economies by Region and Income, FY 2005. World Bank Data. LIC = Low Income, $765 or less GNI per capita in 2005; LMC = Lower Middle Income, $766–3,035; UMC = Upper Middle Income, $3,036–9,385; and HIC = High Income, $9,386 or more.
2. Human Development Report 2007/2008. *Fighting Climate Change: Human Solidarity in a Divided World.* Table 1: Human Development Index, pp. 229–232.
3. Human Development Report 2007/2008. *Fighting Climate Change: Human Solidarity in a Divided World.* Table 28: Gender-related Development Index, pp. 326–329.
4. Human Development Report 2010__20th Anniversary. *The Real Wealth of Nations: Pathways to Human Development.* Table 4: Gender Inequality Index, pp. 156–160.

Table 9.7. Regional Population, Gross Domestic Product (GDP), Total Footprint, Biological Capacity, and Ecological Balance by World Income Regions in 2005

World Regions	*Regional Population (Millions) (2005)*	*GDP PPP (US$) Billions (2005)**	*Total Footprint Million gha (2005)*	*Biological Capacity Global gha (2005)*	*Ecological Balance Million gha (2005)*
Low Income	2370.6	5879.1	2377.2	2089.7	–287.5
% World	36.6	9.7	13.6	15.6	7.0
Middle Income	3097.9	22586.3	6787.0	6684.8	–102.2
% World	47.8	37.3	38.9	50.0	2.5
High Income	971.8	32680.7	6196.0	3561.5	–2634.5
% World	15.0	53.9	35.5	26.7	64.5
World Total	6475.6	60597.3	17443.6	13361.0	–4082.7

*Human Development Report (HDR) 2007/2008. *Fighting Climate Change: Human Solidarity in a Divided World*, table 14, p. 280.

Note: Percentages may not add up to 100 percent due to (i) rounding up; (ii) some countries that were not up to 1 million people in 2005 were not included; and (iii) some countries that had been at war by 2005 were not included.

Source: Global Footprint Network (2008). *The Ecological Footprint Atlas 2008*, Appendix F, table 2, p. 46.

Table 9.8. Public Spending on Health and Education in Asia-Pacific Countries

Countries	*Public Expenditure on Health (% of GDP) 2004*	*Public Expenditure on Education (% of GDP) 2002–2005*
Australia	6.5	4.7
Bangladesh	0.9	2.5
Cambodia	1.7	1.9
China	1.8	1.9
India	0.9	3.8
Indonesia	1.0	3.6
Japan	6.3	3.6
Korea, Rep	2.9	4.6
Lao PDR	0.8	2.3
Malaysia	2.2	6.2
Mongolia	4.0	5.3
Myanmar	0.3	1.3
Nepal	1.5	3.4
New Zealand	6.5	6.5
Pakistan	0.4	2.3
Papua New Guinea	3.0	—
Philippines	1.4	2.7
Singapore	1.3	3.7
Sri Lanka	2.0	—
Thailand	2.3	4.2
Viet Nam	1.5	—

Source: Human Development Report 2007/2008. *Fighting Climate Change: Human Solidarity in a Divided World*. Table 19: Priorities on Public Spending, pp. 294–297.

Table 9.9. Human Development and Affluence on Biological Capacity, Ecological Footprint and Balance in Asia-Pacific Countries in 2005

Countries	Human Development Index (HDI) Value 2005[1]	GDP Per Capita PPP US ($) 2005[2]	Carbon Dioxide Emissions Mil. Tons 2004[3]	Total Footprint Million gha 2005[4]	Bio-Capacity Million gha 2005[4]	Ecological Balance Million gha 2005[4]
Australia	0.962	31794	326.6	157.2	310.9	153.5
Bangladesh	0.547	2053	37.1	81.5	35.6	–45.9
Cambodia	0.598	2727	0.5	13.3	13.1	–0.2
China	0.777	6757	5007.1	2786.8	1132.7	–1654.1
India	0.619	3452	1342.1	986.3	452.1	–534.2
Indonesia	0.728	3843	378.0	211.3	310.1	98.8
Japan	0.953	31267	1257.2	626.6	77.2	–549.4
Korea, Rep	0.921	22029	465.4	178.9	33.4	–145.5
Lao PDR	0.601	2039	1.3	6.3	13.8	7.6
Malaysia	0.811	10882	177.5	61.3	67.8	6.5
Mongolia	0.700	2107	8.5	9.3	38.8	29.5
Myanmar	0.583	1027	9.8	56.0	75.7	19.7
Nepal	0.534	1550	3.0	20.7	10.0	–10.7
New Zealand	0.943	24996	31.6	31.0	56.6	25.6
Pakistan	0.551	2370	125.6	130.2	67.3	–62.9
Papua New Guinea	0.530	2563	2.4	10.0	26.2	16.2
Philippines	0.771	5137	80.5	72.2	45.2	–27.0
Singapore	0.922	29663	52.2	18.0	0.2	–17.9
Sri Lanka	0.743	4595	11.5	21.2	7.8	–13.5
Thailand	0.781	8677	267.9	136.9	62.9	–74.0
Viet Nam	0.733	3071	98.6	106.2	67.7	–38.6
Asia-Pacific[a]	**0.771[a]**	**6604[a]**	**6682.0[a]**	**5758.6**	**2923.3**	**–2835.3**

a. East-Asia and the Pacific.

1. Human Development Report 2007/2008. *Fighting Climate Change: Human Solidarity in a Divided World.* Table 1: Human Development Index, pp. 229–232.
2. Human Development Report 2007/2008. *Fighting Climate Change: Human Solidarity in a Divided World.* Table 1: Human Development Index, pp. 229–232.
3. Human Development Report 2007/2008. *Fighting Climate Change: Human Solidarity in a Divided World.* Table 24: Carbon Dioxide Emissions and Stocks, pp. 310–313.
4. Global Footprint Network (2008). *The Ecological Footprint Atlas 2008, October 28*, tables 2 and 4, pp. 46–50 and pp. 55–58, respectively.

Table 9.10. Population, Ecostructural Factors, and Total Footprint Per Capita in Asia-Pacific Countries in 2005

Countries	*Population (Millions) 2005*[1]	*Population Density People Per Km*^{2} *2003*[2]	*Urban Population % of Total 2005*[3]	*Adult Literacy Rate (%) by 2005*[4]	*Domestic Inequality Gini Index by 2005*[5]	*Total Footprint gha Per Capita 2005*[6]
Australia	20.3	3	88.2	99	35.2	7.8
Bangladesh	153.3	1061	25.1	47.5	33.4	0.6
Cambodia	14	76	19.7	73.6	41.7	0.9
China	1313	138	40.4	90.9	46.9	2.1
India	1134.4	358	28.7	61	36.8	0.9
Indonesia	226.1	118	48.1	90.4	34.3	1
Japan	127.9	349	65.8	99	24.9	4.9
Korea, Rep	47.9	485	80.8	99	31.6	3.7
Lao PDR	5.7	25	20.6	68.7	34.6	1.1
Malaysia	25.7	75	67.3	88.7	49.2	2.4
Mongolia	2.6	2	56.7	97.8	32.8	3.5
Myanmar	48	75	30.6	89.9	—	1.1
Nepal	27.1	172	15.8	48.6	47.2	0.8
New Zealand	4.1	15	86.2	99	36.4	7.7
Pakistan	158.1	193	34.9	49.9	30.6	0.8
Papua New Guinea	6.1	12	13.4	57.3	50.9	1.7
Philippines	84.6	273	62.7	92.6	44.5	0.9
Singapore	4.3	6967	100	92.5	42.5	4.2
Sri Lanka	19.1	297	15.1	90.7	40.2	1
Thailand	63	121	32.3	92.6	42	2.1
Viet Nam	85	250	26.4	90.3	34.4	1.3
Asia-Pacific	**3592.8**[a]	—	**42.8**[b]	**90.7**[b]	—	**1.6**

a. Total minus country of Bhutan that is less than 1 million people in 2005.
b. East-Asia and the Pacific.

1. Human Development Report 2007/2008. *Fighting Climate Change: Human Solidarity in a Divided World*. Table 5: Demographic Trends, pp. 243–246. See also Global Footprint Network (2008). *The Ecological Footprint Atlas 2008, October 28*, pp. 48.
2. World Development Report 2005. *A Better Investment Climate for Everyone*. The World Bank. Table 1: Key Indicators of Development, pp. 256–259.
3. Human Development Report 2007/2008. *Fighting Climate Change: Human Solidarity in a Divided World*. Table 5: Demographic Trends, pp. 243–246. Note: Note: Because data are based on national definitions of what constitutes a city or metropolitan area, cross country comparisons should be made with caution.
4. Human Development Report 2007/2008. *Fighting Climate Change: Human Solidarity in a Divided World*. Table 1: Human Development Index, pp. 229–232.
5. Human Development Report 2007/2008. *Fighting Climate Change: Human Solidarity in a Divided World*. Table 15: Inequality in Income or Expenditure, pp. 281–284. Note: A value of zero (0) represents absolute equality, and a value of hundred (100) absolute inequality.
6. Global Footprint Network (2008). *The Ecological Footprint Atlas 2008, October 28*, table 3, pp. 51–54.

Table 9.11. Descriptive Statistics

	N	*Minimum*	*Maximum*	*Mean*	*Std. Deviation*
Human Development Index	21	.530	.962	.72895	.149720
GDP Per Capita	21	1027	31794	9647.57	10914.993
Carbon Dioxide Emissions	21	.5	5007.1	461.162	1108.9089
Total Footprint	21	6.3	2786.8	272.438	622.1997
Bio-Capacity	21	.2	1132.7	138.338	256.7138
Ecological Balance	21	−1654.1	153.5	−134.119	388.3720
Valid N (listwise)	21				

Table 9.12. Correlations

	Total Footprint	*Human Development Index*	*GDP Per Capita*	*Carbon Dioxide Emissions*	*Bio-Capacity*	*Ecological Balance*
Pearson Correlation						
Total Footprint	1.000	,111	.007	.995	.946	–.977
Human Development Index	.111	1.000	.883	.168	.107	–.108
GDP Per Capita	.007	.883	1.000	.059	–.015	–.022
Carbon Dioxide Emissions	.995	.168	.059	1.000	.935	–.976
Bio-Capacity	.946	.107	–.015	.935	1.000	–.855
Ecological Balance	–.977	–.108	–.022	–.976	–.855	1.000
Sig. (1-tailed)						
Total Footprint		.315	.487	.000	.000	.000
Human Development Index	.315		.000	.234	.321	.321
GDP Per Capita	.487	.000		.400	.473	.462
Carbon Dioxide Emissions	.000	.234	.400		.000	.000
Bio-Capacity	.000	.321	.473	.000		.000
Ecological Balance	.000	.321	.462	.000	.000	
N						
Total Footprint	21	21	21	21	21	21
Human Development Index	21	21	21	21	21	21
GDP Per Capita	21	21	21	21	21	21
Carbon Dioxide Emissions	21	21	21	21	21	21
Bio-Capacity	21	21	21	21	21	21
Ecological Balance	21	21	21	21	21	21

Table 9.13. Variables Entered/Removed[a]

Model	*Variables Entered*	*Variables Removed*	*Method*
1	Carbon Dioxide Emissions	.	Stepwise (Criteria: Probability-of-F-to-enter <= .050, Probability-of-F-to-remove >= .100).
2	Human Development Index	.	Stepwise (Criteria: Probability-of-F-to-enter <= .050, Probability-of-F-to-remove >= .100).

a. Dependent Variable: Total Footprint

Table 9.14. Model Summary[c]

Model	R	R Square	Adjusted R Square	Std. Error of the Estimate	*Change Statistics*					Durbin-Watson
					R Square Change	F Change	df1	df2	Sig. F Change	
1	.995[a]	.990	.989	65.0410	.990	1811.269	1	19	.000	
2	.996[b]	.993	.992	55.7257	.003	7.883	1	18	.012	2.178

a. Predictors: (Constant), Carbon Dioxide Emissions
b. Predictors: (Constant), Carbon Dioxide Emissions, Human Development Index
c. Dependent Variable: Total Footprint

Table 9.15. ANOVA[c]

Model		*Sum of Squares*	*df*	*Mean Square*	*F*	*Sig.*
1	Regression	7662271.930	1	7662271.930	1811.269	.000[a]
	Residual	80376.339	19	4230.334		
	Total	7742648.270	20			
2	Regression	7686751.853	2	3843375.926	1237.660	.000[b]
	Residual	55896.417	18	3105.356		
	Total	7742648.270	20			

a. Predictors: (Constant), Carbon Dioxide Emissions
b. Predictors: (Constant), Carbon Dioxide Emissions, Human Development Index
c. Dependent Variable: Total Footprint

Table 9. 16. Coefficients[a]

	Unstandardized Coefficients		Standardized Coefficients			Correlations			Collinearity Statistics	
Model	*B*	*Std. Error*	*Beta*	*t*	*Sig.*	*Zero-order*	*Partial*	*Part*	*Tolerance*	*VIF*
1 (Constant)	15.031	15.428		.974	.342					
Carbon Dioxide Emissions	.558	.013	.995	42.559	.000	.995	.995	.995	1.000	1.000
2 (Constant)	185.341	62.082		2.985	.008					
Carbon Dioxide Emissions	.564	.011	1.004	49.440	.000	.995	.996	.990	.972	1.029
Human Development Index	–237.034	84.423	–.057	–2.808	.012	.111	–.552	–.056	.972	1.029

a. Dependent Variable: Total Footprint

Table 9.17. Descriptive Statistics

	N	*Minimum*	*Maximum*	*Mean*	*Std. Deviation*
Population	21	2.6	1313.0	170.014	356.7377
Population Density	21	2	6967	526.90	1494.616
Adult Literacy	21	47.5	99.0	81.857	18.4073
Urban Population	21	13.4	100.0	45.657	27.1396
Publ Exp. Educ.	20	.0	40.2	5.235	8.3862
Publ Exp. Health	21	.3	6.5	2.343	1.9268
Gini Index	20	24.9	50.9	38.505	6.9328
Total Footprint	21	.6	7.8	2.405	2.1671
Valid N (listwise)	19				

Table 9.18. Correlations

	Total Footprint	*Population*	*Urban Population*	*Population Density*	*Adult Literacy*	*Publ Exp. Educ.*	*Publ Exp. Health*	*Gini Index*
Pearson Correlation								
Total Footprint	1.000	–.208	.801	.132	.609	–.047	.902	–.242
Population	–.208	1.000	–.197	–.101	–.147	–.146	–.228	.166
Urban Population	.801	–.197	1.000	.428	.638	–.180	.588	–.102
Population Density	.132	–.101	.428	1.000	.066	–.045	–.177	.131
Adult Literacy	.609	–.147	.638	.066	1.000	.172	.601	–.041
Publ Exp. Educ.	–.047	–.146	–.180	–.045	.172	1.000	.054	.098
Publ Exp. Health	.902	–.228	.588	–.177	.601	.054	1.000	–.326
Gini Index	.242	.166	–.102	.131	–.041	.098	–.326	1.000
Sig. (1-tailed)								
Total Footprint		.107	.000	.295	.003	.424	.000	.159
Population	.197		.209	.340	.274	.275	.174	.159
Urban Population	.000	.209		.034	.002	.231	.004	.339
Population Density	.295	.340	.034		.394	.428	.234	.296
Adult Literacy	.003	.274	.002	.394		.240	.003	.434
Publ Exp. Educ.	.424	.275	.231	.428	.240		.413	.345
Publ Exp. Health	.000	.174	.004	.234	.003	.413		.087
Gini Index	.159	.248	.339	.296	.434	.345	.087	
N								
Total Footprint	19	19	19	19	19	19	19	19
Population	19	19	19	19	19	19	19	19
Urban Population	19	19	19	19	19	19	19	19
Population Density	19	19	19	19	19	19	19	19
Adult Literacy	19	19	19	19	19	19	19	19
Publ Exp. Educ.	19	19	19	19	19	19	19	19
Publ Exp. Health	19	19	19	19	19	19	19	19
Gini Index	19	19	19	19	19	19	19	19

Table 9.19. Variables Entered/Removed[a]

Model	*Variables Entered*	*Variables Removed*	*Method*
1	Publ Exp. Health	.	Stepwise (Criteria: Probability-of-F-to-enter <= .050, Probability-of-F-to-remove >= .100).
2	Urban Population	.	Stepwise (Criteria: Probability-of-F-to-enter <= .050, Probability-of-F-to-remove >= .100).

a. Dependent Variable: Total Footprint Per Capita

Table 9.20. Model Summary[c]

					Change Statistics					
Model	*R*	*R Square*	*Adjusted R Square*	*Std. Error of the Estimate*	*R Square Change*	*F Change*	*df1*	*df2*	*Sig. F Change*	*Durbin-Watson*
1	.902[a]	.814	.803	1.0019	.814	74.165	1	17	.000	
2	.962[b]	.925	.916	.6531	.112	24.013	1	16	.000	2.054

a. Predictors: (Constant), Publ Exp. Health
b. Predictors: (Constant), Publ Exp. Health, Urban Population
c. Dependent Variable: Total Footprint Per Capita

Table 9.21. ANOVA[c]

Model	*Sum of Squares*	*df*	*Mean Square*	*F*	*Sig.*
1 Regression	74.452	1	74.452	74.165	.000[a]
Residual	17.066	17	1.004		
Total	91.518	18			
2 Regression	84.694	2	42.347	99.289	.000[b]
Residual	6.824	16	.427		
Total	91.518	18			

a. Predictors: (Constant), Publ Exp. Health
b. Predictors: (Constant), Publ Exp. Health, Urban Population
c. Dependent Variable: Total Footprint Per Capita

Table 9.22. Coefficients[a]

Model	Unstandardized Coefficients		Standardized Coefficients			Correlations			Collinearity Statistics	
	B	*Std. Error*	*Beta*	*t*	*Sig.*	*Zero-order*	*Partial*	*Part*	*Tolerance*	*VIF*
1 (Constant)	.011	.370		.030	.977					
Publ Exp. Health	1.035	.120	.902	8.612	.000	.902	.902	.902	1.000	1.000
2 (Constant)	–.964	.313		–3.083	.007					
Publ Exp. Health	.756	.097	.659	7.811	.000	.902	.890	.533	.655	1.527
Urban Population	.034	.007	.413	4.900	.000	.801	.775	.335	.655	1.527

a. Dependent Variable: Total Footprint Per Capita

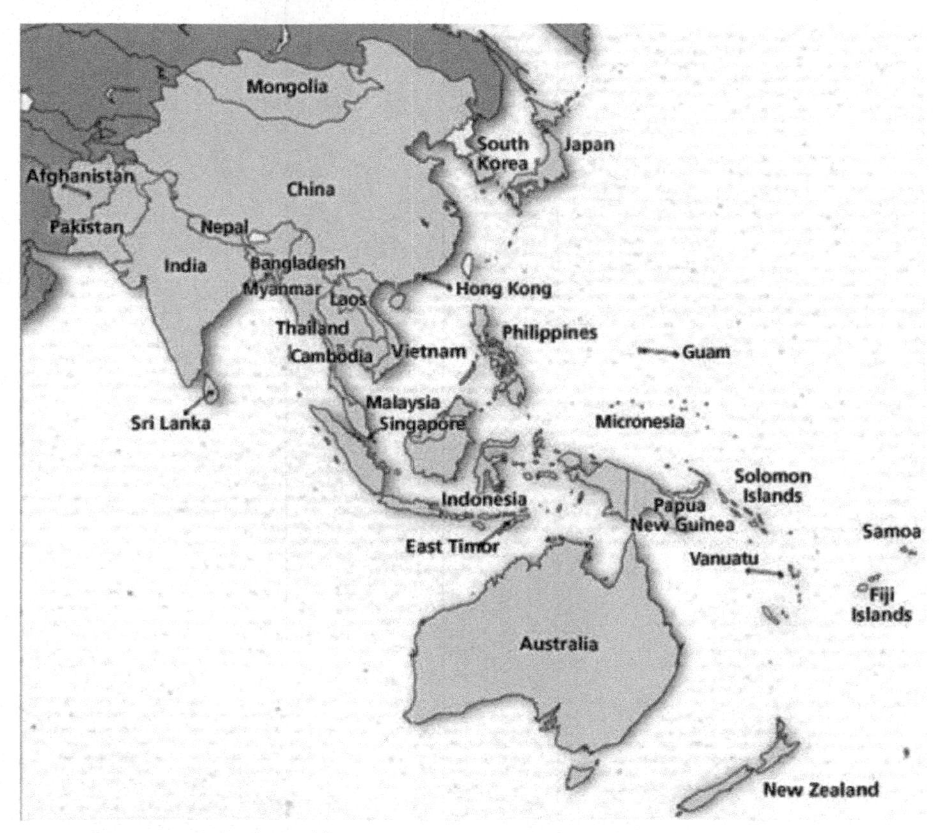

Figure 9.1. Map of Asia-Pacific

Chapter Ten

World Regional and Cultural Footprints and Environmental Sustainability

A Summary of Socio-Economic Determinants

INTRODUCTION

The earth is home to a variety of plants and animals and other living things referred to as biodiversity. Biodiversity provides many of the ecological resources and services for the anthropocentric economy, thus it is threatened by economic sectors in the aggregate (Czech et al, 2000). A fundamental conflict exists between economic growth and biodiversity conservation. To sustain livelihoods and to reduce poverty whilst conserving the earth's resources are major global challenges. Human use of the environment is the largest contributor to habitat modification and ecosystem loss (Goudie, 2000; World Resources Institute, 2000). As the human economy expands, natural capital (resources) is reallocated from non-human uses to the human economy (Czech, 2000a). The manufacturing and service sector of the human economy are responsible for the habitat and ecosystem losses. In the absence of anthropogenic threat, all natural capital is available as habitat for non-human species. Economic growth and development as a measure of gross domestic product (GDP) reflects an increase in the production and consumption of goods and services. The proportion of natural capital allocated to human economy increases with GDP, while the proportion of national capital associated to non-human habitats decreases with GDP (The Wildlife Society, 2003). Growth, according to Zovanyi (2005), constitutes unsustainable behavior because it is incapable of being continued or sustained indefinitely. Growth is blamed for such costly destructive development patterns as urban sprawl, loss of prime agricultural land and ecosystem loss, inefficient provision of public facilities and services, escalating housing prices, pervasive environmental degradation, and loss of community character. Therefore, "smart growth" as alternative to "dumb growth" (Chen, 2000; Lorentz and Shaw, 2000) should be managed to simultaneously confront social sustainability, economic sustainability, and environmental sustainability.

Sustainability refers to the ability of a system to continue and maintain a production level or quality of life for future generations. It means living in material comfort and peacefully with each other within the means of nature (Wackernagel and Rees, 1996). Sustainable development implies that a society should balance social equity, economic prosperity, and environmental integrity (Krizek and Power, 1996). Sustainable development is development that meets the needs of the present without compromising the ability of future generations to meet their needs (World Commission on Environment and Development, 1987). To balance economic and environmental goals, economic growth and development should be sustainable and imply full consideration of environmental factors.

Human societies and their places are the products of environmental, economic, political, and cultural processes, which may work across scales from the local to global. The history of civilization is also the history of ecological degradation and crisis (Chew, 2001). Of course, the world is currently utilizing nature beyond its capacity to renew and regenerate indefinitely. In fact, humankind has consistently remained in a state of "obligate dependence" on the productivity and life support services of the ecosphere, despite the mediating technological sophistication (Rees, 1990). The global modes of production and accumulation are intimately linked to environmental degradation. Market expansion through recent globalization is threatening human race with environmental disasters, and also creating conflict among three essential aims: prosperity, equity, and ecological sustainability (Jorgenson and Kick, 2003). Perhaps, this is what Anderson (2006) called "global ethical trilemma," which are production, consumption, and their consequent environmental degradation.

The pressing issues of our time are the population growth and increasing economic growth, as they interact with resource consumption rates, pushing the world toward global ecological collapse. Thus, population and consumption are the major environmental degradation factors, and resource consumption rates have increased at a faster pace in some world regions than population size in recent decades. In fact, affluence and population size (growth) have been hypothesized to be the primary drivers of human-caused environmental stressors or impacts (Earth Observatory News, 2007). Dietz et al (2007) found that increased affluence exacerbates environmental impacts and, when combined with population growth will substantially increase the human footprint on the planet. Ecological footprint is a measure of the amount of nature it takes to sustain a given population over a course of one year. It is a national-level measurement that quantifies how much land and water are required to produce the commodities consumed and assimilate the waste by them. It actually measures how consumption may affect the environment, thus the concept of ecological footprint analysis (Wackernegel and Rees, 1996; Wackernagel and Silverstein, 2000; Wackernagel et al, 2000; Jorgenson, 2003; Redefining Progress, 2005; Earth Observatory News, 2007). Ecological footprint analysis addresses the issue of human biological metabolism and humanity's industrial metabolism (Rees, 2012); and compares the humanity's ecological footprint with the earth's available capacity. It attempts to operationalize the concept of carrying capacity and sustainability. Carrying capacity is the maximum number of individuals of a defined species that a given environment can support over the long term (Hardin, 1991), or the environment's maximum persistently supportable load (Catton, 1986). In a modern capitalist world system economy, different countries, regions, and cultures have different consumption patterns, different ecological footprints, and different ecological balances.

PURPOSE OF STUDY

Limited studies exist that dealt with humanity's ecological footprint analysis tied to world's biological diversity crisis, and with major emphasis on regional, cultural, and national footprints socioeconomic determinants. Such limited footprints studies without major regional and national determinants analysis include, among others, Burns et al, 2001; Rees, 1990; Wackernagel and Rees, 1996; and Jorgenson, 2003. This study analyzed the ecological footprints and their structural causes in different nations, regions, and cultures for ecological balance; and provided a summary of their socioeconomic determinants that were fleshed out through rigorous multiple regression analyses of their recursive variables. In effect, the study examined and explained the cross-cultural and national variations in total footprints and per capita footprints (personal planetoids) associated with ecostructural factors (socioeconomic processes) and world system human development hierarchy. Ecostructural factors are socioeconomic processes within nations that impact footprints, such as urbanization, literacy rate, gross domestic product per capita (affluence), domestic inequality (Gini Index), government social (public) spending, gender inequality index, and gender-related development index. Progress towards sustainable development can be assessed using human development index (HDI) as an indicator of well-being, and ecological footprint as a measure of demand on biosphere. The process exposes the inequitable distribution of the world's ecological footprints and at the same time heightens the concern about ecological imbalances and overshoots. The expectation is that the empirical analysis of the data and findings should yield some policy recommendations and suggestions on how to ameliorate the impacts of the detected socioeconomic determinants. The question is what factors are responsible for or are driving the total and per capita footprints among nations, regions, and cultures around the world? What type of secure environments should meet the needs of both people and natural environments in the above entities? Thus, the study proffered some solutions on how to reduce the footprints of nations and regions, which also will help to protect the national environment and promote a more equitable and sustainable society.

FOCUS OF STUDY

Nations and regions covered by the study include: African Countries, Organization for Economic Cooperation and Development (OECD) Countries, Latin America and Caribbean (LAC) Countries, Middle East and Central Asia Countries, and Asia-Pacific Countries.

THEORETICAL FRAMEWORK FOR THE STUDY

Within the biosphere, everything is interconnected, including humans, thus sustainable development requires a good knowledge of human ecology. The human life-support functions of the ecosphere are maintained by nature's biocapacity which runs the risks of being depleted, the prevailing technology notwithstanding. Regardless of the humanity's mastery over the natural environment, it still remains a creature of the ecosphere and always in a state of *obligate dependency* on numerous biological goods and services (Rees, 1992, p. 123). Despite the above *dependency* thesis, the prevailing economic mythology assumes a world in which carrying capacity is indefinitely expandable (Daly, 1986; Solow, 1974). The human species has continued to deplete, draw-down, and confiscate nature's bio-capacity with reckless abandon. York et al (2003), indicated that population and affluence account for 95 percent of the variance in total footprints of countries. Thus, large human population all over the world and their excessive consumption of the scarce natural resources are responsible for the national footprint of nations. Thus, total human impact on the ecosphere is given as: population × per capita impact (Ehrlich and Holdren, 1971; Holdren and Ehrlich, 1974); Hardin (1991). In other words, population size and affluence are the primary drivers of environmental impacts (Dietz et al, 2007). The footprints of nations provide compelling evidence of the impacts of consumption, thus the need for humans to change their lifestyles and conserve scarce natural capital. According to Palmer (1998), there are in order of decreasing magnitude, three categories of consumption that contribute enormously to our ecological footprints: wood products (53 percent), food (45 percent), and degraded land (2 percent). Degraded land includes land taken out of ecological availability by buildings, roads, parking lots, recreation, businesses, and industries. Palmer also indicated that about 10 percent or more of earth's forests and other ecological land should be preserved in more or less pristine condition to maintain a minimum base.

The environmental impacts of urban areas should be considered because a rapidly growing proportion of world's population lives in cities, and more than one million people are added to the world's cities each week and majority of them are in developing countries of Africa, Latin America, and Southeast Asia (Wackernagel and Rees, 1996). The reality is that the populations of urban regions of many nations had already exceeded their territorial carrying capacities and depend on trade for survival. Of course, such regions are running an unaccounted ecological deficit; their populations are appropriating and meeting their carrying capacity from elsewhere (Pimentel, 1996; Wackernagel and Rees, 1996; Girardet, 1996, accessed Online on 12/14/07; Rees, 1992; The International Society for Ecological Economics and Island Press, 1994; Vitouset et al, 1986; Wackernagel, 1991; WRI, 1992, p. 374). Undoubtedly, the rapid urbanization occurring in many regions and the increasing ecological uncertainty have implications for world development and sustainability. Cities are densely populated areas that have high ecological footprints which leads to the perception of these populations as "parasitic," since these communities have little intrinsic biocapacity, and instead, must rely upon large hinterlands. Land consumed by urban regions is typically at least an order of magnitude greater than that contained within the usual political boundaries or the associated built-up areas (The International Society for Ecological Economics and Island Press, 1994; Rees, 1992).

According to Rees (1992), every city is an entropic black hole drawing on the concentrated material resources and low-entropy production of a vast and scattered hinterlands many times the size of the city itself. In the same vein, Vitouset et al (1986) asserted that high density settlements "appropriate" or augment their carrying capacity from all over the globe, in the past and the future (see also Wackernagel, 1991). In modern industrial cities, resources flow through the urban system without much concern either about their origins, or about the destination of their wastes, thus, inputs and outputs are considered to be unrelated. The cities' key activities such as transportation, provision of electricity supply, heating, manufacturing and the provision of socio-economic services depend on a ready supply of fossil fuels, usually from far-flung hinterlands than within their usual political boundaries or their associated built-up areas. Cities are not self-contained entities, and their concentration of intense economic processes and high levels of consumption both increase and stimulate their demands on resources. Cities occupy only 2 percent of the world's land surface, but use some 75 percent of the world resources, and release a similar percentage of waste (Girardet, 1996).

Like urbanization, energy footprint, created from energy use and carbon dioxide emissions, is not subject to area constraints. Energy footprint is the area of forest that would be needed to sequester the excess carbon (as carbon dioxide) that is being added to the atmosphere by the burning of fossil fuels to generate energy for travel, heating, lighting, manufacturing, recreation, among other uses. Actually, the demand for energy defines modern cities more than any other single factor. Cities contain enormous concentration of economic activities that consume enormous quantities of energy. The natural global systems of forests and oceans for carbon sequestration are not handling the human carbon contributions fast enough, thus the Kyoto Conference of early 1998. According to Suplee (1998), only half of the carbon humans generate burning fossil fuels can be absorbed in the oceans and existing terrestrial sinks. The oceans absorb about 35 percent of the carbon in carbon dioxide (Suplee, 1998), equivalent of 1.8 giga tons of carbon every year (IPCC, 2001), while the global forests under optimum management of existing forests could absorb about 15 percent of the carbon in the CO_2 produced from the burning of fossil fuels world-wide (Brown et al, 1996). The energy footprint is caused by the un-sequestrated 50 percent in the atmosphere with the potentially troubling ecological consequences, such as rapid global warming and other environmental stresses, including climate change. Carbon dioxide in the atmosphere will continue to increase unless humanity finds alternative energy sources of sufficient magnitude. It is in the humanity's best interest to get off its petroleum addiction (control and minimize fossil fuel use) and develop sustainable consumption habits.

Literacy affects the consumption of natural capital resources. Highly literate groups concentrate in urban areas where they consume more than their fair shares of biospheric resources. Literate populations generally have lower rates of domestic inequality and tend to consume more resources than their illiterate counterparts due to their higher incomes and higher standards of urban living. Furthermore, higher levels of literacy correspond with higher incomes, which allow for greater consumption (Jorgenson, 2003). This is because literate populations are subject to increased consumerist ideologies and contextual images of good life through advertising (Princen et al, 2002); what Leslie Sklair (2001) and Jennifer Clapp (2002) labeled "cultural ideology of consumerism/consumption."

METHODOLOGY FOR THE STUDY

Unit of Analysis

The units of analysis are "country" and "region" from different cultures of the world.

Samples

To test for the national variations in total footprints and per capita footprints associated with human development index, as well as their socioeconomic determinants, Comparative Model Analysis and Stepwise Regression Analysis were used as the tools of analysis. Using comparative model, samples of eligible countries out of many countries from different regions and cultures were analyzed (see Global Footprint Network. *The Ecological Footprint Atlas 2008*). Step-wise Regression Analyses were also conducted using the samples. Each table represents countries in each region of the world studied, and as also represented by Global Footprint Network, *The Ecological Footprint Atlas 2008*. These tables do not include countries that were less than one million people in 2005, as well as the dependent territories without political independence. The larger population of at least one million is chosen for the study because; population is a variable which affects the consumption rates and levels, therefore the ecological footprint. The regional tables used in the analyses are not included in this summary presentation. They were included in previous regional studies by this author published elsewhere, such as *The International Journal of Environmental, Cultural, Economic, & Social Sustainability*; and *The National Social Science Proceedings* of The National Social Science Association (NSSA).

Mode of Analysis

This is an explanatory study using Comparative Model that employs descriptive statistics (such as matrices, totals, averages, ratios, and percentages) as well as Step-wise Regression. Regression analysis was performed to flesh out factors that highly impact regional, national, and per capita footprints; and also to strengthen the results from comparative analysis. Potential technical problems were diagnosed that might affect the validity of the results of various analysis mode components, such as the missing data in some countries or the exclusion or omission of some member countries of a particular region not meeting the stipulated selection criteria such as minimum population threshold; non-political sovereignty (dependent territory status) during the study period; etc. The aforementioned problems did not alter the substantive conclusions of the study, especially regarding the ecological footprint accounts and environmental sustainability of different countries, regions, and cultures.

The national ecological footprints and ecological balance as dependent variables are recursively explained by the country's (region's) various independent variables. The independent variables mediated in explaining the varying levels of consumptions and wastes among different nations of the world capitalist economy. It is hypothesized that Human Development positions and Ecostructural Factors of different countries are likely to be responsible for the variations in the National and Regional Ecological Footprints and Balances. Moreover, the carbon-dioxide emission levels of different countries are included in this study to depict and emphasize the biological and industrial metabolisms (consumption and environmental impacts) of different countries, regions, and cultures. The levels reflect their ecological footprints and balances.

HYPOTHESIS FOR THE STUDY

Null Hypothesis (H_0)

None of the independent variables predicts the national and regional footprints effects. The independent variables do not influence the footprints.

Alternative Hypothesis (H_1)

Some (if not all) of the independent variables predict the national and regional footprints effects.

Variables in the Study

The *Dependent* and *Independent* variables are selected on the basis of the theoretical themes and underpinnings which indicate that national footprint (and regional footprint) as dependent variable is explained by the country's (region's) human development (HDI) hierarchical category, population size, population density, urbanization level, government social (public) spending (as percent of GDP), GDP per capita, domestic inequality (Gini Index), gender inequality index (GII), gender-related development index (GDI), and literacy rate, as the ecostructural or independent variables.

VARIABLE DEFINITIONS

Human Development Index (HDI): The HDI is a summary measure of human development (human welfare). It measures the average achievements of a country in three basic dimensions of human development (Global Footprint Network: *Africa's Ecological Footprint-2006 Factbook*, p. 89): a long and healthy life, as measured by life expectancy at birth; knowledge, as measured by the adult literacy rate (with two-thirds weight) and the combined primary, secondary and tertiary gross enrolment ratio (with one-third weight); and a decent standard of living, as measured by GDP per capita (PPP US$). Purchasing power parity (PPP) is a rate of exchange that accounts for price difference across countries, allowing international comparisons of real output and incomes. At the PPP US$ (as used in this study), PPP US$1 has the same purchasing power in the domestic economy as $1 has in the United States of America.

Gross Domestic Product (GDP) per capita (PPP US$): GDP is converted to US dollars using the average official exchange rate reported by the International Monetary Fund (IMF). GDP alone does not capture the international relational characteristics as does the human development hierarchy of the world economy, which accounts for a country's relative socio-economic power and global dependence position in the modern world system. It is suggested elsewhere that GDP per capita is an inadequate measure of world-system position but a more appropriate indicator of domestic affluence or internal economic development (Burns, Kentor, and Jorgenson, 2003; Jorgenson, 2003; Dietz and Rosa, 1994). The Gross Domestic Product (GDP) per capita data for this study is taken from *Human Development Report 2007/2008*, table 1, and pp. 229–232. See also *World Development Report 2005*.

Domestic Income Inequality (Gini Index): The Gini index measures domestic income inequality of different countries, which had remained stable over a time with its impacts on other variables in the study (Bergesen and Bata, 2002; Jorgenson, 2003). Gini index measures the extent to which the distribution of income (or consumption) among individuals or households within a country deviates from a perfectly equal distribution (*Human Development Report, 2004*, p. 271)). It measures inequality over the entire distribution of income or consumption. A value of zero (0) represents perfect equality, and a value of hundred (100) represents perfect inequality. Data for domestic income inequality measured by Gini index are taken from World Bank, *World Development Report* (2005), table 2, pp. 258–259 and United Nations Development Report, *Human Development Report* 2007/2008, 15, pp. 281–284.

Gender Inequality Index (GII): The index shows the loss in human development due to inequality between female and male achievements in reproductive health, empowerment, and labor market. It reflects women's disadvantage in those three dimensions. The index ranges from zero (0), which indicates that women and men fare equally, to one (1), which indicates that women fare as poorly as possible in all measured dimensions.

The health dimension is measured by two indicators: maternal mortality ratio and the adolescent fertility rate.

The empowerment dimension is measured by two indicators: the share of parliamentary seats held by each sex and by secondary and higher education attainment levels.

The labor market dimension is measured is measured by women's participation in the work force.

Gender Inequality Index is designed to measure the extent to which national achievements in these aspects of human development are eroded by gender inequality; also to provide empirical foundations for policy analysis and advocacy efforts. It can be interpreted as a percentage loss to potential human development due to shortfalls in the dimensions included. Countries with unequal distribution of human development also experience high inequality between women and men, and countries with high gender inequality also experience unequal distribution of human development (UNDP *Human Development Reports*). The Gender Inequality Index data for this study is taken from UNDP's *Human Development Report 2010___20th Anniversary Edition. The Real Wealth of Nations: Pathways to Human Development*. Table 4: Gender Inequality Index, pp. 156–160.

Gender-Related Development Index (GDI): The index is one of the indicators of human development developed by the United Nations. The GDI is considered a gender-sensitive extension of the Human Development Index (HDI), which addresses gender-gaps in life expectancy, education, and income. It highlights inequalities in the areas of long and healthy life, knowledge, and decent standard of living between women and men. It measures achievement in the same basic capabilities as the HDI, but takes note of inequality in achievement between women and men (*Human Development Reports*). The methodology used imposes a penalty for inequality, such that the GDI falls when the achievement levels of both women and men in a country go down or when the disparity in basic capabilities, the lower a country's GDI compared with its HDI. The GDI is simply the HDI discounted, or adjusted downwards, for gender inequality. Thus, if GDI goes down, HDI goes down, while GII goes up. The methodology used to construct the GDI could be used to assess inequalities not only between men and women, but also between other groups such as rich and poor, young and old, etc. (*Human Development Reports*). The Gender-Related Development Index data for this study is taken from UNDP's *Human Development Report 2010___20th Anniversary Edition. The Real Wealth of Nations: Pathways to Human Development*. Table 4: Gender Inequality Index, pp. 156–160.

Urbanization Level (Urban Population as percent of Total Population): Cities are not self-contained entities and their concentration of intense economic processes and high levels of consumption both increase and stimulate their demands on resources. The cities have limited intrinsic biocapacity which undoubtedly must rely upon large hinterlands. Land consumed by urban regions is typically at least an order of magnitude greater than that contained within the usual political boundaries or the associated built-up area (The International Society for Ecological Economics and Island Press, 1994; Rees, 1992). The data are taken from *Human Development Report 2007/2008*, table 5, and pp. 243–246.

Literacy Rate: This variable refers to the percent of a nation's population over the age of fifteen (15) that can read and write in any language of their choice. Literate population generally has low domestic inequality and tends to consume more resources than their illiterate counterpart due to their higher income and urban living. High literate groups concentrate in urban areas where they consume more than their fair shares of biospheric resources. Higher levels of literacy correspond with higher incomes, which allow for greater consumption (Jorgenson,

2003). This is because literate populations are subject to increased consumerist ideologies and contextual images of good life through advertising (Princen, 2002), what Leslie Sklair (2001) and Jennifer Clapp (2002) labeled "cultural ideology of consumerism/consumption." The data for literacy rate is taken from the *Human Development Report 2007/2008*, table 1, pp. 229–232.

Population and Population Density: Apart from consumption, many have attributed to population as driving most of the sustainability problems (Palmer, 1998; Pimentel, 1996). Likewise, Dietz et al (2007) concluded that population size and affluence are the primary drivers of environmental impacts. Population growth and increases in consumption in many parts of the world have increased humanity's ecological burden on the planet. York et al (2003) indicated that population and affluence account for 95 percent of the variance in total footprints of countries. Others also see the ensuing human impact or footprint as a product of population, affluence (consumption), and technology (i.e. I = PAT (Ehrlich and Holdren, 1971; Holdren and Ehrlich, 1974; Hardin, 1991). Population as a variable in this study is taken from, *World Development Report 2005*, table 1, and pp. 256–259.

Government Social Spending Per Capita: Spending by a government (federal, state, and local), municipality, or local authority, which covers such things as spending on healthcare, education, pensions, defense, welfare, interest, and other social services, and is funded by tax revenue, seigniorage, or government borrowing. Public expenditure exerts an effect on economic growth rate through the positive externality in the productivity of the capital stock. Investment in education and health is also investment in people and in the future (See, Education: Crisis Reinforces Importance of a Good Education, says OECD. Accessed Online on 9/23/2011, at http://www.oecd.org/document/21/0,3746. Data for this study were taken from: Human Development Report 2007/2008. *Fighting Climate Change: Human Solidarity in a Divided World.* Table 28: Gender-related Development Index, pp. 326–329; and Human Development Report 2010___20th Anniversary Edition. *The Real Wealth of Nations: Pathways to Human Development*. Table 4: Gender Inequality Index, pp. 156–160.

COMPARATIVE MODEL ANALYSIS

The results and findings from this model of analysis in this study are discussed under the subtitles that include, "Per Capita Ecological Footprint, Biological Capacity, and Ecological Balance by World Regions," "Human Development and Affluence on Biological Capacity, Ecological Footprint and Balance by Regional Countries," and "Ecostructural Factors on Total Footprints Per Capita by Regional Countries." Please see individual national analyses and their results.

REGRESSION ANALYSIS

The tables used in the Comparative Model Analysis are also used for the Regression Analysis. Regression analysis allows the modeling, examining, and exploring of relationships and can help explain the factors behind the observed relationships or patterns. It shows the factors or independent variables that have strong correlation or association with dependent variable. The mathematical formula when applied to the explanatory variables is best used to predict the dependent variable that one is attempting to model. Each independent variable or explanatory variable is associated with a regression coefficient describing the strength and the sign of that variable relationship to the dependent variable.

In this study, the rows of the Unstandardized Coefficients of the "B" column of the Regression Model table is used to show the variables and their coefficients that exhibit strong correlation with total and per capita footprints of the nations and their regions. A regression equation might look like the one given below where Y is the dependent variable, the Xs are the explanatory variables, and the Bs are regression coefficients:

$$Y = B_0 + B_1 X_1 + B_2 X_2 + \text{-----} B_n X_n + E \text{ (Random Error Term/Residuals)}$$

The row of "unstandardized coefficients" or "Bs" gives us the necessary coefficient values for the multiple regression models or equations. See below for the Regression Analysis Terms.

REGRESSION ANALYSIS TERMS

Correlation or co-relation: refers to the departure of two variables from independence or they are non-independent or redundant (D'Onofrio, A., 2001/2002; Richard Lowry, 1999–2008).

Collinearity: Refers to the presence of exact linear relationships within a set of variables, typically a set of explanatory (predictor) variables used in a regression-type model. It means that within the set of variables, some of the variables are (nearly) totally predicted by the other variables [(Sundberg, R. 2002). *Encyclopedia of Environmetrics*, edited by Abdel H. El-Shaarawi and Walter W. Piegorsch (Chichester: John Wiley & Sons, Ltd), Volume 1, pp. 365–366].

Partial Correlation Coefficients (r): When large, it means that there is no mediating variable (a third variable) between two correlated variables (D'Onofrio, A., 2001/2002).

Pearson's Correlation Coefficient (r): This is a measure of the strength of the association between two variables. It indicates the strength and direction of a linear relationship between two random variables. Value ranges from - to +1; –1.0 to –0.7 Strong negative association; –0.7 to –0.3 Weak negative association; –03 to +0.3 Little or no association; +0.3 to +0.7 Weak positive association; +0.7 to 1.0 Strong positive correlation (Luke, B., "Pearson's Correlation Coefficient," Learning *From The Web.net*. Accessed Online on 5/30/2008).

Multiple "R": Indicates size of the correlation between the observed outcome variable and the predicted outcome variable (based on the regression equation).

"R^2" or Coefficient of Determination: Indicates the amount of variation (%) in the dependent scores attributable to all independent variables combined, and ranges from 0 to 100 percent. It is a measure of model performance, summarizing how well the estimated Y values match the observed Y values.

"Adjusted R^2": The best estimate of R^2 for the population from which the sample was drawn. The Adjusted R-Squared is always a bit lower than the Multiple R-Squared value because it reflects model complexity (the number of variables) as it relates to the data.

R^2 and the *Adjusted R^2* are both statistics derived from the regression equation to quantify model performance (Scott and Pratt, 2009. *ArcUser*).

Standard Error of Estimate: Indicates the average of the observed scores around the predicted regression line.

Residuals: These are the unexplained portion of the dependent variable, represented in the regression equation as the random error term (E). The magnitude of the residuals from a regression equation is one measure of model fit. Large residuals indicate poor model fit. Residual = Observed – Predicted.

ANOVA: Decomposes the total sum of squares into regression (= explained) SS and residual (= unexplained) SS.

F-test in ANOVA represents the relative magnitude of explained to unexplained variation. If F-test is highly significant (p = .000), we reject the null-hypothesis that none of the independent variables predicts the effect (scores) in the population.

The "constant" represents the intercept in the equation and the coefficient in the column labeled by the independent variables.

REGRESSION ANALYSIS RESULTS AND MODEL EQUATIONS FOR DIFFERENT REGIONS AND CULTURES

Multiple (step-wise) regression analysis was performed on these two tables, "Human Development and Affluence on Biological Capacity, Ecological Footprint and Balance" and "Population, Ecostructural Factors and Total Footprint Per Capita," from countries of each world region and culture for model equations of the total and per capita footprints socioeconomic determinants. For this summary, only the *coefficients*[a] tables (that contain the "B" columns of the Unstandardized Coefficients) of two tables were used for the model equation constructions of the footprints socioeconomic determinants for different world regions and cultures. The other relevant analyses tables' results were included elsewhere in previous world regional studies by the author as explained above. Tables 10.1a and 10.1b; 10.2a and 10.2b; 10.3a and 10.3b; 10.4a and 10.4b; and 10.5a and 10.5b are associated with the regression analysis results, as well as the model socioeconomic determinants for different world regions and cultures.

AFRICAN REGION FOOTPRINT SOCIOECONOMIC DETERMINANTS MODEL

The Footprint Socioeconomic Determinants in Africa are: *GDP Per Capita, Carbon-Dioxide Emissions, Urban Population, and Literacy Rates.*

ORGANIZATION FOR ECONOMIC COOPERATION AND DEVELOPMENT (OECD) REGION FOOTPRINT SOCIOECONOMIC DETERMINANT MODEL

The Footprint Socioeconomic Determinants in Organization of Economic Cooperation and Development (OECD) are: *Carbon Dioxide Emissions, Ecological Balance, Bio-Capacity, Public Expenditure, Population Density, and Urban Population.*

LATIN AMERICA AND CARIBBEAN REGION FOOTPRINT SOCIOECONOMIC DETERMINANTS MODEL

The Footprint Socioeconomic Determinants in Latin America and Caribbean Region are: *Carbon-Dioxide Emissions, Bio-Capacity, Ecological Balance, and Urban Population.*

MIDDLE EAST AND CENTRAL ASIA REGION FOOTPRINT SOCIOECONOMIC DETERMINANTS MODEL

The Footprint Socioeconomic Determinants in Middle East and Central Asia Region are: *Bio-Capacity, Ecological Balance, Population Density, Population, GDP Per Capita, Gender Inequality Index, Gender Related Development Index, and Carbon Dioxide Emissions.*

ASIA-PACIFIC REGION FOOTPRINT SOCIOECONOMIC DETERMINANTS MODEL

The Footprint Socioeconomic Determinants in Asia-Pacific Region are: *Carbon-Dioxide Emissions, Human Development Index, Public Expenditure on Health, and Urban Population.*

POLICY RECOMMENDATIONS

Regions and nations are designated as high income, medium income, low income, or mix income economies by the United Nations Development Reports. In other words, there are high, low, and mix consumption countries and regions. Therefore, policy recommendations in this study addressed and reflected environmental issues in high, medium, and low income countries, regions, and cultures. Mix income economy regions reflected mix national footprints, environmental problems, recommendations, and solutions.

The above analyses supported the theorization and assertion that the global environmental stress is caused primarily by the increase in resource consumption (demand) and population size and growth in different countries, regions, and cultures. The study showed that the higher the national populations, urban populations, and GDP per capita (affluence), the higher the imbalanced consumption of resources (bio-capacity or natural capital), as well as national and regional footprints. The analyses also indicated that large human population all over the world and their excessive consumption of the scarce natural resources or capital are responsible for the footprints. Thus, a country's or region's total footprint, ecological balance, and carbon-dioxide emissions (waste) are a function of its human development (welfare) position and affluence in the world economy. As human development and income go up, the environmental performance goes down. Absolute population growth in many countries and regions, especially in low income developing countries and regions, as well as their rapid growing largest cities, are exerting tremendous pressures on natural resources. According to Rees (2012), for equity considerations, the way additional population (especially in developing regions) or material growth and consumption (especially in developed regions) can be sustained without ravaging biodiversity and ultimately destroying the ecological basis of human life, is through cutbacks in resource or natural capital consumption by the wealthy nations and regions. Rees further stated that the above action is necessary to vacate the ecological space necessary to justify growth in low income countries and regions.

Carbon-dioxide is a major footprint contributor, and its concentration in the atmosphere will continue to increase unless humanity finds alternative energy sources of sufficient magnitude. It is in the humanity's best interest to get off its petroleum addiction (control and minimize fossil fuel use) and develop sustainable habits. In the countries and regions with footprint intensities like in OECD countries, the amount of resources used in the production of goods and services can be significantly reduced, from energy efficiency in manufacturing and in the home, through minimizing waste and increasing recycling and reuse, to fuel efficient cars. Carbon sequestration can be improved by: reducing the rate of deforestation in the regions, necessary for carbon sinking; and more efficient agriculture and manufacturing, and use of adequate technology, for example, in Africa and Latin America and Caribbean regions with the highest rates of deforestation (biodiversity destruction) in the world.

Population, population density, carbon dioxide emissions and urban population are major factors of national and regional footprints in Africa, Middle East, and Central Asia. Thus, population management is essential and necessary in these regions. For example, urban population could be managed through redistribution to smaller centers to reduce pressures in the burgeoning large cities that are becoming inefficient black holes, in terms of natural resources and energy usage. According to Rees (1992), every city is an entropic black hole drawing on the concentrated material resources and low-entropy production of a vast and scattered hinterlands many times the size of the city itself. In the same vein, Vitouset et al (1986) asserted that high density settlements "appropriate" or augment their carrying capacity from all over the globe, as well as from the past and the future (see also Wackernagel, 1991). Population management could also be achieved through education and empowerment of women, who are major players (especially in developing countries and regions) in reproduction, economy, environmental sustainability, and biodiversity conservation (Aka, 2006). In countries and regions where gender inequality is high and gender development is low, as well as public (social) expenditure low in health and education, for example, in Middle East and Central Asia and African regions, adequate investments in health and education should be in order. This is informed from the fact that investment in health and education in low income countries and regions is also investment in people and in the future, with their positive externalities in the productivity of the capital stock, e.g., human resources.

CONCLUSIONS

This was an explanatory and exploratory study that used Comparative Model that employed descriptive statistics, as well as Step-wise Regression Analysis. Regression analysis was performed to flesh out factors that highly impact regional, national, and per capita and total footprints; and also to strengthen the results from comparative analysis. The study analyzed the ecological footprints and their structural causes in different nations, regions, and cultures for ecological balance; and provided a summary of their socioeconomic determinants that were fleshed out through rigorous multiple regression analyses of the recursive variables. The process exposed the inequitable distribution of the world's ecological footprints and at the same time heightens the concern about ecological imbalances and overshoots. Data analyses results indicate that footprint socioeconomic determinants in different countries, regions, and cultures included: GDP Per Capita, Carbon-Dioxide Emissions, Urban Population, and Literacy Rates (Africa); Carbon-Dioxide Emissions, Ecological Balance, Public (Social) Expenditure, Bio-Capacity, Population Density, and Urban Population (OECD); Carbon-Dioxide Emissions, Ecological Balance, Bio-Capacity, and Urban Population (Latin America and Caribbean); Population Density, Population, Gender Inequality Index, Gender Related Development Index, Public Expenditure on Health and Education, and carbon dioxide emissions (Middle East and Central Asia); and Carbon-Dioxide Emissions, Public Expenditure on Health, and Urban Population (Asia Pacific). Thus, *the study concluded from the above that the major drivers of environmental stress in all the regions and cultures are carbon-dioxide emissions, population, and urban population.*

The above analyses support the theorization and assertion that the global environmental stress is caused primarily by the increase in resource consumption (demand) and population size and growth in different countries, regions, and cultures. Finally, to achieve regional ecological sustainability, emphasis should be through: gender empowerment and access to better education, healthcare, and economic opportunities; population management, especially through redistribution to smaller centers; and the already rich countries and regions to reduce their consumptive appetites and eco-footprints to create the ecological space needed for a justifiable growth in the impoverished countries, regions, and cultures.

REFERENCES

Aka, Ebenezer (2006). "Gender Equity and Sustainable Socio-Economic Growth and Development," *The International Journal of Environmental, Cultural, Economic & Social Sustainability*, Volume 1, Number 5, pp. 53–71.

Andersson, Jan Otto (2006). "International Trade in a Full and Unequal World." Presented at the workshop, "Trade and Environmental Justice," Lund, 15th – 16th of February.

Bergesen, Albert and Michelle Bata (2002). "Global and National Inequality: Are They Connected?" *Journal of World-System Research*, 8: 130–44.

Brown, S., Jayant, S., Cannell, M., and Kauppi, P. (1996). "Mitigation of Carbon Emissions to the Atmosphere by Forest Management," *Commonwealth Forestry Review*, 75, 79–91.

Burns, Thomas J., Byron L. Davis, Andrew K. Jorgenson, and Edward L. Kick (2001). "Assessing the Short- and Long- Term Impacts of Environmental Degradation on Social and Economic Outcomes." Presented at the Annual Meetings of the American Sociological Association, August, Anaheim, CA.

Burns, Thomas J., Jeffery Kentor, and Andrew K. Jorgenson (2003). Trade Dependence, Pollution, and Infant Mortality in Less Developed Countries," Pp. 14–28 in *Crises and Resistance in the 21st Century World-System*, edited by Wilma A. Dunaway (Westport, CT: Pager).

Catton, W. (1986). "Carrying Capacity and the Limits to Freedom." Paper prepared for Social Ecology Session 1, XI, World Congress of Sociology, New Delhi, India (August 18).

Chen, D.T. (2000). "The Science of Smart Growth." *Scientific American*, 283 (6): 84–91.

Chew, Sing C. (2001). *World Ecological Degradation: Accumulation, Urbanization, and Deforestation 3000B.C-A.D 2000* (Walnut Creek, CA: Alta Mira).

Clapp, Jennifer (2002). "The Distancing of Waste: Over-consumption in a Global Economy," in *Confronting Consumption*, edited by T. Princen, M. Maniates, and K. Conca (Cambridge, MA: MIT Press) pp. 155–76.

Czech, B. (2000a). "Economic Growth as the Limiting Factor for Wildlife Conservation," *Wildlife Society Bulletin*, 28 (1): 4–15.

Czech, B.; P .R. Krausman; and P. K. Devers (2000). "Economic Associations Among Causes of Species Endangerment in the United States." *BioScience*, 50: 593–601.

Daly, H.E. (1986). *Beyond Growth: The Economics of Sustainable Development* (Boston, Massachusetts, USA: Beacon Press).

Dietz, Thomas and Eugene Rosa (1994). "Rethinking the Environmental Impacts of Population, Affluence, and Technology," *Human Ecology Review*, 1: 277–200.

Dietz, Thomas, Eugene Rosa, Richard York (2007). "Driving the Human Ecological Footprint." *Frontiers in Ecology and the Environment* (February).

D'Onofrio, Antonia (2001/2002). "Partial Correlation." Ed 710 Educational Statistics, Spring 2003. Accessed Online on 6/3/2008 at http://www2.widener.edu/.

Earth Observatory News (February 2007). "Human's Ecological Footprint in 2015 and Amazonia Revealed." http://earthobservatory.nasa.gov/Newsroom/MediaAlerts/2007. Accessed Online on 2/23/2007.

Education: Crisis Reinforces Importance of a Good Education, says OECD. Accessed Online on 9/23/2011, at http://www.oecd.org/document/21/0,3746.

Ehrlich and Holdren (1971). "Impacts of Population Growth," *Science*, 171, 1212–7.

Girardet, Herbert (1996). "Giant Footprints," Accessed Online on 12/14/07, at http:// www.gdrc .org/uem/footprints/girardet.html.

Global Footprint Network: Africa's Ecological Footprint—2006 Factbook. Global Footprint Network, 1050 Warfield Avenue, Oakland, CA 94610, USA. http://www.footprintnetwork.org/ Africa.

Global Footprint Network (2008). *The Ecological Footprint Atlas 2008*, October 28.

Goudie, A. (2000). *The Human Impact on the Natural Environment. Fifth Edition* (Cambridge, Massachusetts, USA: Harvard University Press).

Hardin, G. (1991). "Paramount Positions in Ecological Economics," in R. Costanza (ed.). *Ecological Economics: The Science and Management of Sustainability* (New York: Columbia University Press) pp. 47–57.

Holdren and Ehrlich (1974). "Human Population and The Global Environment," *American Scientist*, 62: 282–92.

IPCC (2001). Intergovernmental Panel on Climatic Change.

Jorgenson, Andrew K. (2003). "Consumption and Environmental Degradation: A Cross-National Analysis of Ecological Footprint." *Social Problems*. Volume 50, No. 3, pp. 374–394.

Jorgenson, Andrew K. and Edward L. Kick (eds.) (2003). Special Issue: Globalization and the Environment. *Journal of World-System Research*, Volume IX, Number 2, (summer).

Krizek, K.J., Power, J. (1996). *A Planner's Guide to Sustainable Development*. Planning Advisory Service Report Number 467 (Chicago, IL: American Planning Association), 66p.

Lorentz, A., and Shaw, K. (2000). "Are You Ready to Bet on Smart Growth?" *Planning*, 66 (1): 4–9.

Lowry, Richard (1999–2008). "Subchapter 3a. Partial Correlation." Accessed Online on 6/3/08 at http://faculty.vassar.edu/lowry/cha3a.html.

Luke, Brian T. "Pearson's Correlation Coefficient." Learning From The Web.net. Accessed Online on 5/30/2008.

Palmer, A.R. (1998). "Evaluating Ecological Footprints," *Electronic Green Journal*. Special Issue 9 (December).

Pimentel, David (1996). "Impact of Population Growth on Food Supplies and Environment." American Association for the Advancement of Science (AAAS) (February, 9). See also GIGA DEATH, Accessed Online on 12/18/07, at http://dieoff.org/page13htm.

Princen, Thomas (2002). "Consumption and Its Externalities: Where Economy Meets Ecology," in *Confronting Consumption*, edited by T. Princen, M. Maniates, and K. Conca (Cambridge, MA: MIT Press) pp. 23–42.

Redefining Progress (2005). *Footprints of Nations*. 1904 Franklin Street, Oakland, California 94612. See http://www.RedefiningProgress.org.

Rees, W. (1990). Sustainable Development and the Biosphere. Teilhard Studies Number 23. American Teilhard Association for the Study of Man, or: "The Ecology of Sustainable Development." *The Ecologist*, 20 (1), 18–23.

Rees, William E. (1992). "Ecological Footprint and Appropriated Carrying Capacity: What Urban Economics Leaves Out," Environment and Urbanization, Vol. 4, No. 2, (October). Accessed Online on 12/14/07, at http://eau.sagepub.com.

Rees, William E. (2012). "Ecological Footprint, Concept of," in Simon Levin (ed.). *Encyclopedia of Biodiversity (2nd Edition)*.

Scott, Lauren and Monica Pratt (2009). "An Introduction to Using Regression Analysis With Spatial Data." *ArcUser. The Magazine for ESRI Software Users* (spring), pp. 40–43. Lauren Scott and Monica Pratt are ESRI Geo-processing Spatial Statistics Product Engineer and ArcUser Editor respectively.

Sklair, Leslie (2001). *The Transnational Capitalist Class* (Oxford, UK: Blackwell Press).

Solow, R. M. (1974). "The Economics of Resources or the Resources of Economics," *American Economics Review*, Vol. 64, pp. 1–14.

Sundberg, Rolf (2002). *Encyclopedia of Environmetrics*, edited by Abdel H. El-Shaarawi and Walter W. Piegorsch (Chichester: John Wiley & Sons, Ltd), Volume 1, pp. 365–366.

Suplee, D. (1998). "Unlocking the Climate Puzzle." *National Geographic*, 193 (5), 38–70.

The international Society for Ecological Economics and Island Press (1994). "Investing in Natural Capital: The Ecological Approach to Sustainability." Accessed Online on 12/18/07, at http://www.dieoff.org/page13.htm.

The Wildlife Society (TWS) (2003). *The Relationship of Economic Growth to Wildlife Conservation*, Technical Review 03-1 (Bethesda, MD: Wildlife Society).

United Nations Development Program (2004). *Human Development Report 2004 (HDR). Cultural Liberty in Today's Diverse World* (New York, N.Y: UNDP).

United Nations Development Program (2007/2008). *Human Development Report 2007/2008. Fighting Climate Change: Human Solidarity in a Divided World.*

United Nations Development Program (2010). *Human Development Report 2010. The Real Wealth of Nations: Pathway to Human Development*, p. 187.

Vitousek, P., P. Ehrlich, A. Ehrlich and P. Matson (1986). "Human Appropriation of the Products of Photosynthesis," *Bioscience*, Vol. 36, pp. 368–374.

Wackernagel, M. (1991). "Using 'Appropriated Carrying Capacity' as an Indicator: Measuring the Sustainability of a Community." Report for the UBC Task Force on Healthy and Sustainable Communities. UBC School of Community and Regional Planning, Vancouver, Canada.

Wackernagel, Mathis and William Rees (1996). *Our Ecological Footprint: Reducing Human Impact on the Earth* (Gabriola Island, B.C., Canada: New Society Publishers).
Wackernagel, Mathis, Alejandro C. Linares, Diana Deumling, Maria A.V. Sanchez, Ina S.L. Falfan, and Jonathan Loh (2000). *Ecological Footprints and Ecological Capacities of 152 Nations: The 1996 Update* (San Francisco, CA: Redefining Progress).
Wackernagel, Mathis and Judith Silverstein (2000). "Big Things First: Focusing on the Scale Imperative with the Ecological Footprint." *Ecological Economics*, 32: 391–4.
World Bank (2005). *World Development Report 2005. A Better Investment Climate for Everyone* (New York, N.Y: A Co-publication of the World Bank and Oxford University Press).
World Commission on Environment and Development (WCED) (1987). *Our Common Future* (Oxford: Oxford University Press).
World Resources Institute (WRI) (1992). *World Resources, 1992–1993* (New York: Oxford University Press).
World Resources Institute (2000). *World Resources, 2000–2001: People and Ecosystems; The Fraying Web of Life* (Washington, D.C., USA: World Resources Institute).
York, Richard, Eugene A. Rosa, and Thomas Dietz (2003). "Footprints on the Earth: The Environmental Consequences of Modernity." *American Sociological Review*, 68: 279–300.
Zovanyi, G. (2005). "Urban Growth Management and Ecological Sustainability Confronting the 'Smart Growth Fallacy,'" in *Policies for Managing Urban Growth and Landscape Change. A Key to Conservation in the 21st Century*. Published by North Central Research Station, Forest Service U.S. Department of Agriculture, St. Paul, MN 55108.

Table 10.1a. Human Development and Affluence on Biological capacity, ecological Footprint and Balance in African Countries

Coffiecients[a]

Model	*Unstandardized Coefficients*		*Standardized Coefficients*		
	B	*Std. Error*	*Beta*	*t*	*Sig.*
1 (Constant)	.426	.812		.524	.603
GDP Per Capita	2.428E-03	.000	.869	11.099	.000
2 (Constant)	.852	.511		1.667	.104
GDP Per Capita	1.735E-03	.000	.621	10.693	.000
Carbon Dioxide Emissions	6.250E-02	.008	.462	7.947	.000

a. Dependent Variable: Total Footprint

Table 10.1b. Population, Ecostructural Factors, and Total Footprint Per Capita in African Countries

Coefficients[a]

Model		Unstandardized Coefficients		Standardized Coefficients		
		B	*Std. Error*	*Beta*	*t*	*Sig.*
1	(Constant)	−2.384	3.134		−.761	.453
	Urban Population	.247	.078	.520	3.167	.004
2	(Constant)	−9.535	4.495		−2.121	.044
	Urban Population	.229	.074	.482	3.092	.005
	Literacy Rates	.131	.062	.329	2.109	.045

a. Dependent Variable: Total Footprint Per Capita

Table 10.2a. Human Development and Affluence on Biological Capacity, Ecological Footprint and Balance in Organization for Economic Cooperation and development (OECD) Countries

Coefficients[a]

Model		Unstandardized Coefficients		Standardized Coefficients		
		B	*Std. Error*	*Beta*	*t*	*Sig.*
1	(Constant)	21.029	10.029		2.097	.046
	Carbon Dioxide Emissions	.483	.009	.996	55.687	.000
2	(Constant)	21.888	9.090		2.408	.024
	Carbon Dioxide Emissions	.445	.017	.917	26.510	.000
	Ecological Balance	−.165	.064	−.089	−2.587	.016
3	(Constant)	−.131	.138		−.951	.351
	Carbon Dioxide Emissions	−4.472E-5	.001	.000	−.036	.971
	Ecological Balance	−1.000	.002	−.544	−409.409	.000
	Bio-Capacity	1.000	.003	.564	365.988	.000
4	(Constant)	−.129	.124		−1.039	.309
	Ecological Balance	−1.000	.001	−.544	−1944.095	.000
	Bio-Capacity	1.000	.000	.564	2015.521	.000

a. Dependent Variable: Total Footprint

Table 10.2b. Population, Ecostructural Factors, and Total Footprint Per Capita in Organization for Economic Cooperation and Development (OECD) Countries

Coefficients[a]

	Unstandardized Coefficients		Standardized Coefficients		
Model	*B*	*Std. Error*	*Beta*	*t*	*Sig.*
1 (Constant)	3.683	.689		5.347	.000
Public Expenditure	.000	.000	.455	2.608	.015
2 (Constant)	4.363	.708		6.159	.000
Public Expenditure	.000	.000	.431	2.651	.014
Population Density	–.004	.002	–.366	–2.249	.034
3 (Constant)	.677	1.635		.414	.682
Public Expenditure	.000	.000	.302	1.918	.067
Population Density	–.005	.002	–.436	–2.885	.008
Urban Population	.058	.023	.391	2.455	.022

a. Dependent Variable: Total Footprint Per Capita

Table 10.3a. Human Development and Affluence on Biological Capacity, Ecological Footprint and Balance in Latin America and Caribbean Countries

Coefficients[a]

	Unstandardized Coefficients		Standardized Coefficients		
Model	*B*	*Std. Error*	*Beta*	*t*	*Sig.*
1 (Constant)	1.314	9.632		.136	.893
Carbon Dioxide Emissions	.938	.075	.942	12.501	.000
2 (Constant)	–1.325	5.075		–.261	.797
Carbon Dioxide Emissions	.700	.051	.703	13.682	.000
Bio-Capacity	.148	.020	.375	7.307	.000
3 (Constant)	.225	.276		.815	.426
Carbon Dioxide Emissions	–.006	.009	–.006	–.694	.496
Bio-Capacity	1.007	.011	2.546	93.564	.000
Ecological Balance	–1.008	.013	–1.813	–80.199	.000
4 (Constant)	.193	.268		.718	.481
Bio-Capacity	1.000	.003	2.528	377.208	.000
Ecological Balance	–1.000	.004	–1.798	–268.217	.000

a. Dependent Variable: Total Footprint

Table 10.3b. Population, Ecostructural Factors, and Total Footprint Per Capita in Latin America and Caribbean Countries

Coefficients[a]

Model		Unstandardized Coefficients		Standardized Coefficients		
		B	*Std. Error*	*Beta*	*t*	*Sig.*
1	(Constant)	–.744	.799		–.932	.363
	Urban Population	.045	.011	.667	3.897	.001

a. Dependent Variable: Total Footprint Per Capita

Table 10.4a. Human Development and Affluence on Biological Capacity, Ecological Footprint and Balance in Middle East and Central Asia Countries

Coefficients[a]

Model		Unstandardized Coefficients		Standardized Coefficients		
		B	*Std. Error*	*Beta*	*t*	*Sig.*
1	(Constant)	–.630	1.220		–.516	.624
	Human Development Index	–5.219	17.394	–.009	–.300	.774
	GDP Per Capita	3.167E-6	.000	.000	.149	.886
	Gender-Related Development Index	5.571	16.918	.010	.329	.753
	Gender Inequality Index	.721	1.791	.001	.403	.701
	Carbon Dioxide Emissions	.000	.001	–.001	–.446	.671
	Bio-Capacity	1.000	.002	.620	439.371	.000
	Ecological Balance	–1.001	.003	–.465	–398.205	.000

a. Dependent Variable: Total Footprint

Table 10.4b. Population, Ecostructural Factors, and Total Footprint Per Capita in Middle East and Central Asia Countries

Coefficients[a]

Model		Unstandardized Coefficients		Standardized Coefficients		
		B	*Std. Error*	*Beta*	*t*	*Sig.*
1	(Constant)	.829	.334		2.479	.038
	Population Density	.012	.003	.847	4.499	.002
2	(Constant)	.312	.245		1.275	.243
	Population Density	.013	.002	.922	7.917	.000
	Population	.021	.005	.444	3.809	.007

a. Dependent Variable: Total Footprint Per Capita

Table 10.5a. Human Development and Affluence on Biological Capacity, Ecological Footprint and Balance in Asia-Pacific Countries

Coefficients[a]

Model		*Unstandardized Coefficients*		*Standardized Coefficients*		
		B	*Std. Error*	*Beta*	*t*	*Sig.*
1	(Constant)	15.031	15.428		.974	.342
	Carbon Dioxide Emissions	.558	.013	.995	42.559	.000
2	(Constant)	185.341	62.082		2.985	.008
	Carbon Dioxide Emissions	.564	.011	1.004	49.440	.000
	Human Development Index	–237.034	84.423	–.057	–2.808	.012

a. Dependent Variable: Total Footprint

Table 10.5b. Population, Ecostructural Factors, and Total Footprint Per Capita in Asia-Pacific Countries

Coefficients[a]

Model		*Unstandardized Coefficients*		*Standardized Coefficients*		
		B	*Std. Error*	*Beta*	*t*	*Sig.*
1	(Constant)	.011	.370		.030	.977
	Publ Exp. Health	1.035	.120	.902	8.612	.000
2	(Constant)	–.964	.313		–3.083	.007
	Publ Exp. Health	.756	.097	.659	7.811	.000
	Urban Population	.034	.007	.413	4.900	.000

Dependent Variable: Total Footprint Per Capita

Epilogue

FORCES THAT HINDER THE PROGRESS AND PROMOTION OF SUSTAINABILITY AROUND THE WORLD

Many factors have so far hindered sustainability progress around the world. According to Cullingworth and Caves (2014, p. 92), those factors include, among others:

1. Lack of political consensus on what to do in achieving sustainability, especially political differences between developed and developing countries.
2. World system economic ranks of nations and areas have hindered sustainability progress. Some areas seek economic development benefits for some individuals at the expense of the environment and other people; the so called the 'rich get richer' dilemma.
3. Some areas may simply not want to change their current ways of doing things.
4. Some areas pursue their self-interest and ignore the ramifications of their actions on people and neighboring jurisdictions.
5. Failure of some areas to grasp the simple idea that what we do today can greatly affect future generations. Thus, until this idea is acknowledged and recognized by all countries, regions, and cultures, any hopes of achieving sustainability will remain a dream and an illusion.
6. We are really the enemy as human beings and the natural environment are on a collision course, through activities which inflict harsh and often irreversible damage on the environment (Cortese, 1999, p, 1). Fundamental changes are urgent, world over, to avoid the impending calamities of mammoth proportions.

THE FUTURE OF SUSTAINABILITY

There are three major occurring events that currently constitute existential threats to the survival of planet earth, which require urgent attention and solutions. In fact, the future of sustainability is far more threatened by: the ongoing global demographic change, shift, and trends or what Goldstone (2012) called the "New Population Bomb"; the irrefutable climate change that is making the planet Earth hotter, waterier, and wilder (Borenstein, 2015), which constitutes a major threat to environmental sustainability; and the present Islamic State of Iraq and Syria (ISIS), Taliban, and Al-Qaida extensions (e.g. Boko Haram in Nigeria, Niger, Chad, and Cameroon) all over the globe, which constitute security threats to regimes and cultures, as well as danger to the overall survival and sustainability of different nations and governments. The current global terrorism is impossible to fully assess.

The future of sustainability is far more threatened by the "New Population Bomb" enunciated by Jack Goldstone, when compared to the warnings 50 years ago of Paul Ehrlich in "Old Population Bomb" and Meadows and others (1972) "*Limits to Growth*" combined (Goldstone, 2012, pp. 13–17). According to Goldstone, sustainability, no matter how it is defined, in this 21st century depends largely on how the global population is composed and distributed than on the absolute number of people inhabiting the world as depicted earlier by Paul Ehrlich and Meadows et al. The current global demographic trends as noted by Howe and Jackson (2011) are likely to produce greater disruption in the future. For example, that the industrialized world, with the exception of the United States, will have an aging and declining population in coming decades, while the developing world's population will be passing through demographic transition. According to them, these trends will have a profound impact on economic growth and productivity as well as on security and stability. In the same vein with Howe and Jackson, Goldstone emphasized that the current declining fertility rates will stabilize world population in the middle of twenty-first century, and the shifting demographics will bring about significant changes in both rich and poor countries, however. In fact, the industrial countries will account for less of the world's population, their economic influence will diminish, and they will need more migrant workers; and meanwhile, most of the world's population growth will take place in the developing world, especially the poorest countries, and those populations will be increasing urban.

Thus, the social, economic, cultural, and environmental sustainability and whatever gains thereof, are on the verge of being seriously compromised by, where the populations are declining and where they are growing, which countries and regions are relatively older and which are more youthful, and how demographics will influence population movements across regions and cultures. The vital demographics revealed by Goldstone exposed four historic shifts or megatrends that would fundamentally change the world over the next coming decades if adequate measures to mitigate them are not taken. The historic shifts include: the relative demographic weight of the world's developed countries will drop by nearly 25 percent, shifting economic power to developing nations; the developed countries' labor forces will substantially age and decline, constraining economic growth in the developed world raising the demand for immigrant workers; most of the world's expected population growth will increasingly be concentrated in today's poorest, youngest and heavily Muslim countries, which have a dangerous lack of quality education, capital, and employment opportunities; and for the first time in history, most of the world's population will become urbanized, with the largest urban centers being in the world's poorest countries, where policing, sanitation, and health care are often scarce (Goldstone, p. 13). In other words, the megatrends of Europe's reversal of fortunes, Europe's aging pains, youth and Islam in developing countries, and urban sprawl in low-income countries of Africa, Asia, Middle East, and South America, will undoubtedly pose greater sustainability and ecological challenges than those noted by Ehrlich and Meadows et al. Defusing and averting the potential dangers posed by the new population bomb will require global efforts that will also require a major reconsideration of the world's basic global governance structures, especially reconsidering the restructure of various global institutions policing the global environments. Nonetheless, if we are lucky though not guaranteed, the transition in the long run might usher in a new era in which demographic trends would promote global stability, according to "demographic peace" thesis propounded by Howe and Jackson (2012, p. 121). They noted that as demographic transition progresses, population growth slow, medium ages rise, and child dependency burdens fall. Demographic peace thesis predicts that economic growth and social and political stability will follow, as attested in developed regions of the world. The thesis emphasized that societies with rapidly growing populations and young age structures are often mired in poverty and prone to civil violence and state failure; while those with no or low population growth and older age structures tend to be more affluent and stable. Relatively, the jury is still out, especially in developing countries, such as the poorest and least stable countries of Sub-Saharan Africa, where the fulfillment of the transition thesis is yet to be confirmed since the transition has not run its course.

As mentioned above, there should be major socioeconomic and sustainability challenges in decades to come, which must be collectively tackled posthaste to avoid serious problems that will affect many world regions and cultures. The challenges include, among others, the shift in disproportionate consumption of natural capital resources in favor of the more populous developing countries that will become the main driver of global economic expansion. In the coming decades, the gross domestic product (GDP) growth is headed for a sharp reversal from Europe, United States, and Canada to newly industrialized countries of developing world, which will have more global middle class capable of purchasing durable consumer products, such as cars, appliances, and electronics. The developed countries will become economically dynamic because their populations will be substantially older, having increasingly large proportion of retirees and increasingly small proportions of workers of prime working age (15–59 years old), which will have dramatic impact on economic growth, health care, and military strength in the developed world, thereby shifting the world capitalist economic and political systems and their ranks.

Contrary to the industrialized countries of Europe, North America, and Northeast Asia, the fast-growing countries of Africa, Latin America, the Middle East, and Southeast Asia will have exceptionally youthful populations. These countries are likely to have few ways of providing employment to their young, fast-growing populations who are likely to be increasingly attracted to the labor markets of the aging developed countries of Europe, North America, and Northeast Asia. According to Goldstone (p. 15), the forces of immigration will act strongly on the Muslim world, where many economically weak countries will continue to experience dramatic population growth in the decades ahead. For example, worldwide, of the 48 fastest-growing countries today—those with annual population growth of 2 percent or more,—28 are majority Muslim or have Muslim minorities of 33 percent or more (p. 15).

The world is urbanizing to an unprecedented degree, and for the first time in history, a majority of the world's people live in cities rather than in the countryside. The cities are the black-hole of excessive consumption patterns, and most of the world's largest urban agglomerations are found in low-income countries of Asia, Africa, and Latin America. Such urban agglomerations include, Mumbai (20.1 million), Mexico City (19.5 million), New Delhi (17 million), Shanghai (15.8 million), Calcutta (15.6 million), Karachi (13.1 million), Cairo (12.5 million), Manila (11.7 million), Lagos (10.6 million), and Jakarta (9.7 million); and most countries have multiple cities with over one million residents each, such as Pakistan (8), Mexico (12), and China (more than 100) (Goldstone, p. 15). It has been concluded also elsewhere by other demographers that countries with younger populations and heavily urbanized are prone to civil unrest and anarchic violence, and less able to create or sustain democratic institutions; as well as having sprawling and impoverished cities that are vulnerable to crime lords, gangs, and petty rebellions; and even in 19th century Europe, cyclical employment, inadequate policing, and limited sanitation and education often spawned wide spread labor strife, periodic violence, and sometimes even revolution (Cincotta and Leahy, 2006/2007; Goldstone, 2012).

The four megatrends will surely alter the ongoing debates and disagreements between developed and developing countries about what region and culture is to be blamed for greater wastes, footprints, global warming, and damage to the environment, as well as a major driver of environmental stress.

In any case, the anthropocentric world is to be overly blamed, unless urgent measures are taken posthaste to stem the impending catastrophe resulting from human and industrial metabolisms.

Since the Kyoto, Japan, Protocol in 1997, the Planet Earth has become hotter, waterier, and wilder causing a great deal of havoc around the world. As carbon levels in the atmosphere break new records and also greenhouse gases hit record levels around the world, the Planet Earth has entered "uncharted territory" due to global warming, according to AFP (French Press Agency), November 10, 2015. Spewing large amount of carbon dioxide into atmosphere is no more hitherto prerogative of the advanced industrial nations, but today fingers are being pointed at emerging giants, such as China and India, as they spew carbon dioxide burning coal to power ever expanding populations and economies. According to Britain's Weather Office on November 9, 2015, the earth has heated up by one degree Celsius (1.6 degrees Fahrenheit) which is threatening environmental security with its catastrophe climate change. This announcement that local mean surface temperature of above one Celsius was above the pre-industrial level for the first time, half way to the 2 Celsius threshold scientists said humanity must not cross. According to World Metrological Organization (WMO) Chief, Michel Jarraud, "life on Earth is threatened as we are now really in uncharted territory for the human race." WMO reported that "the level of climate-altering gases in the air pushed through the psychological barrier of 400 parts per million; and that concentrations of greenhouse gases in the atmosphere are now reaching levels not seen on Earth for more than 800,000 years, maybe even one million years" (Yahoo News on November 10, 2015; see also AFP, 11/10/15).

There are many events happening today that are as a consequence of changing climate due to global warming. For example, on November 9, 2015, NOAA released its data for the Pacific Ocean temperatures that cause El Nino, which indicated the biggest El Nino on record and the strongest in history. El Nino happens when ocean water temperatures rise above normal across the central and eastern Pacific, near the equator. Before 2015, the largest weekly reading of El Nino occurred during the week of November 26 in 1997, but was recently surpassed in November this year, 2015. For example, in 1997, the El Nino index set a record of 2.8, but on November 9, 2015, the Pacific Ocean in region 3.4 (where El Nino is measured) hit a record of 3.0 (Krasting, 2015). According to Krasting, this means in the months to come, there will be some hellacious weather in the US Pacific/Texas, drought in Australia and Indonesia, and other parts of the globe will also feel the consequences of the mega Nino. In the U.S., the effects will leave the Northeast warmer than usual, the Midwest drier, and the West and the South wetter. If history is the gauge, the coming La Nina should bring a record hurricane season in the summer and fall of 2016 and a return to the crushing droughts in Pacific West, especially in some cities in US. such as,

Phoenix, Las Vegas, San Francisco, and Los Angeles that are highly dependent on the Colorado River/Lake Mead.

The Earth has changed a lot since 1997 Kyoto Protocol and the speed of climate change is proceeding faster than expected, e.g. Arctic sea ice loss, forest die-off due to drought, etc. People are really experiencing climate change through the frequent extreme temperatures we have been witnessing ever since, causing large downpours, floods, mudslides, the deeper and longer droughts, rising sea levels from melting ice, forest fires. Studies by climate scientists have shown that man-made climate change contributed in a number of recent weather disasters of significant proportions, such as: the 2003 European heat wave that killed 70,000 people in the deadliest such disaster in a century; Hurricane Sandy, worsened by sea level rise, which caused more than $67 billion in damage and claimed 159 lives; the 2010 Russian heat wave that left more than 55,000 dead; the drought is still gripping California; Typhoon Haiyan, which killed more than 6,000 in the Philippines in 2013; and the heavy flooding in the Southern State of Tamil Nadu, India in December 2015 after the heaviest cloudburst in over a century that claimed over 270 lives, among others (Borenstein, 2015). The Earth is really getting a lot more dangerous than it was used to be and the urgency to embark on mitigation measures is much clearer than before. Given below are also some drastic alarming numbers of events that have occurred since 1997 due to global warming (Borenstein, 2015):

1. The West Antarctic and Greenland ice sheets have lost 5.5 trillion tons of ice, or 5 trillion metric tons, according to Andrew Shephard at the University of Leeds, who used NASA and European satellite data.
2. The five-year average surface global temperature for January to October has risen by nearly two-thirds of a degree Fahrenheit or 0.36 degrees Celsius, between 1993–97 and 2011–15, according to the U.S National Oceanic and Atmospheric Administration (NOAA). In 1997, Earth set a record for the hottest year, but did not last. Records were set in 1998, 2005, 2010, and 2014, and it is sure happen again in 2015 when the results are in from the year, according to NOAA.
3. The average glacier has lost about 39 feet, or 12 meters of ice thickness since 1997, according to Samuel Nussbaumer at the World Glacier Monitoring Service in Switzerland.
4. With 1.2 billion more people in the world, carbon dioxide emissions from the burning fossil fuels climbed nearly 50 percent between 1997 and 2013, according to the U.S Department of Energy. The world is spewing more than 100 million tons of carbon dioxide a day now.
5. The seas have risen nearly 2 ½ inches, or 6.2 centimeters, on average since 1997, according to calculations by the University of Colorado.

6. At its low point during the summer, the Arctic sea ice is on average 820,000 square miles smaller than it was 18 years ago, according to the National Snow and ice Data Center. That's a loss equal in area to Texas, California, Montana, New Mexico, and Arizona combined.
7. The five deadliest heat waves of the past century—in Europe in 2003, Russia in 2010, India and Pakistan in 2015, Western Europe in 2006, and Eastern Asia in 1998—have come in the past 18 years, according to the International Disaster Database run by the Centre for Research on the Epidemiology of Disaster in Belgium.
8. The number of weather and climate disasters worldwide has increased 42 percent, though deaths are down 58 percent. From 1993 to 1997, the world averaged 221 weather disasters that killed 3,248 people a year. From 2010 to 2014, the yearly average of weather disasters was up to 313, while deaths dropped to 1,364, according to the disaster database.

As some regions get drier due to global warming and climate change, the growing water scarcity will likely generate new levels of tension at local, national, and even international levels, because water sharing among countries and states with limited water availability often increases the risk of friction, social tensions, and conflict. In fact, in the coming decades, fresh water availability should constitute the key geopolitical resource as oil is currently. The increasing drought and scarcity of water due to climate change will in a number of regions generate serious dislocation in water supply, thereby making consumption patterns not sustainable. Water possesses major challenges in addressing issues of public health, economic development, gender equity, humanitarian crises, environmental degradation, and global security. According to Peterson and Posner (2014), water has meant the difference between life and death, health and sickness, prosperity and poverty, environmental sustainability and degradation, progress and decay, stability and insecurity. Furthermore, societies subject to flood-drought cycles and without the capacity to manage water flows, have found themselves confronting tremendous social and economic challenges in development. This is especially true in developing countries of South Asia and Sub-Saharan Africa, in the so-called "poverty belt" of the world, which is home to 2.3 billion people that account for one-third of the global population. Water problem affects the earning power of women, especially in poor countries where women and girls spend hours caring for sick family members; and spend hours each day walking to collect water for daily drinking, cooking, and washing.

The effect of global warming and climate change is making distribution of the fraction of water suitable for human consumption more unequal, inefficient, and unsustainable. In the decades to come, the capacity of governments, policy makers, and development experts around the world to address the constellation of challenges that relate to water access, sanitation, ecosystems, infrastructure, adoption of technologies, and the mobilization of resources, will mean the difference between rapid economic development and continued poverty. Adequate management of freshwater sources should represent means of preventing and mitigating conflict between countries and regions with shared water resources. Thus, to avoid impending conflicts and hardships, governments in water stressed regions should effectively and transparently mediate and negotiate concerns and demands of various constituencies within and between countries, regions, and cultures.

The debate on who is to blame for the prevailing sustainability issues, for example, the spewing of enormous carbon dioxide and other greenhouse gases in the atmosphere should no longer be defined by rich and poor political divide or global North and South or developed and developing countries, but all should step up together to shoulder the responsibility, and with less foot-dragging in negotiations, because every country wants sustainable life and living environment. The debacle between developed and developing countries over the years has stymied progress. The present environmental negotiations going on since November 2015 in Paris, France, after Kyoto, Japan negotiations 18 years ago in 1997 is an attempt in good direction, as the urgency for practicable, workable, and enforceable agreement is much clearer now than before. The main goal is to half average warming at 3.6 degrees Fahrenheit. It is hoped that binding measures that hinge on environmental security should be created from the agreement in Paris to curb dangerous levels of global warming. Indeed, in December 2015, the delegates from nearly 200 nations, large and small, in Paris agreed to cut carbon emissions. To be more confident that the planet should be in better shape than before, this historic and landmark agreement stipulated that every nation should set its own target for a more sustainable environment. The U.S. President, Barak Obama, saw the agreement as the "best chance we have to save the one planet that we have." He also noted that low carbon pursuit by different nations may likely create the much needed jobs in the clean energy investments. Nevertheless, since verification has remained a major problem of previous agreements enforcement among nations, probably creating an environmental security council may be necessary at this time.

A major natural solution to greenhouse gas effect on the environment is to wean ourselves from fossil fuels that emit heat-trapping gases. Without gainsaying, the present global improvements in technology are pointing to the possibility of a world weaned from fossil fuels, which will make Planet Earth greener. For example, some current local sustainability efforts in some countries and regions have included: promoting climate protection through climate action plans; promoting the use of clean energy; promoting alternative modes of transportation; promoting recycling and composting; and developing a sustainable buildings and sites policy, i.e. Green building regulations. As local governments strive to balance what we have and want for today and tomorrow, they have focused on some issues surrounding air, community development (e.g. recycling program), energy (e.g. energy efficiency promotion), transportation (e.g. offering mass transportation to residents), and waste resources management (e.g. yard composting program); they are consistent with Brundtland Commission's definition of sustainability. Moreover, chapter 12 of the Kyoto Protocol (1997) required cities to play a pivotal role in reducing global warming pollution level by taking a variety of actions at the local level to help meet or exceed the targets set. Thus, local governments are required to encourage environmentally friendly or environmentally responsible building in the design of homes and buildings in cities by incorporating healthy, resource- and energy- efficient materials and building methods, e.g. the Leadership in Energy and Environmental Design (LEED) standard /code developed by the U.S Green Building Council (Meisal, 2010; Retzlaff, 2009). Although a shift away from dirty old fuels is the only path toward reducing several of the greatest ecological security threats the planet faces, nonetheless, we must thread or step carefully. This is because a greener world might not be a more peaceful world, as Rothkopf (2009) noted that as the world seeks alternative energy sources, there is a distinct possibility that a greener world will not necessarily be a more peaceful one. He opined that trade disputes, resource scarcity, and the dangers of alternative energy sources threaten to make the shift to more environmentally sound energy production a security challenge for both industrialized and developing countries.

The third currently occurring event, apart from the new population bomb and demographic shifts and extreme climate changes due to global warming, is the Islamic fundamentalists' uprising and radicalism that have metamorphosed into ongoing intractable terror attacks around the globe. There are many Islamic radical terrorist groups operating in different world regions under different names, such as Taliban, Al-Qaida, ISIS, Boko Haram, among others. The above Jihadists are in collision course with world's freedom, liberty, and security to the extent that public gathering is threatened. They operate in different parts of the globe killing and maiming hundreds and thousands individuals in Iran, Saudi Arabia, Syria, Yemen, Somalia, Sudan, Tunisia, Morocco, Egypt, Lebanon, Iraq, Afghanistan, Pakistan, India, Turkey, Indonesia, Kenya, Nigeria, Libya, Niger, Chad, Cameroon, Mali, USA, Belgium, France, and many other countries.

The fundamentalists attack their targets both locally and internationally, with recent examples between October and December 2015 at Radisson Western Hotel in Mali killing about 22 people; downing of Russian Airline killing 224 people; Paris music hall attack that killed above 130 people; and the San Bernardino, California massacre of 14 people that also wounded 21 people on December 2, 2015, which an FBI official termed "an act of terrorism" by the Muslim extremists; and the list goes on. According to the recent U.S. counterterrorism guide, Islamic extremism is rearing its heads across the globe, especially in Africa. Al-Qaida is growing in Africa at the same time like ISIS, which has been behind the beheadings and other atrocities in North Africa. Many off-shoots of Al Qaida and ISIS are pledging allegiance to them, such as: Nigerian-based Boko Haram, under Abubakar Shekau, has offered allegiance to ISIS with its fealty or fidelity horrific deeds. Since 2009, Boko Haram insurgency has killed some 20,000 people and forced about 2.3 million from their homes; the Somalia-based terror group, Al-Shabaab and its mujahedeen fighters affiliated to al-Qaida, has also recently offered allegiance to ISIS; and the North African jihadist group, Al-Murabitun, has also declared its allegiance to Al Qaida organization in Islamic Maghreb. It was Al-Murabitun that claimed responsibility for the November 2015 attack at Radisson Blu Hotel in Bamako, Mali that claimed many lives.

A major threat is the human (both men and women) suicide bombers with a lot of wires on them penetrating through national borders, especially European and U.S borders that are relatively open and porous, including their maritime or sea borders. The bombers' activities have caused many wars and human displacements and dislocations, that thousands and millions of people are fleeing the war-torn areas, creating unbelievable hardships. Hence, many refugees are seeking asylums, especially in Western European countries (e.g. Spain, Italy, Germany, Belgium, France, and Britain) and United States of America. In fact, the more wars we have on this planet, and the longer they last, the more population dislocations we have, and the more crowded our cities become and less the resources are devoted to basic necessities of life, thus making environmental sustainability and overall sustainability virtually impossible. The aftermaths of the ongoing Islamic insurgency and extremism are disproportionately affecting the poor and low-income communities in different countries, regions, and cultures. How to prevent these attacks is a major challenge to different governments and policy makers. To prevent, stem, and deal adequately with the effects of the present Islamic insurgency and the heightened security of terror, will undoubtedly require a great deal of resources, sharing of intelligence, and working together; else the overall future global sustainability shall be in jeopardy and not guaranteed. The above three megatrends, the new population bomb; global warming and the irrefutable climate change that devastate individuals and communities; and Islamic fundamentalists' insurgency and terrorism that main, kill, and dislocate individuals worldwide, should be confronted headlong in order to guarantee our environmental, social, cultural, and economic survival and sustainability, especially for the future generations.

REFERENCES

Borenstein, Seth (2015). "Earth is a Wilder, Warmer Place since Last Climate Deal Made." *Associated Press (AP)/Yahoo News*, 11/30.2015.

Cincotta, R.P and E. Leahy (2006/2007). "Population Age Structures and the Risk of Civil Conflict: a Metric," *Environment Change & Security Project Report*, 12: pp. 55–58. http://www.wilson center.org/topics/pubs/PopAgeStructures&CivilConflict12-pdf

Cortese, Anthony (1999). "Education for Sustainability: The Need for a New Human Perspective." *Second Nature, Inc.* Boston Massachusetts.

Cullingworth, Barry and Roger W. Caves (2014). *Planning in the USA: Policies, Issues, and Processes, 4th Edition* (London and New York: Routledge. Taylor & Francis Group).

Ehrlich and Holdren (1971). "Impacts of Population Growth," *Science*, 171, 1212–7. See also Ehrlich, P.R. and A.H. Ehrlich (1990). *The Population Explosion* (New York, USA: Simon and Schuster).

Goldstone, Jack A. (2010). "The New Population Bomb: The Four Megatrends That Will Change the World," *Foreign. Affairs*, January/February, 2010. See also Jack A. Goldstone (2012). "The New Population Bomb: The Four Megatrends That Will Change the World," in *Annual Editions: Developing World, 12/13, Twenty-Second Edition*. Edited by Robert J. Griffiths, pp. vi, 13–17.

Howe, Neil and Richard Jackson (2011). "Global Aging and the Crisis of the 2020s," *Current History*, (January). See also Neil Howe and Richard Jackson (2012). "Global Aging and the Crisis of the 2020s, in *Annual Editions: Developing World, 12/13, Twenty-Second Edition*. Edited by Robert J. Griffiths, pp. x, 119–123.

Krasting, Bruce (2015). "My Take on Financial Events," *Business Insider*, November 18, 2015, 6:43 pm.

Meadows, D.H, D.L. Meadows, J. Randers, and W.W. Behrens (1972). *The Limits to Growth* (New York, NY: Universe Books).

Meisal (2010). *LEED Materials: A Resource Guide to Green Building*. See, Retzlaff (2009). "The Use of LEED in Planning and Development Regulation: An Explanatory Analysis." See also discussions on the Leadership in Energy and Environmental Design (LEED) and BRE Environmental Assessment Method (BREEAM) can be found at: www.new.usgbc.org and http://www .breeam.org.

Peterson, Erik R. and Rachel A. Posner (2014). "The World's Water Challenge," *Annual Editions: The Developing World, 24e*, Edited by Robert J. Griffiths, pp. 165–167. See also *Current History*, January 2010, pp. 31–34.

Rothkopf, David J. (2009). "Is a Green World a Safer World? Not Necessarily," *Foreign Policy*, September/October, 2009. See also "Is a Green World a Safer World? Not Necessarily," in *Annual Editions: Developing World, 12/13, Twenty-Second Edition*. Edited by Robert J. Griffiths, pp. xii, pp. 186–188.

Index

CPSIA information can be obtained
at www.ICGtesting.com
Printed in the USA
BVOW03*2319140417
481127BV00003B/3/P

9 780761 868644